Elements of Optical Coherence Theory

Elements of Optical Coherence Theory

ARVIND S. MARATHAY
Optical Sciences Center
University of Arizona

1807 175 YEARS OF PUBLISHING 1982

John Wiley & Sons

New York • Chichester • Brisbane • Toronto • Singapore

Library of Congress Cataloging in Publication Data:

Marathay, Arvind S., 1933-
Elements of optical coherence theory.

(Wiley series in pure and applied optics, ISSN 0277-2493)
Bibliography: p.
Includes index.
1. Coherence (Optics) I. Title. II. Series.

QC476.C6M37 535'.2 82-2707
ISBN 0-471-56789-2 AACR2

Printed in the United States of America

10 9 8 7 6 5 4 3 2 1

To my Mother and Father

Preface

The theory of partial coherence has been evolving over the past three decades or more. During this time, numerous papers and several research monographs have appeared, covering a wide range of topics at a relatively advanced level.

There is now a definite need for a textbook to introduce the student to the basic elements of optical coherence theory. The present book is intended to fill that need. The book presupposes a basic knowledge of interference and diffraction of light and a familiarity with Fourier theory. The required level of knowledge of interference and diffraction is that of a good general physics course at the undergraduate level. Some knowledge of statistics would be helpful, but it is not absolutely necessary since some basic discussion on this topic is included in the Appendixes to Chapter 3. The book may be used for a senior undergraduate or a first-year graduate course. It can also be used by scientists and engineers working in other fields of research. They can apply the elements and methods of coherence theory to their own fields of work.

The approach taken in writing this book is that of "back to basics." A good foundation will prepare the reader for applications and uses of coherence theory in the area of interest, without outside help. Homework problems are included wherever possible. For the most part, they are meant to further the understanding or to develop a new aspect of the topics covered in the text. Some references are supplied for the same purpose and also to give the reader sources for further research into the topic of interest. There are a few problems that call for short numerical calculations, in order to provide an understanding of the order of magnitude of the quantities involved in the theory.

Insofar as possible, without sacrificing clarity, the book is kept to a limited size so that a major portion of it may be comfortably covered during a one semester course. The selection of topics was a difficult task. Instead of covering a large number of topics inadequately, it was decided to limit the topics to what may be called "elements" of coherence theory, including basic properties of sources and the light they emit. As a result, some topics that might be expected to be covered in a book on coherence theory are not included here.

As the title suggests, the subject matter of the book is theoretical. Wherever appropriate, however, basic experiments are brought into the discussion so that the reader may appreciate the usefulness of theoretical concepts and quantities in relation to elementary optical experiments. Theory inevitably entails derivation of results. This can be lengthy but an attempt has been made to make it painless. In some cases the details of the derivation are given in an appendix; in others the "route" from one intermediate step to another is detailed in words. In the discussion of the diffraction of light in the language of the mutual coherence function (MCF), Appendixes 5.1 and 5.2 are used extensively to arrive at the result of the generalized van Cittert–Zernike theorem, which is the working equation for the rest of Chapter 5.

Chapter 1 discusses briefly the history of coherence theory. Chapter 2 details the complex analytic signal representation, and in Chapter 3 the MCF is introduced in this language, with a discussion of the field statistics. Chapter 4 develops mathematical familiarity with the MCF, along with the introduction of the concept of noncoherence. It is customary to discuss noncoherence in terms of sharply peaked narrow functions and an attempt has been made to introduce noncoherence with emphasis on the property of the constant or "uniform" spatial frequency spectrum and its interpretation in parallel with the "white" noise sources of electrical engineering. In this way, the properties of the noncoherent source follow more easily.

In the field of optics the term *intensity* is used loosely: Not properly defined, it has no place in the radiometric scheme. In order not to confuse it with the term *radiant intensity* defined in radiometry, intensity is always qualified with an appropriate adjective. Thus, when referring to the time average of the square of the field variable, we use the term *optical intensity* and denote it by I or by $I(x)$ to display the space coordinate(s). Frequently, it is necessary to distinguish optical intensity from its spectral version, which we call *optical spectral intensity* (OSI) and denote by $\hat{I}(x, \nu)$. We follow the Système International for units and nomenclature, and for the purpose of this book we display the radiometric symbols in sans serif type. Thus, irradiance is denoted by E to distinguish it from the electric field $\mathbf{E}$. By using a scaling factor C with appropriate units, we use $\mathsf{E} = \mathsf{C}I$ for irradiance, with units of W m^{-2}. Similarly, $\hat{\mathsf{E}} = \mathsf{C}\hat{I}$ is used for spectral irradiance [W m^{-2} Hz^{-1}]. Although this approach entails dealing with a wide range of symbols and terms, the author feels it is best to follow this scheme in order to clearly specify what is meant in each particular situation.

As much as possible, an attempt has been made to keep the notation uniform throughout the book. A list of symbols, notation, and abbreviations is also given.

Chapter 5 deals with the propagation of the mutual coherence function (MCF) and related topics. The generalized van Cittert–Zernike theorem and the van Cittert–Zernike theorem with noncoherent sources are discussed. By way of example, it is shown how the partially coherent-source result approaches the noncoherent-source result by going to the limit of the constant spatial frequency spectrum. The Thompson and Wolf experiment, the Michelson stellar interferometer, and the two-beam interferometer are studied to explain the spatial and temporal coherence of light. The book is concerned exclusively with the second-order statistics, namely, the mutual coherence function. The Hanbury-Brown and Twiss interferometer is mentioned only briefly, to make the reader aware of the relatively recent advances. The chapter continues with a discussion of how the beam energy is distributed (spread out) in the right half-space after it leaves the source. The measurement of the mutual coherence function is discussed at the end.

The propagation of the MCF is used to discuss image formation in Chapter 6. The discussion makes use of the object and image space spatial frequencies defined with the object and image distances measured from the entrance and exit pupil planes, respectively. In theoretical formulations, there is a tendency to omit the "constants" that occur outside of the integrals or to collect several constants under one common constant as the development proceeds. This "snowball" effect has been avoided. As a result, the relationships in the chapter may be verified for dimensional balance at every stage of the development. In Chapter 6, Hopkins' effective source is introduced in order to study the influence of partial coherence on optical imaging. The chapter ends with a summary and a brief discussion of resolution criteria. In particular, it is shown how the various resolution criteria are special cases of the one that may be formulated by use of the spatial coherence function and the van Cittert–Zernike theorem.

The last chapter, Chapter 7, is on radiometry. After a brief review of conventional radiometry, a study is made of the properties of the noncoherent source. The study is based on the well-established diffraction calculation. It is shown that the idealization of noncoherence in the theory of partial coherence is the same as the idealization of a Lambertian source of conventional radiometry. Two approaches to generalized radiometry applicable to sources of any state of coherence are presented. The results of the special cases derived in the first approach (largely unpublished) are tabulated for ease of reference. The second approach, pioneered by Walther, Marchand, and Wolf, is also discussed along with the results that follow for some of the special cases. Because the discussion is brief, the original paper of Marchand and Wolf is reproduced in Appendix 7.2. No attempt is made (at least not intentionally) to upgrade or downgrade one or the other approach; only the results as they follow for the special cases are displayed for ease of comparison by the reader. A word of caution to the reader at this point: the first approach is given in order to observe what the results would be if that path were followed; the approach that is widely used throughout the literature is the one due to Walther, Marchand, and Wolf.

I offer sincere thanks to my former professors, H. H. Hopkins and E. L. O'Neill, with whom I have had the pleasure of associating and studying optics. I should also like to express my appreciation to Professor E. Wolf, who has done so much for so long in the field of optics and from whose wide range of publications I learned coherence theory. I am extremely fortunate to associate with the distinguished faculty of the Optical Sciences Center, University of Arizona. In particular, my association with Professors H. H. Barrett, R. V. Shack, R. R. Shannon, P. N. Slater, and J. C. Wyant has proved to be a valuable learning experience through helpful discussions. It is a pleasure to acknowledge helpful discussions on radiometry with Professor W. L. Wolfe, Dr. F. O. Bartell, and Dr. J. M. Palmer. And, of course, in an institution of learning, the students form an important component: I am thankful for their discussions and, in particular, I should like to mention Dr. M. J. Lahart, Dr. V. N. Mahajan, and Dr. R. E. Wagner. Furthermore, I am especially grateful to Professor P. A. Franken, Director of the Optical Sciences Center, for providing an environment conducive to learning and for his constant encouragement.

Many thanks to Don Cowen for the ink drawings of the figures in this book. My special appreciation and sincere thanks to Martha Stockton for reading, editing, typing and reediting and retyping the manuscript. If the reader should find the book readable it is because of her untiring efforts. However, I shall appreciate a brief communication upon discovery of errors, for they are entirely my own.

ARVIND S. MARATHAY

Tucson, Arizona
April 1982

Contents

Symbols, Notation, and Abbreviations

a	radius of lens or entrance pupil
a'	radius of exit pupil
a_0	radius of circular source
a_0, b_0	rectangular source sides
$\mathcal{a}_c$	coherence area for spatial coherence
$A(\mathbf{x} - m\boldsymbol{\xi})$	Fourier transform of the exit pupil function
$\mathscr{A}$	symbol for area
Besinc $(x) \equiv 2J_1(x)/x$	J_1 is the Bessel function of order 1 and x stands for the unitless argument
BFP	back focal plane
c	speed of light in vacuum
c-radiometry	conventional radiometry
$\mathrm{cyl}(r_s/a)$	$= \begin{cases} 1, & r_s < a,\ r_s = (x_s^2 + y_s^2)^{1/2} \\ 0, & r_s > a \end{cases}$
$C(\mathbf{f}'_1, \mathbf{f}'_2, \nu)$	Hopkins' frequency response (HFR) function
C	suitable constant for use with the scalar field ψ: $\mathsf{E} = \mathsf{C}\langle\psi^2\rangle = \mathsf{C}I$
$E\{\ \}$	ensemble average
EP	enclosed power
ESP	enclosed spectral power
$\mathbf{E}$	electric field vector
E	irradiance [W m^{-2}]
$\hat{\mathsf{E}}$	spectral irradiance [W m^{-2} Hz^{-1}]
f, g	pair of spatial frequencies: $f = (1/\lambda)p$, $g = (1/\lambda)q$
$f = \alpha/\lambda s,\ g = \beta/\lambda s$	object-space spatial frequencies

$\mathbf{f} = \hat{\mathbf{i}}f + \hat{\mathbf{j}}g$	two-dimensional vector for object-space frequencies
$f' = \alpha'/\lambda s'$ $g' = \beta'/\lambda s'$	image-space spatial frequencies
$\mathbf{f}' = \hat{\mathbf{i}}f' + \hat{\mathbf{j}}g'$	two-dimensional vector for image-space frequencies
FOV	field of view
FTS	Fourier transform spectrometry
g-radiometry	generalized radiometry
HFR	Hopkins' frequency response
$\mathcal{H}[\cdots]$	Hilbert transform of $[\cdots]$; see Eq. (2.18)
$I = \langle \psi^2 \rangle$	optical intensity, Eq. (2.3)
I	radiant intensity [W sr^{-1}]
$\hat{\mathsf{I}}$	spectral radiant intensity [W sr^{-1} Hz^{-1}]
$J_0, J_1, J_{3/2}$	Bessel functions of order 0, 1, 3/2 respectively
$k = 2\pi/\lambda$	propagation constant
$K(Q, P)$	approximate multiplicative free-space propagator for light from P to Q, Eq. (5-142)
LHS	left-hand side
$\mathsf{L}, \hat{\mathsf{L}}$	radiance [W m^{-2} sr^{-1}] and spectral radiance [W m^{-2} sr^{-1} Hz^{-1}], respectively
$\mathcal{L}(\boldsymbol{\alpha}')$	amplitude and phase transmittance of the system described in the exit pupil coordinates
$(\Delta l)_c$	coherence length
m	image magnification
m_p	pupil magnification
MCF	mutual coherence function
MOI	mutual optical intensity
MSDF	mutual spectral density function
MTF	modulation transfer function
M	radiant exitance [W m^{-2}]
$\hat{\mathsf{M}}$	spectral radiant exitance [W m^{-2} Hz^{-1}]
$(\mathrm{N.A.})_c$	numerical aperture of the condenser
$(\mathrm{N.A.})_o$	numerical aperture of the objective
OI	optical intensity
OPD	optical path difference

OSI	optical spectral intensity
OTF	optical transfer function
p, q, m	direction cosines: $m = \begin{cases} +(1-p^2-q^2)^{1/2}, & p^2+q^2 \le 1 \\ +i(p^2+q^2-1)^{1/2}, & p^2+q^2 > 1 \end{cases}$
PSF	point spread function
QH source	quasihomogeneous source
QM field	quasimonochromatic field
Q	energy in the field [joules] [J]
$r_{12} = (x_{12}^2 + y_{12}^2)^{1/2}$	radius vector in difference coordinates
$\mathbf{r} = \hat{\mathbf{i}}x + \hat{\mathbf{j}}y + \hat{\mathbf{k}}z$ $= \hat{\mathbf{i}}r\sin\theta\cos\phi$ $+\hat{\mathbf{j}}r\sin\theta\sin\phi$ $+\hat{\mathbf{k}}r\cos\theta$	position vector with spherical polar coordinates
$\text{Rect}(x/a_0)$	rectangular function, equals unity for $\|x\| \le a_0$ and zero for $\|x\| > a_0$
RHS	right-hand side
$s(t)$	temporal Fourier transform of $\hat{s}(\nu)$
$\hat{s}(\nu)$	step function, equals $+1$ for $\nu \ge 0$ and zero for $\nu < 0$
$\text{sgn}(\nu)$	signum function, equals $+1$ for $\nu > 0$ and -1 for $\nu < 0$; see Eq. (2.23)
$\mathbf{s} = \hat{\mathbf{i}}x_s + \hat{\mathbf{j}}y_s + \hat{\mathbf{k}}z_s$	point on a surface $\mathcal{S}$
$S_{\text{coh}}, T_{\text{coh}}$	amplitude, impulse response, and transfer function of the lens system, respectively, for the coherent case
$S_{\text{ncoh}}, T_{\text{ncoh}}$	impulse response and transfer function of the lens system, respectively, for the noncoherent case
$S_{\text{pcoh}}, T_{\text{pcoh}}$	impulse response and transfer function of the lens system, respectively, for the partially coherent case
$S_{\text{poly}}, T_{\text{poly}}$	impulse response and transfer function of the lens system, respectively, for the polychromatic case in the noncoherent limit
SFS	spatial frequency spectrum

$\mathrm{Sinc}(x) = (\sin x)/x$	x is a unitless argument of the trigonometric sine function
t	time variable [s]
t_{ncoh}	normalized form of T_{ncoh}, OTF
$\hat{t}_{\mathrm{ob}}(\boldsymbol{\xi}, \nu)$	amplitude transmittance of the object
$(\Delta v)_c$	coherence volume
$\mathcal{V}$	Michelson's visibility function for interference fringes
$V(t)$	$= (1/\sqrt{2})[V^{(r)}(t) + iV^{(i)}(t)]$, analytic signal
$V(\mathbf{r}, t)$	analytic signal for the field at point $\mathbf{r}$
$V^{(i)}(t)$	imaginary part of the analytic signal
$V^{(r)}(t)$	real physical field
$\hat{V}(\nu)$	temporal Fourier transform of $V(t)$
$\hat{V}^{(r)}(\nu)$	complex temporal Fourier transform of $V^{(r)}(t)$
$\tilde{V}(\kappa p, \kappa q, z, t)$	two-dimensional spatial Fourier transform of $V(\mathbf{r}, t)$
$\mathring{V}(\kappa p, \kappa q, z, \nu)$	total (spatial and temporal) Fourier transform of $V(\mathbf{r}, t)$
$\mathcal{V}$	visibility
w	coherence width
$\mathcal{w}_x$	coherence width for spatial coherence along x
$\mathcal{w}_y$	coherence width for spatial coherence along y
$W(\boldsymbol{\alpha}') \equiv W(\alpha', \beta')$	wavefront aberration in the exit pupil coordinates
WMW	Walther, Marchand, and Wolf approach
$\mathcal{w}$	energy density (field) [J m^{-3}]
x, y, z	space coordinates
$x_{12} = x_1 - x_2$, $y_{12} = y_1 - y_2$	difference coordinates
$\mathbf{x} = \hat{\mathbf{i}}x + \hat{\mathbf{j}}y$	two-dimensional vector and coordinates for the image plane
$\boldsymbol{\alpha} = \hat{\mathbf{i}}\alpha + \hat{\mathbf{j}}\beta$	two-dimensional vector and coordinates for the entrance pupil
$\boldsymbol{\alpha}' = \hat{\mathbf{i}}\alpha' + \hat{\mathbf{j}}\beta'$	two-dimensional vector and coordinates for the exit pupil
$\boldsymbol{\alpha}_0 = \hat{\mathbf{i}}\alpha_0 + \hat{\mathbf{j}}\beta_0$	two-dimensional vector and coordinates for the effective source

$\gamma_{11}(\tau)$	complex degree of self-coherence
$\|\gamma_{12}\|$	degree of partial coherence of light from points 1 and 2
$\gamma_{12}(0)$	complex degree of spatial coherence
$\gamma_{12}(\tau) = \|\gamma_{12}\| \exp(+i\phi_{12})$	normalized MCF, complex degree of coherence
$\hat{\gamma}(\nu) = \hat{\gamma}_{11}(\nu)$	when points P_1 and P_2 coincide, normalized spectrum
$\hat{\gamma}_{12}(\nu)$	temporal Fourier transform of $\gamma_{12}(\tau)$
$\Gamma_{11}(\tau)$	self-coherence function; describes temporal coherence
$\hat{\Gamma}_{11}(\nu)$	optical spectral intensity (OSI), spectral density function, temporal Fourier transform of $\Gamma_{11}(\tau)$
$\Gamma_{12}(0)$	MOI for a pair of points P_1 and P_2, $\tau = 0$; describes spatial coherence
$\Gamma_{12}(t_1, t_2) = \Gamma(\mathbf{r}_1, t_1; \mathbf{r}_2, t_2)$	mutual coherence function for the nonstationary case
$\Gamma_{12}(\tau) = \Gamma(P_1, P_2, \tau)$	mutual coherence function
$\hat{\Gamma}_{12}(\nu)$	MSDF for a pair of points P_1 and P_2; it is the temporal transform of $\Gamma_{12}(\tau)$
$\tilde{\Gamma}(f_1, g_1, f_2, g_2, z, \tau)$	spatial Fourier transform of $\Gamma_{12}(\tau)$; (f, g), pair of spatial frequencies
$\mathring{\Gamma}(f_1, g_1, f_2, g_2, z, \nu)$	spatial and temporal Fourier transform of $\Gamma_{12}(\tau)$
$\delta_{\mu\mu'}$	Kronecker delta, equals unity for $\mu = \mu'$, zero otherwise
$\delta(x - x')$	one-dimensional Dirac delta function
$\delta(\mathbf{x} - \mathbf{x}') = \delta(x - x')\delta(y - y')$	two-dimensional delta function using two-dimensional vectors $\mathbf{x}$ and $\mathbf{x}'$
$\delta^{(2)}(\mathbf{s}_1 - \mathbf{s}_2)$	two-dimensional delta function with three-dimensional vectors $\mathbf{s}$, with components x_s, y_s, z_s
$\Delta\nu$	spectral spread of QM fields
ε	enclosed power
$\boldsymbol{\xi} = \hat{\mathbf{i}}\xi + \hat{\mathbf{j}}\eta$	object-space two-dimensional vector
$\boldsymbol{\xi} \cdot \boldsymbol{\xi} = \xi^2 + \eta^2$	square of the vector length
$d\boldsymbol{\xi} \equiv d\xi\, d\eta$	area element
$f(\boldsymbol{\xi}) \equiv f(\xi, \eta)$	function of ξ and η
$\kappa = 1/\lambda$	kappa, reciprocal of the wavelength

λ	wavelength of light in vacuum
ν	temporal frequency [s^{-1}]
$\bar{\nu}$	mean frequency of a QM field
$\rho_j = \lvert \mathbf{r} - \mathbf{s}_j \rvert$	dijtance between P_j and Q, $j = 1, 2$
τ	time delay
τ_c	coherence time
$\mathbf{\Phi}$	average radiant power [W]
$\hat{\mathbf{\Phi}}$	spectral radiant power [W Hz^{-1}]
ψ	scalar field function
$\omega = 2\pi\nu$	circular frequency
$\langle \ \rangle$	angular brackets for time average

1

Coherence

The rectilinear propagation of light has always been more or less self-evident to even the most casual inquirer. Less obvious has been the periodic nature of light. The first experiment to offer a glimpse of the periodicity of light was done by Grimaldi (1613–1663), who used sunlight to observe fringes in the shadow of a hair. The hair was arranged accurately parallel to a narrow vertical slit opening in an otherwise opaque window shade.

Newton (1642–1727) devised a variety of experiments in the study of light, and by the Newton's rings and related experiments he discovered clear evidence of light's periodicity. In explaining his results, Newton found it necessary to associate something periodic—he called it "fits"—with light rays, which until then had been regarded as uniform. He determined the interrelationships among length of period, color of light, and refractive index, and also found the law of the radii of the bright and dark rings in a single color of light. Newton was a thorough experimentalist, whose experiments were much more refined than those of his predecessors. However, he made it perfectly clear that he was not interested in idle hypotheses. At times he offered analogies, like the waves created in a pond by a falling stone, but he refrained from hypothesizing and emphasized the importance of accurately describing the findings. During Newton's lifetime, Huygens (in 1690) had proposed the wave hypothesis for light and had given his method for calculating the future shape of the wave front. But these considerations did not influence Newton's explanations of his own experiments.

YOUNG'S INTERFEROMETER

Almost a century later, two outsiders to the field of optics, Young (in 1807) and Fresnel (in 1868), provided a fresh look at the properties of light. Young, who was conversant with experiments in acoustics, was a physician by profession. His famous experiment, sketched in Fig. 1-1, with two pinholes illuminated by light from a single pinhole, was the cornerstone for the demonstration of the wave nature of light. The experiment was later redone by Fresnel, who was an engineer by profession. He used slit apertures instead of pinholes, making it more convenient to observe the fringes. These experiments, to be sure, were well thought out; they did not come about serendipitously. It is quite possible that Young, in his attempt to see interference for the first time, may have placed two light sources behind the two apertures and found no fringes. He may then have realized, due to his knowledge of acoustics, that there ought to be a single source that supplies light to the two apertures. It was Young who stated explicitly the path-difference condition for the occurrence of the maxima and minima of the light distribution in the plane of observation (see Fig. 1-1).

Subsequently, both Young and Fresnel enumerated the relevant conditions for observing the interference of light. They found, for example, that the interfering beams must be derived from the same source point or must have a common origin. That is, if two beams of light of the same color but from two different sources are brought together with near zero path difference, they may first find themselves in phase and augmenting each other to produce brightness, but an instant later they may be out of phase and neutralizing each other to produce darkness. Fresnel referred to this as the fast shifting of the interference fringes making them imperceptible to the eye. The change of phase of the two light beams can happen so fast that the eye is unable to recognize the changes in brightness. In effect, the eye sees a

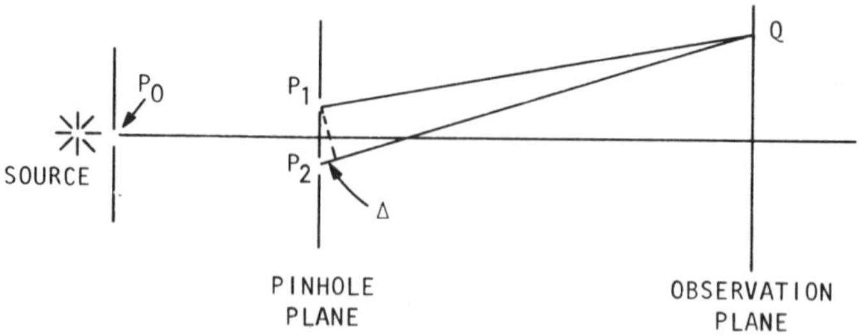

Fig. 1-1. Sketch of Young's pioneering experiment. Pinholes P_1 and P_2 are illuminated by light from a single pinhole P_0. Point of observation is Q. The optical path difference (OPD) is $\Delta = P_2Q - P_1Q$. There is a maximum at Q if OPD is an integral multiple of the wavelength.

certain average over numerous such phase changes. The eye is said to perform a time average of the instantaneous interference of the two beams. This same thing happens for light from two different points of the same source and also for light coming from a single point but at different instants of time. Such noninterfering beams are called *noncoherent*. (The word *incoherent* is also widely used.) In Young's original experiment the conditions of small path difference (especially important for white light interference) and smallness of the source (i.e., with limited spatial extent) were both taken into account to create two beams capable of interfering. The experiment was not easy to do; it took considerable skill and powers of critical observation on his part. The interfering beams were referred to as *coherent*. The terms coherent and noncoherent come about because of the time averaging of the observation process.

From Grimaldi's experiments with red and blue glass, and also from Newton's experiments, it was known that the red fringes are more widely spaced than the blue fringes. Young deduced from his experiments that red light has a wavelength λ of 5.176×10^{-7} m—a very small value indeed. Young also gave a wave theoretic explanation of all the findings of Newton's experiments related to the periodic nature of light.

The reason for our journey through this part of history is to indicate that the conditions of interference or coherence were far less obvious than the wave nature of light. It is inappropriate to blame the difficulties of performing these experiments on merely the smallness of the wavelength of light; the concept of coherence is fundamental. This point should be amply clear nowadays with the advent of the laser-type light sources. Once a coherent light source is obtained, the interference fringes may be seen almost everywhere. They are then truly ubiquitous.

MICHELSON'S TWO-BEAM INTERFEROMETER

In the Newton's ring experiment, Newton observed that in white light only a few colored rings are seen, whereas in light of a single color the light and dark rings may be seen across the entire surface of the lens. He concluded that the rings in white light are composed of many single-color rings of unequal radius. It was Fizeau (in 1849) who made the first attempt, with some success, to go beyond the small-path-difference condition to observe the disappearance (loss of visibility) of interference fringes for a single color of light. Later, Michelson (in 1882) with his two-beam interferometer, sketched in Fig. 1-2, was actually able to reach the limits of visibility by introducing sufficiently long path differences for a single color of light. In this way he studied several light sources. He attributed the loss of visibility

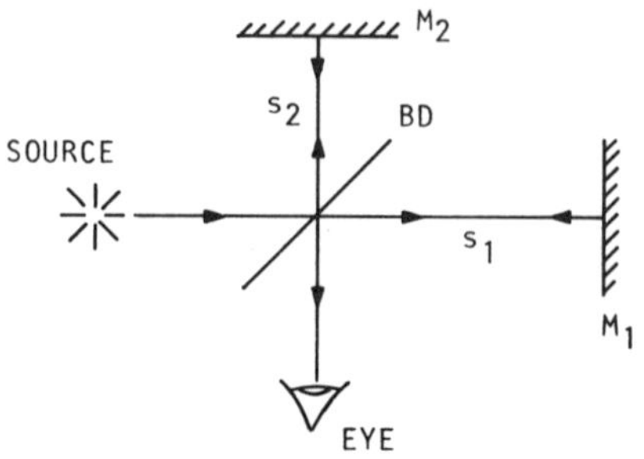

Fig. 1-2. Sketch of Michelson's two-beam interferometer. M_1 and M_2 are two plane mirrors; BD is the beam divider. The observer's eye receives the recombined beams.

to the characteristic structure in terms of the neighboring wavelength components contained in a "single-color" source.

Now, it was Fresnel who realized and explicitly stated that the periodicity of light is not perfectly regular (e.g., like the regular vibrations from a tuning fork) but rather is subject to frequent disturbances. Light from a given source thus contains sudden random phase jumps that lead to loss of interference. To see this, we make use of the so-called "wave train" model of the light generated by the source. In this model we suppose that each point of the source emits light in the form of wave trains of finite length. The resulting light from the source is thus supposed to be composed of the sum total of numerous wave trains, emitted at several different times and proceeding in various directions from each point of the source. Consider this light to be incident on the beam divider of the two-beam interferometer shown in Fig. 1-2. Each wave train suffers a division of amplitude at the beam divider, and the separate parts travel the paths of lengths s_1 and s_2. After their return from the respective mirrors M_1 and M_2, they are recombined at the same beam divider to allow for possible interference.

Now suppose that the lengths s_1 and s_2 are very nearly equal, that is, $s_1 \simeq s_2$. In this case, the superposed beams are at every instant assured to be the two parts of the same wave train, yielding nearly perfect interference. Next, suppose the two paths are of sufficiently different lengths that at any instant the two superposed beams are ones that have come from the same source point but have been emitted at entirely different times. The superposed beams thus do not belong to the same wave train. Because the emission times are random, the phase difference of the superposed beams is sufficiently uncertain so that no interference is perceptible to the eye. Finally, suppose that we begin again with nearly equal path lengths but that we gradually change the length of one or both paths so that the path difference, $s_1 - s_2$, is gradually increased from its zero value. For such a nonzero path difference the overlap of the wave train with itself will be partial. The portion of the wave train that did not overlap with its sister wave train might be expected to overlap (partly) with a wave train emitted by the source point at a slightly later instant. The overlap with the sister wave train will produce good interference, but the overlap with the later train will contribute poorly to interference, since the phases of the two trains will be randomly related. Thus, due to the time-averaging nature of our observation, the interference will be less than perfect whenever the two beams do not have a constant phase difference. This may happen when the path difference is great, as demonstrated with the Michelson two-beam interferometer. It can also happen when the path difference is small, as in Young's experiment, if the source is of sufficiently large size. In that case, the beams coming from the two pinholes are made up of contributions from several independent radiators, and the phase difference no longer remains constant during observation. Between the limits of coherent and noncoherent, the beams that produce less than perfect interference are called *partially coherent*.

OUTLOOK

This book is concerned with the study of partial coherence and its effect on the outcome of optical experiments. Conditions of coherence and noncoherence are treated as limiting cases of the state of partial coherence. To begin our work we need a convenient language to describe light. That is the subject matter of the next chapter.

Several authors over many years have worked on the formulation of coherence theory. In the chapters to follow, we refer frequently to the works of van Cittert, Zernike, Dumontet, Blanc-Lapierre, Hopkins, and Wolf. We shall follow the modern formulation due to Wolf and also discuss Hopkins' formulation, which is an important special case of the general formulation.

REFERENCE

I highly recommend the following delightful little book by Ernst Mach (1838–1916), from which the historical material for this chapter came. Unfortunately, the book is now out of print and available only in libraries.

Mach, Ernst (1926). *The Principles of Physical Optics: An Historical and Philosphical Treatment* (English translation by J. S. Anderson and A. F. A. Young, from earlier German edition), Dover, New York, 324 pp.

2

Field Representation

In optical experiments, we wish to study the passage of light through optical instruments and then perform a measurement of the outcome. Now, it is an experimental fact that detectors respond to the *electric field* **E** [V/m]. There are detectors that reach a steady state for a constant incident beam power with units of watts [W], and there are those like the photographic film that integrate the incident power over a certain time. For a constant beam power the darkening of the photographic film depends on the product of power and exposure time, measured in units of joules [J].

Since detectors inherently take the time average, the quantity of importance in a beam of light is the *average radiant power* $\mathbf{\Phi}$ [W]. Furthermore, light beams have a finite cross-sectional area, so it is meaningful to talk about the average power in the beam per unit area of its cross section measured in square meters [m^2] or square centimeters [cm^2]. In radiometry, which deals with the detection of light, this sort of measurement is called the *irradiance* E [W m^{-2}]; it is the average power density. (We will postpone the study of radiometry until Chapter 7). Each component of the field **E** contributes to the total irradiance, which may be given by the expression

$$\mathsf{E} = \frac{1}{2}\left(\frac{\varepsilon}{\mu}\right)^{1/2} \langle \mathbf{E} \cdot \mathbf{E} \rangle, \tag{2-1}$$

where ε and μ are the (dielectric) permittivity [F/m] and (magnetic) permeability [H/m], respectively. The angular brackets $\langle \quad \rangle$ symbolize the operation of time average.

As is often the case, the vectorial nature of the field may be ignored because the optical instruments (nonpolarizing) do not distinguish among the components. In addition, the detection process gives the total irradiance. Under these circumstances, light beams may be described by the artifice of a single scalar function ψ. In a situation where only one component of the field is present and the others are zero, the scalar may be identified with that component. In general, however, there is no simple relationship available to "derive" a scalar quantity from the components of the vector field; it is only a convenient association found useful in practice. For the most part we shall work with the scalar function ψ and assume that it obeys the wave equation the same way the vector field does. For this scalar ψ, the irradiance may be defined by

$$\mathsf{E} = \mathsf{C}\langle \psi^2 \rangle, \tag{2-2}$$

where C is a suitable constant so that the units of E are watts per square meter [W m^{-2}].

The international agreement on the radiometric symbols and nomenclature is relatively recent; see, for example, the discussion by MacAdam (1967). Throughout the optics literature, however, a quantity has been used called "intensity," generally with the symbol I. It is defined by

$$I = \langle \mathbf{E} \cdot \mathbf{E} \rangle \quad \text{or} \quad I = \langle \psi^2 \rangle \tag{2-3}$$

for the vector or scalar fields, respectively. It has no place in the radiometric scheme but, nevertheless, it is convenient since in theoretical work one frequently uses quadratic time averages of field quantities. We shall use it but call it *optical intensity* so as not to confuse it with the radiometric term (radiant) intensity I with units of watts per steradian [W sr^{-1}]. When a calculation is made with reference to a specific measurement we shall identify the appropriate radiometric quantity such as the irradiance E.

COMPLEX REPRESENTATION

An entirely physical entity like light can be described completely by using only real-valued functions. That is, the $\mathbf{E}$ field and the scalar ψ are real functions of the space coordinates x, y, and z and the time t. However, in dealing with linear systems it is much more convenient to use complex-valued functions instead. We shall regard the function of interest, say ψ, as a function of time, $\psi(t)$, in order to find a complex representation that is convenient to work with and with which to calculate time averages. As with the introduction of a scalar, the complex representation is also a convenient artifice. In order to appreciate the usefulness of this idea, we shall first deal with a simple example of a monochromatic plane wave and then generalize the procedure for more general solutions of the wave equation.

Consider a cosinusoidal monochromatic plane wave propagating in the $+z$ direction as the time t advances:

$$\psi(t) = A\cos(2\pi\nu t - kz + \phi). \tag{2-4}$$

This equation describes a plane wave of amplitude A, temporal linear frequency ν, and propagation constant $k = 2\pi/\lambda$, where λ is the wavelength. The expression in Eq. (2-4) is a solution of the wave equation provided the condition $\nu\lambda = c$ is fulfilled, where c is the speed of propagation of the wave.

When such a wave passes through an optical instrument—or in the simplest case through a glass plate—it will experience a change in its amplitude, its phase, or both. Symbolically the instrument may be denoted by a black box L, as in Fig. 2-1, and it may be thought of as operating on the incoming wave ψ_{in} to produce the outgoing wave ψ_{out}:

$$L\psi_{in} = \psi_{out}. \tag{2-5}$$

If the operation of L on ψ_{in} is interpreted as a simple product, the change in the amplitude of the incoming wave can be brought about by a suitable parameter, say a, used as a multiplicative factor. Its value may be smaller than, equal to, or greater than unity in order to account for absorption, no absorption, or even amplification, respectively.

A simple product relationship cannot produce a phase change, however, In this case, as illustrated in Fig. 2-2, the instrument seems to operate on two quantities at the same time in two different ways and somehow combines the result to produce the phase-changed output. In expressing this

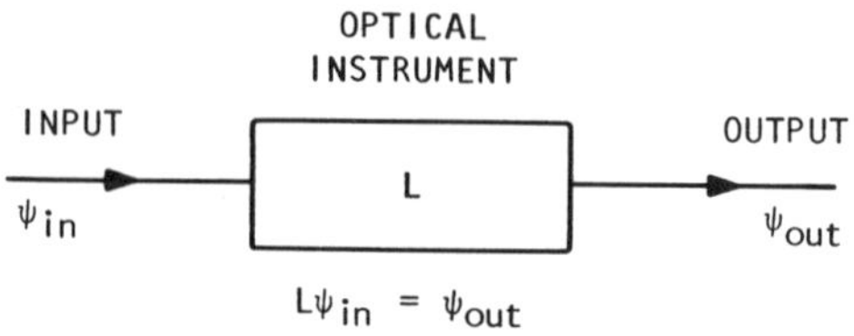

Fig. 2-1. Block diagram of an optical instrument L operating on the input ψ_{in} to produce the output ψ_{out}, Eq. (2-5).

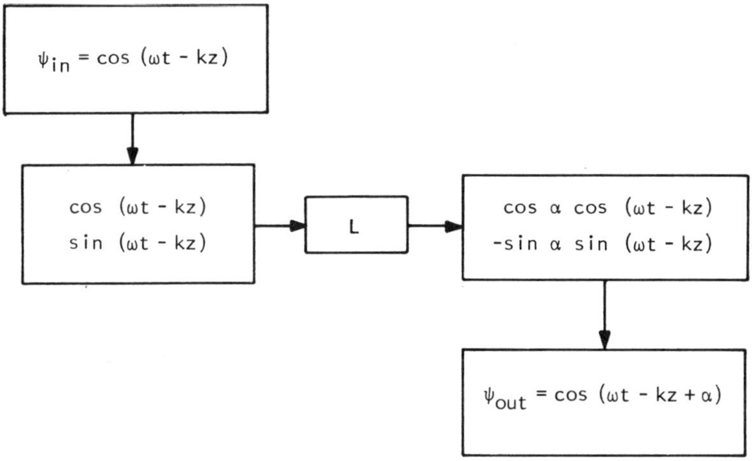

Fig. 2-2. Illustrative example of the inadequacy of the product relationship (see Fig. 2-1) to produce a phase change with only real functions.

mathematically, however, it is convenient to preserve the product relationship of Eq. (2-5). For this reason, we seek a function that contains a combination of cosines and sines and has the property that its multiple products result in the sum of its arguments. The complex exponential is just such a function; it allows us to bring about an amplitude and phase change in the form $L\psi_{in} = \psi_{out}$:

$$[a\exp(-i\alpha)]\left\{\frac{A}{\sqrt{2}}\exp[-i(\omega t - kz + \phi)]\right\} = \frac{aA}{\sqrt{2}}\exp[-i(\omega t - kz + \phi + \alpha)], \tag{2-6}$$

where ω is the circular frequency ($\omega = 2\pi\nu$). The instrument operator is specified by two real parameters, a and α. We agree to use the complex exponential to represent the field $\psi(t)$, with the stipulation that *only the real part has physical significance.* We complete the prescription of associating a complex quantity with the real physical field by noting that its imaginary part is created by shifting its real part in phase by $+\pi/2$. A shift of $-\pi/2$ would be just as good, but we will adopt the above convention. The numerical factor $1/\sqrt{2}$ is useful in the calculation of the time-average irradiance.

In practice, the time interval T for averaging is taken to be very large compared to any of the time periods involved in the process: for the case in Eq. (2-4) we ask for $T \gg 1/\nu$. Even without any mathematics, it is obvious that $\langle \psi^2 \rangle = \frac{1}{2}$; that is, the irradiance is

$$\mathsf{E} = \frac{\mathsf{C}A^2}{2}. \tag{2-7}$$

If the complex representation is used, this same answer is obtained if we agree to use

$$\mathsf{E} = \mathsf{C}\langle \psi\psi^* \rangle = \frac{\mathsf{C}A^2}{2}, \tag{2-8}$$

where ψ^* is the complex conjugate of ψ. If, in addition to the irradiance, we want the output field, it is proportional to the real part of the output from Eq. (2-5).

This procedure for working with the complex representation will enable us to formulate the representation of more general fields. It is important to note that the above prescription and the one below can be *uniquely* associated with the real physical field. Furthermore, due to the product of ψ with ψ^*, the fast-oscillating parts of the field disappear and only the time average of the relatively slowly varying envelope function enters the irradiance calculation.

ANALYTIC SIGNAL

Gabor (1946) introduced the analytic signal representation for general wave fields. It is a natural generalization of the complex representation of the previous section.

Let $V^{(r)}(t)$ be the real physical field. Assume that its Fourier transform exists—

$$V^{(r)}(t) = \int_{-\infty}^{\infty} \hat{V}^{(r)}(\nu) \exp(-i2\pi\nu t)\, d\nu. \tag{2-9}$$

Since $V^{(r)}(t)$ is real, the transform is complex symmetric,

$$\hat{V}^{(r)*}(-\nu) = \hat{V}^{(r)}(\nu),$$

which means that the amplitude $A(\nu)$ is even and the phase $\phi(\nu)$ is an odd function of ν, where $\hat{V}^{(r)}(\nu) = A(\nu)\exp[+i\phi(\nu)]$. Because of these restrictions, Eq. (2-9) becomes

$$V^{(r)}(t) = 2\int_0^{\infty} A(\nu)\cos[2\pi\nu t - \phi(\nu)]\,d\nu. \tag{2-10}$$

Thus all the information about $V^{(r)}(t)$ is contained in the positive-frequency part of the spectrum. Therefore, it should be possible to construct a complex representation that has a temporal frequency spectrum on the positive-frequency axis and that is identically zero for negative frequencies.

Let $V^{(i)}(t)$ be a function obtained from $V^{(r)}(t)$ by shifting each frequency component by $+\pi/2$,

$$V^{(i)}(t) = -2\int_0^{\infty} A(\nu)\sin[2\pi\nu t - \phi(\nu)]\,d\nu. \tag{2-11}$$

To a real physical field $V^{(r)}(t)$, we associate a complex function $V(t)$ in the following way:

$$V(t) = \frac{1}{\sqrt{2}}\left[V^{(r)}(t) + iV^{(i)}(t)\right]. \tag{2-12}$$

This is Gabor's analytic signal representation. By construction, its decomposition in frequencies is

$$V(t) = \sqrt{2}\int_0^{\infty} A(\nu)\exp\{-i[2\pi\nu t - \phi(\nu)]\}\,d\nu, \tag{2-13}$$

which implies that the Fourier transform $\hat{V}(\nu)$ of $V(t)$ is

$$\hat{V}(\nu) = \begin{cases} \sqrt{2}\,A(\nu)\exp[+i\phi(\nu)] = \sqrt{2}\,\hat{V}^{(r)}(\nu), & \nu \geq 0 \\ 0, \quad \nu < 0 \end{cases} \tag{2-14}$$

More insight may be gained by taking an alternative route. We seek a complex function $V(t)$ defined in terms of the real field $V^{(r)}(t)$ such that its Fourier transform $\hat{V}(\nu)$ is one-sided:

$$\hat{V}(\nu) = \sqrt{2}\,\hat{V}^{(r)}(\nu)\,\hat{s}(\nu), \tag{2-15}$$

where $\hat{s}(\nu)$ is a step function defined by

$$\hat{s}(\nu) = \begin{cases} +1, & \nu > 0 \\ 0, & \nu < 0. \end{cases} \tag{2-16}$$

Corresponding to the product relationship in the frequency domain, the convolution in the time domain is

$$V(t) = \sqrt{2}\int_{-\infty}^{\infty} V^{(r)}(t')\, s(t-t')\, dt'. \tag{2-17}$$

The knowledge of the Fourier transform $s(t)$ of the step $\hat{s}(\nu)$ (see Problem 2-3) leads us to

$$V(t) = \frac{1}{\sqrt{2}}\left[V^{(r)}(t) + iV^{(i)}(t)\right],$$

where $V^{(i)}(t)$ is now prescribed as the Hilbert transform of $V^{(r)}(t)$; that is,

$$V^{(i)}(t) = \mathcal{H}\left[V^{(r)}(t)\right] = +\frac{1}{\pi}⨍_{-\infty}^{\infty} \frac{V^{(r)}(t')}{t'-t}\, dt'. \tag{2-18}$$

The bar on the integral indicates that the Cauchy principal value is to be taken at the singularity $t' = t$ of the integrand. There are several ways of obtaining the principal value [see Beran and Parrent (1964) and Roman (1965)]. They are all equivalent to taking the limit

$$⨍_{-\infty}^{\infty} f(t')\, dt' = \lim_{\varepsilon\to 0}\left[\int_{-\infty}^{t-\varepsilon} f(t')\, dt' + \int_{t+\varepsilon}^{+\infty} f(t')\, dt'\right], \tag{2-19}$$

where $f(t)$ is the function with a singularity at $t' = t$. For a table of Hilbert transforms see Alavi-Sereshki and Prabhakar (1972).

It is not immediately obvious that the specification of $V^{(i)}(t)$ as a Hilbert transform of $V^{(r)}(t)$ is equivalent to the previous method of Eq. (2-11). The equivalence is easily established, however, by observing that the Hilbert transform of the cosine function is related to the sine as given by

$$\begin{aligned}\mathcal{H}[\cos 2\pi\nu t] &= +\frac{1}{\pi}⨍_{-\infty}^{\infty} \frac{\cos 2\pi\nu t'}{t'-t}\, dt' \\ &= \begin{cases} +\sin 2\pi\nu t, & \nu < 0 \\ -\sin 2\pi\nu t, & \nu > 0. \end{cases}\end{aligned} \tag{2-20}$$

The details of this work are delegated to Problem 2-4. The Hilbert transform of the cosine of Eq. (2-20) shows that the complex representation of the previous section is also an analytic signal.

The nature of the spectrum $\hat{V}(\nu)$ of the analytic signal $V(t)$ may be understood by calculating the Fourier transform of the imaginary part,

$\hat{V}^{(i)}(t)$, of Eq. (2-18). The left-hand side yields $\hat{V}^{(i)}(\nu)$. For the right-hand side we make the simplifying assumption that the Fourier transform operation may be interchanged with the Hilbert, principal-value integral. Whether the integral is on t' or t, the singularity at $t' = t$ is to be handled according to the Cauchy prescription. In this case, we need the Hilbert transform of the exponential given by

$$\mathcal{H}[\exp(+i2\pi\nu t')] = \begin{cases} -i\exp(+i2\pi\nu t'), & \nu < 0 \\ +i\exp(+i2\pi\nu t'), & \nu > 0. \end{cases} \tag{2-21}$$

The subsequent Fourier integral on t' then leads us to the result

$$\hat{V}^{(i)}(\nu) = \begin{cases} +i\hat{V}^{(r)}(\nu), & \nu < 0 \\ -i\hat{V}^{(r)}(\nu), & \nu > 0 \end{cases}$$

or, in more compact form,

$$i\hat{V}^{(i)}(\nu) = \operatorname{sgn}(\nu)\,\hat{V}^{(r)}(\nu). \tag{2-22}$$

The sgn function gives the sign of the variable ν:

$$\operatorname{sgn}(\nu) = \begin{cases} -1, & \nu < 0 \\ +1, & \nu > 0 \end{cases} \tag{2-23}$$

Now the spectrum $\hat{V}(\nu)$ is made up of two parts,

$$\hat{V}(\nu) = \frac{1}{\sqrt{2}}\left[\hat{V}^{(r)}(\nu) + i\hat{V}^{(i)}(\nu)\right], \tag{2-24}$$

which, due to Eq. (2-22), become

$$\hat{V}(\nu) = \frac{1}{\sqrt{2}}\left[\hat{V}^{(r)}(\nu) + \operatorname{sgn}(\nu)\hat{V}^{(r)}(\nu)\right]. \tag{2-25}$$

It is not surprising that this leads us to the expression of $\hat{V}(\nu)$ given in Eqs. (2-14) and (2-15). The point to be made is that the part of the spectrum contributed by $V^{(i)}(t)$ cancels identically the part due to $V^{(r)}(t)$, for $\nu < 0$. This is true with respect to both the real and imaginary parts of $\hat{V}^{(r)}(\nu)$,

$$\hat{V}^{(r)}(\nu) = A(\nu)\cos\phi(\nu) + iA(\nu)\sin\phi(\nu). \tag{2-26}$$

We shall now consider some examples.

Example 2-1

Consider the real physical field

$$V^{(r)}(t) = \frac{1}{(2\pi\sigma^2)^{1/2}} \exp\left[-\frac{(t-t_0)^2}{2\sigma^2}\right] \cos 2\pi\nu_0 t. \tag{2-27}$$

This function is plotted in Fig. 2-3. That the real and imaginary parts of the spectrum $\hat{V}^{(r)}(\nu)$ are even and odd, respectively, is seen in Figs. 2-4*a* and 2-5*a*. The nature of the spectrum—$i\hat{V}^{(i)}(\nu)$—as expressed in Eq. (2-22) is exemplified in Figs. 2-4*b* and 2-5*b*. For $\nu < 0$, $i\hat{V}^{(i)}(\nu)$ is equal to but opposite in sign from $\hat{V}^{(r)}(\nu)$. For this reason the spectrum $\hat{V}(\nu)$, which is the sum as indicated in Eq. (2-24), is zero for $\nu < 0$. The real and imaginary parts of the spectrum $\hat{V}(\nu)$ of the analytic signal are plotted in Figs. 2-4*c* and 2-5*c*, respectively.

Although it is not evident from Figs. 2-4 and 2-5, the functions $A(\nu)\cos\phi(\nu)$ and $A(\nu)\sin\phi(\nu)$ have long tails that extend over the entire frequency range, $-\infty \leq \nu \leq +\infty$. This fact makes for a rather complicated

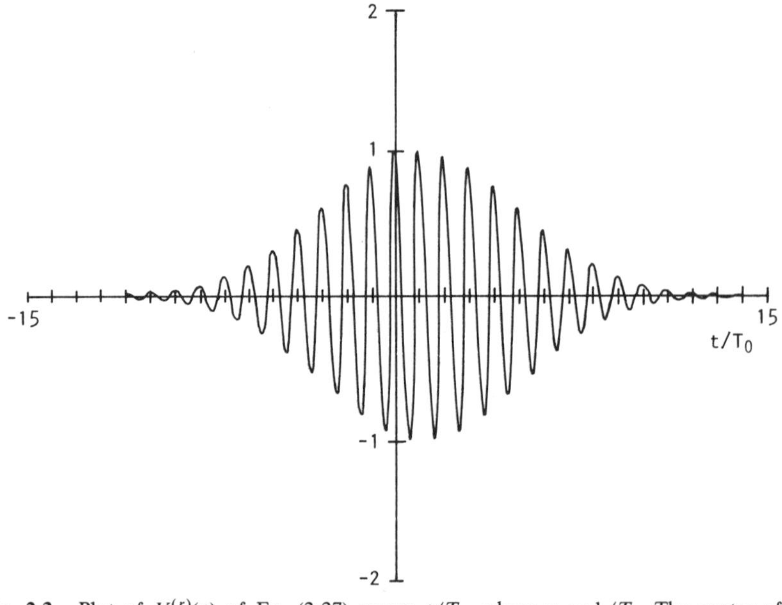

Fig. 2-3. Plot of $V^{(r)}(t)$ of Eq. (2-27) versus t/T_0, where $\nu_0 = 1/T_0$. The center of the Gaussian envelope is shifted by one period, $t_0 = T_0$, and its halfwidth is ten periods, $(2\pi\sigma^2)^{1/2} = 10T_0$.

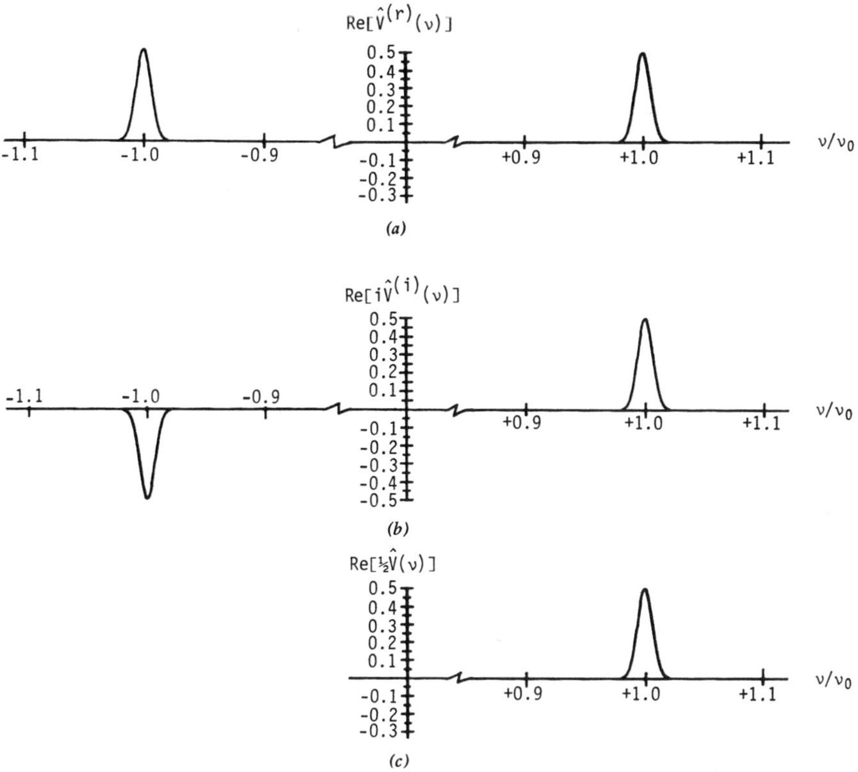

Fig. 2-4. The Re[$\hat{V}^{(r)}(\nu)$] $= A(\nu)\cos\phi(\nu)$ is an even function of ν. Re[$i\hat{V}^{(i)}(\nu)$] $=$ $+$Re[$\hat{V}^{(r)}(\nu)$] for $\nu > 0$, and Re[$i\hat{V}^{(i)}(\nu)$] $= -$Re[$\hat{V}^{(r)}(\nu)$] for $\nu < 0$. The Re[$\hat{V}(\nu)$] is the sum of the contributions in graphs *a* and *b*; hence it is zero for $\nu < 0$.

expression of the analytic signal. The exact expression of $V(t)$ is displayed in Problem 2-5. The mathematical complexity dwarfs the claim of convenience of the analytic signal. In practice, however, an approximate form may be usefully employed. From Figs. 2-4 and 2-5 we observe that the significant portions of the real and imaginary parts of $\hat{V}^{(r)}(\nu)$ are concentrated in the neighborhood of the carrier frequency, $\nu = \pm\nu_0$. The one-sidedness (existing over the half-range $\nu \geq 0$) of $\hat{V}(\nu)$ can be satisfied to a good approximation by noting that $A(\nu)\exp[i\phi(\nu)]$ is for all intents and purposes significant in the neighborhood of $\pm\nu_0$ and negligible for $|\nu \pm \nu_0| > (2\pi\sigma^2)^{-1/2}$. In this way, the expression for the analytic signal may be approximated by

$$V(t) \simeq \frac{1}{2(\pi\sigma^2)^{1/2}} \exp\left[-\frac{(t-t_0)^2}{2\sigma^2}\right] \exp(-i2\pi\nu_0 t). \qquad (2\text{-}28)$$

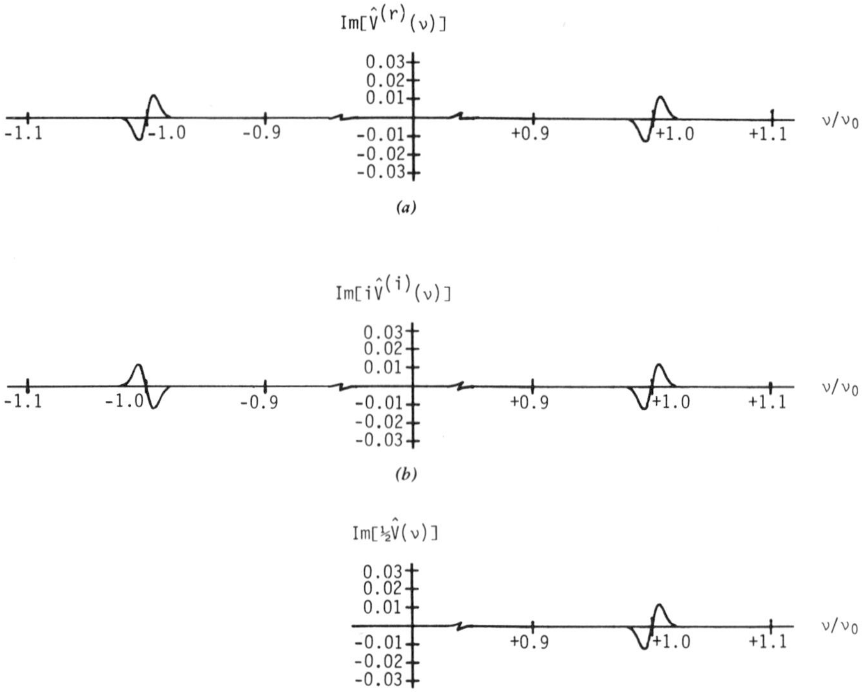

(c)

Fig. 2-5. The $\mathrm{Im}[\hat{V}^{(r)}(\nu)] = A(\nu)\sin\phi(\nu)$ is an odd function of ν. $\mathrm{Im}[i\hat{V}^{(i)}(\nu)] = \mathrm{Im}[\hat{V}^{(r)}(\nu)]$ for $\nu > 0$, and $\mathrm{Im}[i\hat{V}^{(i)}(\nu)] = -\mathrm{Im}[\hat{V}^{(r)}(\nu)]$ for $\nu < 0$. The $\mathrm{Im}[\hat{V}(\nu)]$ is the sum of the contributions in graphs *a* and *b*; hence it is zero for $\nu < 0$.

The approximation gets better as the half-width at the value $e^{-\pi}$ of the envelope becomes much larger than the period T_0 of the carrier, $(2\pi\sigma^2)^{1/2} \gg T_0$.

Example 2-2

As another example, we consider the real field given by

$$V^{(r)}(t) = \frac{1}{\pi\sigma}\mathrm{Sinc}\left(\frac{t - t_0}{\sigma}\right)\cos 2\pi\nu_0 t. \tag{2-29}$$

The band-limited envelope (Sinc) function,† coupled with the discussion of the previous example, indicates that the analytic signal is

$$V(t) = \frac{1}{\sqrt{2}\,\pi\sigma}\,\mathrm{Sinc}\left(\frac{t - t_0}{\sigma}\right)\exp(-i2\pi\nu_0 t), \tag{2-30}$$

† For the definition of Sinc see the List of Symbols, Notation, and Abbreviations.

where $\nu_0 > (2\pi\sigma)^{-1}$. This can be established quite simply as asked for in Problem 2-6.

INSTRUMENT FUNCTION

The analytic signal representation is a natural generalization of the simple complex exponential representation. It permits us to deal with the polychromatic nature of light. In general, optical instruments are frequency dependent; dispersion of light in a glass plate is such an example. We shall consider a linear description of the interaction of an instrument with the incoming field. Each Fourier frequency of the incoming field experiences a different amplitude and phase change, described by the following equation (see Problem 2-7):

$$\hat{L}(\nu)\hat{V}_{\text{in}}(\nu) = \hat{V}_{\text{out}}(\nu), \tag{2-31}$$

where $\hat{L}(\nu)$ is the temporal Fourier transform of the time domain description of the instrument function, $L(t)$. The two parameters a and α that pertained to the function in Eq. (2-6) are now functions of the frequency ν. In the time domain, Eq. (2-31) is replaced by

$$\int_{-\infty}^{\infty} L(t-t')\, V_{\text{in}}(t')\, dt' = V_{\text{out}}(t). \tag{2-32}$$

The linear nature of this relationship assures us that V_{out} is also an analytic signal uniquely associated with the outgoing real field.

IRRADIANCE

We introduced the concept of irradiance in the beginning of this chapter and used it in a qualitative manner on the road to the introduction of the analytic signal. This concept involves the measurement of average power. The time average in it is formally defined by

$$\langle VV^* \rangle = \lim_{T\to\infty} \frac{1}{2T} \int_{-T}^{T} V(t)\, V^*(t)\, dt, \tag{2-33}$$

where V is the complex analytic signal associated with the scalar field. For the real physical field, the integrand in the above definition contains the square of the real field. By this definition, the average power in the field is a constant, or time independent. When applied to fields in the form of isolated pulses that grow and decay, as in the examples of Eqs. (2-28) and (2-30), it leads to zero (see Problem 2-8).

At this stage, an application of the time-average operation will be instructive. To this end, we consider a periodic real physical field $V^{(r)}(t)$, with period T_0,

$$V^{(r)}(t) = A_0 + 2\sum_{\mu=1}^{\infty} A_\mu \cos\left(2\pi\frac{\mu}{T_0}t - \phi_\mu\right). \tag{2-34}$$

It exists over all times t, $-\infty \leq t \leq +\infty$, and its average power is nonzero, as will be verified in what is to follow.

The question of what is the associated analytic signal $V(t)$ for the case of periodic functions and the corresponding integral relationship between the real and imaginary parts is discussed in detail in Appendix 2.1. For the above case, it is found that

$$V(t) = A_0 \exp\left(+i\frac{\pi}{4}\right) + \sqrt{2}\sum_{\mu=1}^{\infty} A_\mu \exp(+i\phi_\mu)\exp\left(-i2\pi\frac{\mu}{T_0}t\right). \tag{2-35}$$

The Fourier expansion coefficients V_μ of $V(t)$ are identically zero for $\mu < 0$, which is analogous to the one-sided nature of the spectrum of the analytic signal.

Returning to the calculation of the time average, we find that

$$\langle V(t)\,V^*(t)\rangle = A_0^2 + 2\sum_{\mu=1}^{\infty}\sum_{\mu'=1}^{\infty} A_\mu A_{\mu'} \exp(i\phi_\mu - i\phi_{\mu'})$$

$$\times \lim_{T\to\infty} \operatorname{Sinc}\left[2\pi(\mu - \mu')\frac{T}{T_0}\right].$$

The indicated limit is equal to the Kronecker delta, $\delta_{\mu\mu'}$. Therefore, the irradiance E for our periodic field is

$$\mathsf{E} = \mathsf{C}\langle VV^*\rangle = \mathsf{C}\left(A_0^2 + 2\sum_{\mu=1}^{\infty} A_\mu^2\right). \tag{2-36}$$

It is equal to the sum of the irradiances of the individual Fourier components $V_\mu^{(r)}$ for all μ; that is, $-\infty \leq \mu \leq +\infty$. It is instructive to verify that the irradiance of this periodic field calculated by using the real field $V^{(r)}(t)$ is the same as that found in Eq. (2-36) (see Problem 2-9).

Since the process of light detection can only measure average power at optical frequencies, the operation of the time average is used frequently through the rest of this book.

ANALYTIC PROPERTY OF THE SIGNAL

Although we do not wish to dwell on the mathematical discussion, we mention a property of the analytic signal $V(t)$ in the complex plane

$$z = t + it',$$

where t is the real physical time and t' is a real quantity. The Fourier integral representation of $V(t)$ is

$$V(t) = \int_0^{\infty} \hat{V}(\nu) \exp(-i2\pi\nu t)\, d\nu.$$

In the complex plane this integral reads

$$\begin{aligned} V(z) &= V(t + it') \\ &= \int_0^{\infty} \hat{V}(\nu) \exp(-i2\pi\nu t) \exp(+2\pi\nu t')\, d\nu. \end{aligned} \tag{2-37}$$

Due to the one-sided nature of the spectrum, the integral is restricted to positive frequencies. For $\nu > 0$, it is evident that $V(z)$ will be an analytic function in the lower half of the complex plane, $t' < 0$. The field $V(t)$ itself is then the boundary value of $V(z)$ as t' approaches zero, $t' \to 0^-$, through its negative values. The name "analytic signal" of the complex representation derives from this property. For a further discussion on this point, the reader is referred to Gabor (1946) and Wolf (1962).

PROBLEMS

2-1. Invent a matrix representation that may be associated with a real, physical *scalar* field and used instead of the analytic signal introduced in this chapter. Find a corresponding representation of the optical instrument. You are allowed to use only *real* quantities for the matrix elements. Describe the interaction of the field and the optical instrument and show how the real field in the output may be recovered from the matrix representation. *Hint*: Do not confuse the Jones matrix calculus used for polarized light with the above problem. One may start with a two-component vector to represent the real scalar field. For example, for the field of Eq. (2-4), use the cosine and sine functions for the first and second components, respectively, of the (matrix) vector. This idea may be expanded and generalized to work through the problem.

2-2. Is the analytic signal representation a solution of the wave equation?

2-3. Show that the Fourier transform $s(t)$ of $\hat{s}(\nu)$ is

$$s(t) = \frac{1}{2}\left[\delta(t) - \frac{i}{\pi t}\right],$$

and verify that $V^{(i)}(t)$ is the Hilbert transform of $V^{(r)}(t)$.

2-4. This problem is concerned with the verification of (a) the equivalence of the Hilbert transform specification, $V^{(i)}(t) = \mathcal{H}[V^{(r)}(t)]$, of Eq. (2-18), with the alternative definition given by Eq. (2-11), namely,

$$V^{(i)}(t) = -2\int_0^\infty A(\nu)\sin[2\pi\nu t - \phi(\nu)]\,d\nu,$$

and (b) verification that the spectrum of $V(t)$ is given by Eq. (2-25), namely,

$$\hat{V}(\nu) = \frac{1}{\sqrt{2}}[1 + \mathrm{sgn}(\nu)]\hat{V}^{(r)}(\nu).$$

The mathematically inclined may also wish to verify that the Hilbert transform of the exponential is [see Eq. (2-21)]

$$\mathcal{H}[\exp(i2\pi\nu t')] = \begin{cases} -i\exp(+i2\pi\nu t'), & \nu < 0 \\ +i\exp(+i2\pi\nu t'), & \nu > 0. \end{cases}$$

In this problem, special attention should be given to the signs.

2-5. Let the real physical field $V^{(r)}(t)$ be given by

$$V^{(r)}(t) = \frac{1}{(2\pi\sigma^2)^{1/2}}\exp\left[-\frac{(t - t_0)^2}{2\sigma^2}\right]\cos 2\pi\nu_0 t.$$

Calculate the spectrum $\hat{V}^{(r)}(\nu)$ and show that its real and imaginary parts are given by

$$A(\nu)\cos\phi(\nu) = \tfrac{1}{2}\left\{\exp\left[-2\pi^2\sigma^2(\nu+\nu_0)^2\right]\cos 2\pi(\nu+\nu_0)t_0 + \exp\left[-2\pi^2\sigma^2(\nu-\nu_0)^2\right]\cos 2\pi(\nu-\nu_0)t_0\right\}$$

and

$$A(\nu)\sin\phi(\nu) = \tfrac{1}{2}\left\{\exp\left[-2\pi^2\sigma^2(\nu+\nu_0)^2\right]\sin 2\pi(\nu+\nu_0)t_0 + \exp\left[-2\pi^2\sigma^2(\nu-\nu_0)^2\right]\sin 2\pi(\nu-\nu_0)t_0\right\}.$$

In the calculation of the exact form of the analytic signal one encounters the complex Fourier transform of a Gaussian over the half-range. It is related to the "degenerate hypergeometric function $\Phi(\alpha, \beta; z)$" as defined by Gradshteyn and Ryzhik (1965). This same function is also referred to as the "confluent hypergeometric function" in the literature; see, for example, Abramowitz and Stegun (1964). Show that the exact expression for the analytic signal is

$$V(t) = \frac{1}{2\sigma\sqrt{\pi}}\left\{1 - i\left[\frac{t-t_0}{(2\pi\sigma^2)^{1/2}}\right]\Phi\left(\frac{1}{2}, \frac{3}{2}; \frac{(t-t_0)^2}{2\sigma^2}\right)\right\} \times \exp\left[-\frac{(t-t_0)^2}{2\sigma^2}\right]\cos 2\pi\nu_0 t.$$

The approximate form is given in the text, Eq. (2-28).

2-6. For a real field with a band-limited envelope function,

$$V^{(r)}(t) = \frac{1}{\pi\sigma}\operatorname{Sinc}\left(\frac{t-t_0}{\sigma}\right)\cos 2\pi\nu_0 t,$$

show that the analytic signal is

$$V(t) = \frac{1}{\sqrt{2}\,\pi\sigma}\operatorname{Sinc}\left(\frac{t-t_0}{\sigma}\right)\exp(-i2\pi\nu_0 t),$$

where $\nu_0 > (2\pi\sigma)^{-1}$.

2-7. Pedagogically, it is instructive to begin with a single-frequency field in the time domain. Since the product relationship belongs in the frequency domain, it is pertinent to ask: Why was it used in describing the passage of the time-varying field through an instrument as in Eq. (2-5)?

Hint: To answer this question, first establish that the product relationship, Eq. (2-31), and the convolution, Eq. (2-32), may be used interchangeably when a single (monochromatic) frequency ν is present.

2-8. Show by calculation that the average power in an isolated pulse is zero. For this purpose, construct your own pulse or choose one from Eq. (2-28) or (2-30) of the text.

2-9. Consider a periodic, real, physical field,

$$V^{(r)}(t) = A_0 + 2\sum_{\mu=1}^{\infty} A_\mu \cos\left(2\pi \frac{\mu}{T_0} t - \phi_\mu\right),$$

with period T_0. Obtain the associated analytic signal $V(t)$. Verify that the irradiance calculated by using this real representation is the same as that found by using the corresponding analytic signal and that it is the same as the irradiance associated with the imaginary part $V^{(i)}(t)$.

REFERENCES

Abramowitz, M., and I. A. Stegun (1964). *Handbook of Mathematical Functions*, National Bureau of Standards, Applied Mathematics Series, Vol. 55, Chap. 13, p. 504.

Alavi-Sereshki, M. M., and J. C. Prabhakar (1972). A tabulation of Hilbert transforms for electrical engineers, *IEEE Trans. Commun.* **Com-20**:1194–1198.

Beran, M. J., and G. B. Parrent, Jr. (1964). *Theory of Partial Coherence*, Prentice-Hall, Englewood Cliffs, NJ, 189 pp.

Gabor, D. (1946). Theory of communication. Part I. The analysis of information, *J. Inst. Elec. Eng. Lond.* **93** (3):429–441.

Gradshteyn, I. S., and I. M. Ryzhik (1965). *Table of Integrals, Series, and Products*, Academic Press, New York, p. 1058, art. 9.2.1.

MacAdam, D. L. (1967). Nomenclature and symbols for radiometry and photometry, *J. Opt. Soc. Am.* **57**:854.

Roman, P. (1965). *Advanced Quantum Theory*, Addison-Wesley, Reading, MA, Appendix 4, Sec. A4-3.

Wolf, E. (1962). Is a complete determination of the energy spectrum of light possible from measurements of the degree of coherence? *Proc. Phys. Soc.* **80** (6):1269–1272.

APPENDIX 2.1. ANALYTIC SIGNAL FOR PERIODIC FUNCTIONS

The theory in what is to follow is the discrete version of the analytic signal discussed in Chapter 2. Consider a real physical field $V^{(r)}(t)$, with period

T_0. The Fourier series representation for it is

$$V^{(r)}(t) = \sum_{\mu=-\infty}^{\infty} V_\mu^{(r)} \exp\left(-i2\pi \frac{\mu}{T_0} t\right), \tag{A2-1}$$

where the expansion coefficients are

$$V_\mu^{(r)} = \frac{1}{T_0} \int_{-T_0/2}^{T_0/2} V^{(r)}(t) \exp\left(+i2\pi \frac{\mu}{T_0} t\right) dt. \tag{A2-2}$$

Since $V^{(r)}(t)$ is real, $V_\mu^{(r)} = V_{-\mu}^{(r)*}$. Therefore, the field may be expressed in the form

$$V^{(r)}(t) = A_0 + 2 \sum_{\mu=1}^{\infty} A_\mu \cos\left(2\pi \frac{\mu}{T_0} t - \phi_\mu\right), \tag{A2-3}$$

where we have put $V_\mu^{(r)} = A_\mu \exp(+i\phi_\mu)$. The A_μ is real and positive, ϕ_μ is real, and we choose $\phi_0 = 0$. One may allow for $\phi_0 = 0$ or π, but that does not affect the discussion presented here.

We define a function $V^{(i)}(t)$ by shifting the argument of the cosine by $+\pi/2$,

$$V^{(i)}(t) = A_0 - 2 \sum_{\mu=1}^{\infty} A_\mu \sin\left(2\pi \frac{\mu}{T_0} t - \phi_\mu\right). \tag{A2-4}$$

Now the analytic signal $V(t)$ is defined by following Eq. (2-12):

$$\begin{aligned} V(t) &= \frac{1}{\sqrt{2}} \left[V^{(r)}(t) + iV^{(i)}(t)\right] \\ &= A_0 \exp\left(+i\frac{\pi}{4}\right) + \sqrt{2} \sum_{\mu=1}^{\infty} A_\mu \exp(+i\phi_\mu) \exp\left(-i2\pi \frac{\mu}{T_0} t\right). \end{aligned} \tag{A2-5}$$

Observe that the expansion coefficients V_μ of the Fourier series of the analytic signal are

$$V_\mu = \begin{cases} \sqrt{2}\, A_\mu \exp(+i\phi_\mu) = \sqrt{2}\, V_\mu^{(r)}, & \mu \geq 1 \\ A_0 \exp(+i\pi/4), & \mu = 0 \\ 0, & \mu < 0. \end{cases} \tag{A2-6}$$

The one-sided Fourier series of Eq. (A2-5) was used in the example of Chapter 2.

For the sake of completeness, we also discuss the integral relationship between the real and imaginary parts,

$$V^{(i)}(t) = \frac{1}{T_0}\int_{-T_0/2}^{T_0/2} f(t, t')\, V^{(r)}(t')\, dt'. \tag{A2-7}$$

We wish to determine $f(t, t')$. It can be shown that f is periodic in both t and t' with period T_0 and that f is real. We assume a double Fourier series expansion on the full range given by

$$f(t, t') = \sum_{\mu_1=-\infty}^{\infty} \sum_{\mu_2=-\infty}^{\infty} f_{\mu_1, \mu_2} \exp\left(-i2\pi \frac{\mu_1 t - \mu_2 t'}{T_0}\right).$$

Using the one-sided series expansion of $V^{(r)}(t)$ and $V^{(i)}(t)$ and matching the coefficients on both sides of the integral relation shows that $f(t, t')$ is a function of only the time difference:

$$f(t - t') = 1 - 2\sum_{\mu=1}^{\infty} \sin\left[\frac{2\pi\mu(t - t')}{T_0}\right]. \tag{A2-8}$$

There exists an inverse relationship, namely,

$$V^{(r)}(t) = \frac{1}{T_0}\int_{-T_0/2}^{T_0/2} f^{-1}(t - t')\, V^{(i)}(t')\, dt', \tag{A2-9}$$

where

$$f^{-1}(t - t') = 1 + 2\sum_{\mu=1}^{\infty} \sin\left[\frac{2\pi\mu(t - t')}{T_0}\right]. \tag{A2-10}$$

The product of the function f and its inverse integrated over a period T_0 yields

$$\frac{1}{T_0}\int_{-T_0/2}^{T_0/2} f(t - t')\, f^{-1}(t' - t'')\, dt'$$

$$= \sum_{\mu=-\infty}^{\infty} \exp\left[\frac{-i2\pi\mu(t - t'')}{T_0}\right].$$

The right-hand side (RHS) is in fact the Fourier series of a Dirac comb function with period T_0, namely,

$$T_0 \sum_{n=-\infty}^{\infty} \delta(t - t'' - nT_0) = \sum_{\mu=-\infty}^{\infty} \exp\left[\frac{-i2\pi\mu(t - t'')}{T_0}\right].$$

The functions f and f^{-1} occur under an integral over a period, so if both t and t'' are limited to the interval $-T_0/2$ to $T_0/2$, then

$$\frac{1}{T_0}\int_{-T_0/2}^{T_0/2} f(t - t') f^{-1}(t' - t'')\, dt' = T_0\delta(t - t''). \qquad \text{(A2-11)}$$

Now although the series for $f(t - t')$ is divergent, it may be summed by using a limiting procedure,

$$f(t - t') = 1 - 2 \lim_{\lambda \to 0}\left[\frac{\sin\beta \exp(-\lambda)}{1 - 2\exp(-\lambda)\cos\beta + \exp(-2\lambda)}\right], \qquad \text{(A2-12)}$$

where we have put

$$\beta \equiv \frac{2\pi(t - t')}{T_0}.$$

To obtain the above relationship we used the formula

$$\begin{aligned}\sum_{\mu=0}^{\infty} \exp(i\beta\mu) &= \lim_{\lambda \to 0} \sum_{\mu=0}^{\infty} [\exp(-\lambda + i\beta)]^{\mu} \\ &= \lim_{\lambda \to 0} [1 - \exp(-\lambda + i\beta)]^{-1}.\end{aligned}$$

It is important that the limit be taken *after* the indicated integral of Eq. (A2-7) is performed. If the warning is ignored, one may deliberately put $\lambda = 0$ in Eq. (A2-12) and find that

$$f(t - t') = 1 - \cot\left[\frac{\pi(t - t')}{T_0}\right].$$

This is unacceptable, however, because for $t = t'$ the RHS is infinite whereas from Eq. (A2-8) and also from Eq. (A2-12), before taking the limit, we know that $f(0) = 1$.

Finally, we mention the eigenfunctions of the kernel $f(t - t')$. We start with the equation

$$\frac{1}{T_0}\int_{-T_0/2}^{T_0/2} f(t - t')\,\psi(t')\,dt' = \lambda\psi(t). \tag{A2-13}$$

If $\psi(t)$ is an eigenfunction with the eigenvalue λ, then obviously ψ^* is also an eigenfunction with λ^* as its eigenvalue. The eigenfunctions are

$$\psi_\mu(t) = \exp\left(-i2\pi\frac{\mu}{T_0}t\right), \tag{A2-14}$$

where the integer μ has the full range $-\infty \leq \mu \leq +\infty$ and

$$\lambda_\mu = \begin{cases} +i, & \mu < 0 \\ +1, & \mu = 0 \\ -i, & \mu > 0. \end{cases}$$

There is degeneracy for $\mu < 0$ and for $\mu > 0$. Thus, for example, for $\mu > 0$, a linear combination $\Sigma_{\mu=1}^{\infty} c_\mu \psi_\mu(t)$ is also an eigenfunction with the eigenvalue $\lambda = -i$, where c_μ are arbitrary (possibly complex) constants. In terms of its own eigenfunctions, the self-representation of the kernel is

$$f(t - t') = 1 + i\sum_{\mu=1}^{\infty}\left[\psi_\mu^*(t)\,\psi_\mu(t') - \psi_\mu(t)\,\psi_\mu^*(t')\right]. \tag{A2-15}$$

3

Mutual Coherence Function

It is not easy to demonstrate coherence with just any type of light source. This is because the phase of the two waves V_1 and V_2 arriving at the observation point must be identical. Two such waves are called coherent, and their sum is referred to as *coherent superposition*,

$$V = V_1 + V_2, \tag{3-1}$$

which leads to the optical intensity,

$$I = \langle VV^* \rangle = I_1 + I_2 + 2(I_1 I_2)^{1/2} \cos \phi_{12}. \tag{3-2}$$

Here ϕ_{12} is the constant phase difference between the two waves; the term $2(I_1 I_2)^{1/2} \cos \phi_{12}$ is called the *interference term*. With a laser source, the generated waves have a constant phase difference relative to each other, rendering easy observation of interference. (Waves from two independently operated lasers, however, do not exhibit this coherence.)

In contrast to laser sources, thermal sources almost completely lack coherence. Before the laser was invented, light sources used for experiments were thermal in nature. Examples of thermal sources are flames—or flames fed by an element like sodium—mercury vapor lamps, and the blackbody.

With thermal sources, the interference term changes often. During observation it is, in general, as many times negative as positive, so its average is zero or negligible compared to the sum of the optical intensities; therefore

$$I = I_1 + I_2. \tag{3-3}$$

This is *noncoherent superposition.*

To understand why thermal sources produce noncoherent superposition, it is necessary to understand the nature of thermal radiation. Thermal radiation is completely randomized. We can imagine it as being produced by a large collection of isotropic, independent, "elementary" radiators. These might be individual atoms or perhaps a collection of atoms—their particular makeup is unimportant for our present discussion. For each radiator, the time, duration, and temporal form of emission is highly irregular. The interval between emissions might change as often as 10^8 times per second. [This number comes from the "natural linewidth" for the transition of the outer electrons in an atom, which are responsible for radiation in the visible region. For example, for mercury vapor (Hg) lamps the width is on the order of 10^8 Hz. The corresponding time interval is 10^{-8} s—beyond which the phase is significantly disturbed. For blackbody sources the bandwidth is much larger, approaching the order of 10^{14} Hz.] The total light from such a source would be a messy mixture (linear superposition) of light from all the radiators. So at any point or on a plane of observation irradiated by such a source, the light amplitude A and the phase ϕ would be completely irregular or random. Even if the amplitude were regarded as nearly constant for a uniformly lit plane, the phase ϕ would be completely uncertain over the range 0 to 2π. Furthermore, its value might change just as often as the change in the emission times (on the order to 10^8 times per second).

Now, even with the fastest available detectors, the duration of observation is very long compared to this time of 10^{-8} s over which the phase remains unchanged. Thus, although the linearity of the wave equation assures linear superposition and hence interference *on an instantaneous basis*, yet on the time scale of the observation no interference is evident. (Even with such sources, interference *can* be demonstrated, but only with special experimental arrangements, as described in Chapter 1. With such arrangements, while the phase of the wave V_1 arriving at the observation point is changing irregularly and rapidly, the phase of a second wave V_2 arriving at that point is changing in a like manner.)

The two limiting cases described above (coherent superposition and noncoherent superposition) and the intermediate states of partial coherence mentioned in Chapter 1 may be brought into a single framework by introducing the *mutual coherence function*. We shall see how it occurs naturally in the study of interference.

YOUNG'S TWO-PINHOLE INTERFERENCE REVISITED

Imagine a thermal source feeding light to the two pinholes of the screen shown in Fig. 3-1. We wish to calculate the optical intensity at a point Q in the plane of observation.

Let $V(P_s, t)$ be the optical field wave amplitude at the pinhole P_s (where $s = 1, 2$) at time t due to the source Σ. The determination of the amplitude at Q in terms of that at the pinhole P_s involves an integral relationship. But as we shall show in Chapter 5, we may use a multiplicative free-space propagator $K(Q, P_s)$ to a good approximation. It may be interpreted as the response at Q due to a disturbance of unit wave amplitude and zero phase at P_s. Later (in Chapter 5) we discuss the existence of such a factor and show that when *retarded* time arguments are used (as is done in what is to follow), the factor is purely imaginary; it depends on geometry and is independent of time.

The instantaneous amplitude $V(Q, t)$ at Q and at time t is obtained by a linear superposition of the amplitude at the pinholes at earlier times

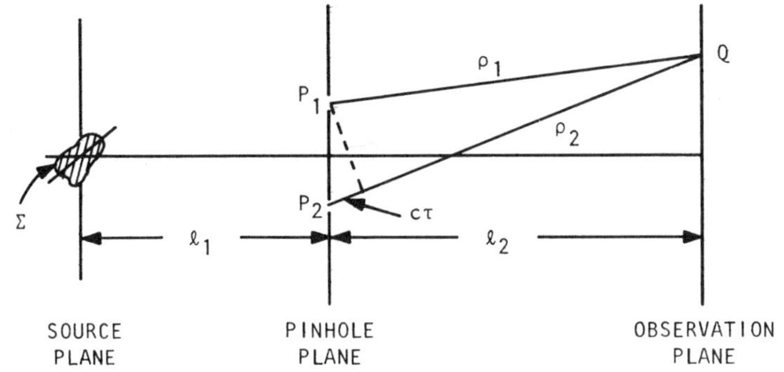

Fig. 3-1. Young's interference experiment. The length $c\tau = \rho_2 - \rho_1$ is the extra path required for light to travel from P_2 to the observation point Q compared to that from P_1 to Q. The plane of the paper contains the two pinholes and the normal to the pinhole plane. To a good approximation the position of the point Q may be labeled by the values of the time delay τ when it lies on the line of intersection of the paper and the observation plane.

$t - \rho_s/c$, for $s = 1$ and 2:

$$V(Q, t) = K(Q, P_1)\,V\!\left(P_1, t - \frac{\rho_1}{c}\right) + K(Q, P_2)\,V\!\left(P_2, t - \frac{\rho_2}{c}\right), \tag{3-4}$$

where ρ_s is the distance from P_s to Q.

The optical intensity $I(Q) = \langle VV^* \rangle$ at Q is given by four terms:

$$\begin{aligned} I(Q) = &|K(Q, P_1)|^2 \left\langle V\!\left(P_1, t - \frac{\rho_1}{c}\right) V^*\!\left(P_1, t - \frac{\rho_1}{c}\right)\right\rangle \\ &+|K(Q, P_2)|^2 \left\langle V\!\left(P_2, t - \frac{\rho_2}{c}\right) V^*\!\left(P_2, t - \frac{\rho_2}{c}\right)\right\rangle \\ &+K(Q, P_1)\,K^*(Q, P_2)\left\langle V\!\left(P_1, t - \frac{\rho_1}{c}\right) V^*\!\left(P_2, t - \frac{\rho_2}{c}\right)\right\rangle \\ &+K^*(Q, P_1)\,K(Q, P_2)\left\langle V^*\!\left(P_1, t - \frac{\rho_1}{c}\right) V\!\left(P_2, t - \frac{\rho_2}{c}\right)\right\rangle. \end{aligned} \tag{3-5}$$

By definition, the third angular bracket is

$$\begin{aligned} &\left\langle V\!\left(P_1, t - \frac{\rho_1}{c}\right) V^*\!\left(P_2, t - \frac{\rho_2}{c}\right)\right\rangle \\ &\qquad = \lim_{T\to\infty} \frac{1}{2T}\int_{-T}^{T} V\!\left(P_1, t - \frac{\rho_1}{c}\right) V^*\!\left(P_2, t - \frac{\rho_2}{c}\right) dt. \end{aligned}$$

A change of variable, $t' = t - \rho_2/c$, shows that this angular bracket actually depends on the time difference

$$\tau = \frac{\rho_2 - \rho_1}{c}$$

rather than the separate times ρ_1/c and ρ_2/c. It denotes the extra time required for light to reach Q from P_2 compared to P_1. Therefore, it is convenient to define a function $\Gamma(P_1, P_2, \tau)$ of two space points and the time delay by

$$\begin{aligned} \Gamma(P_1, P_2, \tau) &= \left\langle V\!\left(P_1, t - \frac{\rho_1}{c}\right) V^*\!\left(P_2, t - \frac{\rho_2}{c}\right)\right\rangle \\ &= \lim_{T\to\infty} \frac{1}{2T}\int_{-T}^{T} V(P_1, t' + \tau)\,V^*(P_2, t')\,dt'. \end{aligned} \tag{3-6}$$

It follows that the fourth angular bracket, which is the complex conjugate of the third bracket of Eq. (3-5), is

$$\Gamma(P_2, P_1, -\tau) = \left\langle V^*\left(P_1, t - \frac{\rho_1}{c}\right) V\left(P_2, t - \frac{\rho_2}{c}\right)\right\rangle$$
$$= \Gamma^*(P_1, P_2, \tau) \tag{3-7}$$

and that the first two are the zero ordinates of Γ:

$$\Gamma(P_s, P_s, 0) = \left\langle V\left(P_s, t - \frac{\rho_s}{c}\right) V^*\left(P_s, t - \frac{\rho_s}{c}\right)\right\rangle \tag{3-8}$$

for $s = 1$ and 2. If one of the pinholes is temporarily closed, the optical intensity at Q due to the other is

$$I_s(Q) = |K(Q, P_s)|^2\, \Gamma(P_s, P_s, 0) \tag{3-9}$$

for $s = 1$ or 2.

The notation can be simplified by using

$$\Gamma_{12}(\tau) \equiv \Gamma(P_1, P_2, \tau), \tag{3-10}$$

and the physical interpretation of the optical intensity at Q is facilitated by defining the normalized function

$$\gamma_{12}(\tau) = \frac{\Gamma_{12}(\tau)}{[\Gamma_{11}(0)\, \Gamma_{22}(0)]^{1/2}} \tag{3-11}$$

and writing it in the form

$$\gamma_{12}(\tau) \equiv |\gamma_{12}(\tau)| \exp[+i\phi_{12}(\tau)]. \tag{3-12}$$

The normalization is such that

$$0 \leq |\gamma_{12}(\tau)| \leq 1, \tag{3-13}$$

as may be established by Schwartz's inequality; see Problem 3-1.

Now, since the factor $K(Q, P_s)$ is purely imaginary, $K(Q, P_1)\, K^*(Q, P_2)$ is a real number. This fact, coupled with the definitions introduced above, permits us to express the optical intensity at Q of Eq. (3-5) in the form

$$I(Q) = I_1(Q) + I_2(Q) + 2[I_1(Q)I_2(Q)]^{1/2}\,|\gamma_{12}(\tau)|\cos\phi_{12}(\tau). \tag{3-14}$$

Except for the extra factor $|\gamma_{12}(\tau)|$ appearing in it, the expression is similar to the optical intensity, Eq. (3-2), for coherent superposition. Evidently this extra factor governs the quality of the fringes, which is quantified through Michelson's visibility function, v:

$$v = \frac{I_{max} - I_{min}}{I_{max} + I_{min}}. \tag{3-15}$$

The I_{max} and I_{min} are the optical intensities at the neighboring maxima and minima in the plane of observation.

Consider an experimental arrangement such that, to a good approximation,

$$I_1(Q) \simeq I_2(Q) = I_0,$$

for all Q in a portion of the plane where the fringes are observed. The neighboring maxima and minima located at Q_+ and Q_- are

$$I_{max}(Q_+) = 2\,I_0[1 + |\gamma_{12}(\tau_+)|]$$

and

$$I_{min}(Q_-) = 2\,I_0[1 - |\gamma_{12}(\tau_-)|].$$

If $\gamma_{12}(\tau)$ is slowly varying compared to the optical intensity function $I(Q)$, so that

$$|\gamma_{12}(\tau_+)| \simeq |\gamma_{12}(\tau_-)|,$$

then the concept of visibility is meaningful and is given by

$$v = |\gamma_{12}(\tau)| \tag{3-16}$$

(see Problem 3-2). If $|\gamma_{12}(\tau)|$ is slowly varying, consider a portion of the observation plane in the neighborhood of zero path difference, $c\tau \simeq 0$. In one limit, where $|\gamma_{12}(0)| \simeq 1$, the fringes are of good quality and the pinholes P_1 and P_2 are said to be *mutually coherent*. In the other limit, where

$|\gamma_{12}(0)| \simeq 0$, the poor quality or the lack of fringes is attributed to the *mutual noncoherence* of the light from the pinholes. In general, the $|\gamma_{12}(0)|$ is said to describe the *degree of partial coherence* of the light from the pinholes.

The argument $\phi_{12}(0)$ of $\gamma_{12}(0)$ dictates the location of the zero-order fringe. For example, if $\phi_{12}(0) = 0$, the optical intensity maximum is located at zero path difference, $c\tau = 0$; when $\phi_{12}(0) \neq 0$, the maximum is shifted from this location. Thus, by making irradiance measurements

$$\mathsf{E}(Q) = \mathsf{C}I(Q)$$

in the plane of observation, we can derive the state of coherence of the light from the pinholes.

Now suppose the experiment is repeated all over with the same source Σ at the same position but for a new pair of locations of the pinholes P_1 and P_2. The state of coherence of the light from the new pinholes will, in general, be different. A repetition of this experiment for all pairs of points in the screen illuminated by the source Σ will reveal the state of coherence of the illumination in space. The quantity $\gamma_{12}(0)$ is said to describe the *complex degree of spatial coherence* of the illumination in the screen plane. The function $\Gamma(P_1, P_2, \tau)$ was introduced by Wolf (1954, 1955) and named the *mutual coherence function* (MCF). It is the cornerstone in the study of partial coherence of the light field. It offers a general framework for this study and goes beyond the simple intuitive notions of coherence and noncoherence.

As we shall see in what is to follow, the mutual coherence function, $\Gamma_{12}(\tau)$, plays a central role in the description of optical phenomena. It is basically a *statistical quantity* defined through the time-averaging procedure and is a function of two space points P_1 and P_2 and the time delay τ. It is said to describe the *cross correlation* of the optical fields emanating from the two pinholes. The question of whether or not there is field correlation is the same as the question of coherence in optics, which in turn is answered through the presence or the absence of the fringes.

The quantity $\Gamma_{12}(\tau)$ depends only on the time difference and not on the separate times ρ_1/c and ρ_2/c. It is therefore independent of time origin, and so is the outcome of the two-pinhole interference experiment that it correctly describes. The outcome of the experiment does not depend on *when* it is carried out, as long as the light source Σ is kept turned on and everything else is left unchanged. The particular values assumed by the light amplitude $V(P, t)$ for one run of the experiment may be quite different, in general, from those assumed by it for some other run of the experiment. However, the time-averaged (statistical) quantity $\Gamma_{12}(\tau)$ will remain the same for all the runs. The underlying statistics of the light field are said to be *stationary*. This point will be discussed in more depth in the next section.

FIELD STATISTICS

Wolf introduced the MCF as a statistical quantity. In this section, we discuss the "big picture" of what is involved in the statistical description of the field. A brief qualitative discussion of the statistical theory is given in Appendixes 3.1 and 3.2 at the end of this chapter and it may be regarded as a prerequisite to this section.

Statistical Average

We consider the optical field variable V as a random variable. That is, it is *not* known in the usual sense as a function of space and time, $V(\mathbf{r}, t)$. The random variable V stands for an entire ensemble of (class of) functions. Each member of the ensemble is a possible realization of, and shares all the attributes of, the optical field under study; for example, it satisfies the wave equation.

The random variable V is thought to signify the entire ensemble, and one can offer only a probabilistic description of it. It is known only through the knowledge of a hierarchy of probability density functions: $W_1, W_2, \ldots, W_n$. For example, the density W_2 depends on V_1 and V_2, and possibly on times t_1 and t_2, and is written as $W_2(V_1, V_2; t_1, t_2)$. The joint probability that the random variable V has a value between V_1 and $V_1 + dV_1$ at time t_1 and that it has a value between V_2 and $V_2 + dV_2$ at time t_2 is given by $[W_2(V_1, V_2; t_1, t_2)\, dV_1\, dV_2]$. The individual density functions are all deterministic; they collectively describe the random variable V. Equivalently, V may also be described by means of the hierarchy of correlation functions of all orders.

The above description of the random variable is as complete as possible. From the point of view of predicting the outcome of optical experiments, we are not short-changed in any way. The experimental observation (outcome) is related to the statistical (ensemble) average of some deterministic function $f(V)$ of this random variable, and it is found through

$$E\{f(V)\} = \int_{-\infty}^{\infty} f(V)\, W_1(V; t)\, dV. \tag{3-17}$$

In particular, the mean value of V itself is

$$E\{V\} = \int_{-\infty}^{\infty} V W_1(V; t)\, dV, \tag{3-18}$$

which is the first moment of the density W_1. If the experimental quantity should depend on two field values, V_1 and V_2, at two times, t_1 and t_2, then the second density W_2 is used. The second-order field correlation defined as an ensemble average is

$$E\{V_1V_2\} = \iint_{-\infty}^{\infty} V_1V_2W_2(V_1, V_2; t_1, t_2)\, dV_1\, dV_2 \tag{3-19}$$

and is denoted by $R_V(t_1, t_2)$. If the field values at two different space–time points are correlated, we may write $R_{12}(t_1, t_2)$ and drop the subscript V as understood. It is this correlation function that comes close to the MCF, $\Gamma_{12}(\tau)$, except that it is sensitive to the time origin since it depends on the separate times t_1 and t_2, instead of just the difference, $\tau = t_2 - t_1$.

The statistics are said to be stationary in time when the entire hierarchy of functions, $W_1, W_2, \ldots, W_n$ are all invariant under arbitrary (linear) time displacements. Thus W_1 can be made independent of time, W_2 depends on $t_2 - t_1$, W_3 on $t_2 - t_1$ and $t_3 - t_1$, and so forth for all the functions. Thus, under the conditions of time stationarity, the second-order correlation function defined through the ensemble average will depend on the time difference, bringing it still closer to the definition of the MCF.

We shall regard the optical field statistics as stationary, and, by appealing to the ergodic hypothesis, we shall further assume that its correlation functions defined through the ensemble (statistical) average are respectively the same as those defined through the time-average operation. (See Appendix 3.1, the section entitled Time Average and Statistical Average.) Thus the MCF defined by using the time average is the same as the one that may be defined by using the statistical average. In this context the mean value of the field becomes time independent, and since the real optical field is as many times positive as negative, or since the phase of the complex representation is uniformly distributed, its mean value is zero. The equivalence of the time and statistical average is important; in some cases the statistical average definition of the MCF is more convenient to use (see Beran and Parrent, 1964; Marathay et al. 1970).

For the sake of simplicity, we have regarded the random variable optical field V as a real quantity. For the complex analytic signal V, we need to define the probability density functions W_n for both the real and imaginary parts or equivalently for both V and V^*. If $V(\mathbf{r}, t)$ is a member of the ensemble, then $V^*(\mathbf{r}, t)$ is also a valid member since both $V(\mathbf{r}, t)$ and $V^*(\mathbf{r}, t)$ are solutions of the wave equation.

Gaussian statistics is an important special case in which the first and the second moments completely determine the first probability density function W_1. All its higher moments are expressible in terms of the first two moments. Furthermore, each one of the higher-order correlation functions may be expressed in terms of the first moment and the second-order correlation function of the random variable. In other words, the knowledge of the first two probability density functions *or* the knowledge of the first moment and the second-order correlation function is necessary and sufficient to know all that there is to be known of the Gaussian random variable. This fact, coupled with the applicability of the Gaussian case to a wide range of problems, makes this special case attractive from the point of view of theory. A working knowledge of the Gaussian statistics is indispensable. Instead of providing only a brief review in the appendix, we refer the reader to two of several excellent source books: Blackman (1966) and Davenport and Root (1958).

It turns out that the light field $V(\mathbf{r}, t)$ from a thermal source is Gaussian distributed. This can be argued theoretically and has been shown to be the case experimentally. Several publications discuss the theoretical aspect: A few early ones are van Cittert (1934), Blanc-Lapierre and Dumontet (1955), Janossy (1957, 1959), Wolf (1957), Glauber (1963), and Mandel (1964); also see a review article by Mandel and Wolf (1965), wherein still further references may be found. A particularly elegant theoretical treatment is found in a book by Klauder and Sudarshan (1968). Evidence of the Gaussian nature is revealed experimentally through the measurement of the higher-order moments. The reader may start by referring to the earlier papers, such as by Hanbury-Brown and Twiss (1956, 1957), Martienssen and Spiller (1964), Arecchi et al. (1966), and Morgan and Mandel (1966).

The autocorrelation function $\Gamma_{11}(\tau)$ is a contracted form of the MCF $\Gamma_{12}(\tau)$ for $P_1 = P_2$. It is pointed out in Appendix 3.2 that it makes no sense to talk about the (temporal) Fourier transform of the individual member function of the ensemble. These members, for stationary statistics, continue forever and, in general, do not possess a Fourier transform. By allowing the member functions to possess infinite energy over an infinite interval but a finite energy over a finite interval, we find that their total power is nonzero. The autocorrelation function is assumed to exist. The zero ordinate of the autocorrelation function equals the (finite) total power, and its Fourier transform is the *power spectral density*. The autocorrelation function is regarded as something common to all the member functions of the ensemble. They all have the same power spectrum too. Now, by the ergodic hypothesis, the operation of the time average is taken to be equivalent to the statistical average. It is pointed out in Appendix 3.2 that the Fourier transform (power spectral density) of the autocorrelation function defined via the statistical average (i.e., the average "across" the ensemble) is real and positive. Thus, with either definition of the autocorrelation function, its Fourier transform is the true power spectral density. The statement that the autocorrelation function of a random function and its power spectral density form a Fourier transform pair is the Wiener–Khintchine theorem.

The autocorrelation function $\Gamma_{11}(\tau)$ enters naturally in the study of Michelson's two-beam interferometer discussed in Chapter 1. The displacement of one mirror relative to the other changes the path difference $c\tau$, and it correlates the field with itself. In the optical context, we shall reserve the word "power" only to a properly defined radiometric quantity and refer to the Fourier transform of $\Gamma_{11}(\tau)$ as the *optical spectral intensity*,† $\hat{\Gamma}_{11}(\nu)$, abbreviated OSI. The power at frequency ν within the interval $d\nu$ is proportional to $\hat{\Gamma}_{11}(\nu)\,d\nu$. Michelson studied the spectral lineshapes in this way. We will return to the details of this discussion in Chapter 5.

Special Forms

Let us gather together the definitions and special forms of the functions related to the mutual coherence function as used in optical theory. The time-average definition of MCF is

$$\Gamma_{12}(\tau) = \langle V_1(t+\tau) V_2^*(t) \rangle. \tag{3-20}$$

With the assumed equivalence of the time and ensemble average, we also have

$$\begin{aligned} \Gamma_{12}(\tau) &= E\{V_1 V_2^*\} \\ &= \iint_{-\infty}^{\infty} V_1 V_2^* W_2(V_1, V_2^*, \tau)\, d^{(2)}V_1\, d^{(2)}V_2^*, \end{aligned} \tag{3-21}$$

where the superscript (2) on the differentials indicates that the integration is intended over both the real and imaginary parts.

The optical spectral intensity function $\hat{\Gamma}_{11}(\nu)$ or the more general mutual spectral density function (MSDF) $\hat{\Gamma}_{12}(\nu)$ for two points may be defined (see Appendix 3.2) by

$$\hat{\Gamma}_{12}(\nu) = \lim_{T\to\infty} E\left\{ \frac{\hat{V}_T(\mathbf{r}_1, \nu)\hat{V}_T^*(\mathbf{r}_2, \nu)}{2T} \right\}, \tag{3-22}$$

where the subscript T on $\hat{V}$ indicates that it is the spectrum of the field $V(\mathbf{r}, t)$ limited to the time interval $-T$ to T.

For the nonstationary case the MCF is defined by

$$\Gamma_{12}(t_1, t_2) = E\{V_1(t_1) V_2^*(t_2)\} \tag{3-23}$$

†In the literature this quantity is referred to as power spectral density.

and the mutual spectral density function in the ensemble average sense is

$$\hat{\Gamma}_{12}(\nu_1, \nu_2) = E\{\hat{V}_1(\nu_1)\hat{V}_2^*(\nu_2)\}. \tag{3-24}$$

The subscripts 1 and 2 on Γ and V stand for the two different space points $\mathbf{r}_1$ and $\mathbf{r}_2$, respectively. The temporal Fourier transform will be indicated by a caret over the symbol, for example,

$$\Gamma_{12}(\tau) = \int_0^\infty \hat{\Gamma}_{12}(\nu)\exp(-i2\pi\nu\tau)\,d\nu. \tag{3-25}$$

Further discussion on the properties of the transform is given in Chapter 4.

The arguments of the MCF may be displayed in one of several ways. Thus,

$$\Gamma_{12}(\tau) = \Gamma(\mathbf{r}_1, \mathbf{r}_2, \tau) = \Gamma(P_1, P_2, \tau),$$

or when the points are in a plane $z_1 = z_2 = z$, we write

$$\Gamma_{12}(\tau) = \Gamma(x_1, y_1, x_2, y_2, z, \tau). \tag{3-26}$$

In addition to being time stationary, the statistics may be homogeneous or spatially stationary (also called shift invariant). In that case,

$$\Gamma_{12}(\tau) = \Gamma(x_1 - x_2, y_1 - y_2, z, \tau) \tag{3-27}$$

wherein frequently we shall use the notation $x_{12} = x_1 - x_2$ and $y_{12} = y_1 - y_2$ for short. As a still further special case, the coherence may be independent of angle or direction in the x, y plane; then

$$\Gamma_{12}(\tau) = \Gamma\left((x_{12}^2 + y_{12}^2)^{1/2}, z, \tau\right), \tag{3-28}$$

where the field statistics are said to be homogeneous and isotropic. A feature common to these special cases is that the associated optical intensity ($x_1 = x_2$, $y_1 = y_2$, and $\tau = 0$) is uniform over the plane $z =$ constant,

$$I = \Gamma(0, 0, z, 0). \tag{3-29}$$

Spatial stationarity is useful in theoretical work. It may be approximated to a high degree of accuracy in practical applications if the apertures involved are very large compared to the distance over which Γ_{12} remains almost unchanged and has a significant value. When reference is made to an irradiance measurement in a plane $z =$ constant, we shall use the symbol

$$\mathsf{E}(x, y, z) = \mathsf{C}\Gamma(x, y, x, y, z, 0) \tag{3-30}$$

as we did in Chapter 2. For spectral irradiance in the above situation we shall write

$$\hat{\mathsf{E}}(x, y, z, \nu) = \mathsf{C}\hat{\Gamma}(x, y, x, y, z, \nu) \tag{3-31}$$

with units of watts per square meter per hertz [W m^{-2} Hz^{-1}]. Frequently the z dependence will be dropped in E and $\hat{\mathsf{E}}$ if it is clear from the context.

The autocorrelation $\Gamma_{11}(\tau)$ with coincident space points is sometimes called the *self-coherence* function since it correlates the field with itself. It describes the temporal coherence "along" the beam-propagation direction, such as in the Michelson two-beam interferometer. Similarly, the *spatial coherence function* $\Gamma_{12}(0)$ describes the coherence "across" the beam. In the literature, $\Gamma_{12}(0)$ is referred to as the mutual intensity, but we shall call it the *mutual optical intensity* (MOI) function to set it apart from the radiometric terminology.

The spatial Fourier transform of the MCF will be denoted by a tilde over the symbol,

$$\begin{aligned}\tilde{\Gamma}_{12}(f_1, g_1, f_2, g_2, z, \tau) = \iiiint_{-\infty}^{\infty} & \Gamma(x_1, y_1, x_2, y_2, z, \tau) \\ & \times \exp[-i2\pi(f_1x_1 + g_1y_1 - f_2x_2 - g_2y_2)] \\ & \times dx_1\, dy_1\, dx_2\, dy_2,\end{aligned} \tag{3-32}$$

where f and g are the respective spatial frequencies with units of reciprocal length [m^{-1}] and they take on values over the full range $-\infty$ to $+\infty$. The total Fourier transform over both time and space will be denoted by an open circle over the symbol, namely, $\mathring{\Gamma}(f_1, g_1, f_2, g_2, z, \nu)$.

In conclusion, we mention that when Gaussian statistics are applicable—as, for example, for light from a thermal source—the mutual coherence function $\Gamma(P_1, P_2, \tau)$ is all that is required to describe the related optical phenomena. If P_1 is set equal to P_2 and τ is set equal to 0, the MCF equals the optical intensity at P_1. Its contracted form, $\Gamma_{11}(\tau)$, is important in the study of spectra. Furthermore (as we shall see in what is to follow), $\Gamma_{12}(\tau)$, unlike the field itself, is in principle determinable by experiment (it is observable); it also propagates like a Huygens wave. The mutual coherence function $\Gamma_{12}(\tau)$ thus claims a unique position in optical theory.

PROBLEMS

3-1. Show that

$$\Gamma(P_1, P_2, \tau) = \Gamma^*(P_2, P_1, -\tau)$$

and establish the inequality

$$0 \leq |\gamma_{12}(\tau)| \leq 1,$$

by using Schwartz's inequality,

$$\left|\int fg^*\,dt\right|^2 \leq \int |f|^2\,dt \int |g|^2\,dt,$$

where f and g are complex-valued functions of the variable t.

3-2. Obtain an expression for the optical intensity in the plane of observation for the two-pinhole experiment of Fig. 3-1 and show that the visibility of fringes in general has the form

$$v = \frac{2(I_1 I_2)^{1/2}}{(I_1 + I_2)} |\gamma_{12}(\tau)|\,.$$

REFERENCES

Arecchi, F. T., E. Gatti, and A. Sona (1966). Time distribution of photons from coherent and Gaussian sources, *Phys. Lett.* **20**:27–29.

Blackman, N. M (1966). *Noise and Its Effect on Communication*, McGraw-Hill, New York.

Blanc-Lapierre, A., and P. Dumontet (1955). La notion de coherence en optique, *Rev. Opt.* (*Théor. Instrum.*) **34**:1–21.

Beran, M. J., and G. B. Parrent, Jr. (1964). *Theory of Partial Coherence*, Prentice-Hall, Englewood Cliffs, NJ, 189 pp.

Davenport, W. B., Jr., and W. L. Root (1958). *Random Signals and Noise*, McGraw-Hill, New York, 393 pp.

Glauber, R. J. (1963). Photon correlations, *Phys. Rev. Lett.* **10**:84–86.

Hanbury-Brown, R., and R. Q. Twiss (1956). Correlation between photons in two coherent beams of light, *Nature* **177**:27–29.

Hanbury-Brown, R., and R. Q. Twiss (1957). Interferometry of the intensity fluctuations in light. I. Basic theory: the correlation between photons in coherent beams of radiation, *Proc. R. Soc. Lond. Ser. A* **242**:300–324.

Janossy, L (1957). On the classical fluctuation of a beam of light, *Nuovo Cimento* **6**:111–124.

Janossy, L. (1959). The fluctuations of intensity of an extended light source, *Nuovo Cimento* **12**:369–384.

Klauder, J. R., and E. C. G. Sudarshan (1968). *Fundamentals of Quantum Optics*, Benjamin, New York.

Mandel, L. (1964). Some coherence properties of non-Gaussian light, pp. 101–109 in N. Bloembergen and P. Grivet, eds., *Quantum Electronics*, Vol. I (Proceedings of the Third International Congress on Quantum Electronics, Paris, 1963), Dunod, Paris; Columbia University Press, New York.

Mandel, L., and E. Wolf (1965). Coherence properties of optical fields, *Rev. Mod. Phys.* **37**:231–287.

Marathay, A. S., L. Heiko, and J. L. Zuckerman (1970). Study of rough surfaces by light scattering, *Appl. Opt.* **9**:2470–2476.

Martienssen, W., and E. Spiller (1964). Coherence and fluctuations in light beams, *Am. J. Phys.* **32**:919–926.

Morgan, B. L., and L. Mandel (1966). Measurement of photon bunching in a thermal light beam, *Phys. Rev. Lett.* **16**:1012–1015.

van Cittert, P. H. (1934). Die wahrscheinliche Schwingungsverteilung in einer von einer Lichtquelle direkt oder mittels einer Linse beleuchteten Ebene, *Physica* **1**:201–210.

Wolf, E. (1954). Optics in terms of observable quantities, *Nuovo Cimento* **12**:884–888.

Wolf, E. (1955). A macroscopic theory of interference and diffraction of light from finite sources. II: Fields with a spectral range of arbitrary widths, *Proc. R. Soc. Lond. Ser. A* **230**:246–265.

Wolf, E. (1957). Intensity fluctuations in stationary optical fields, *Phil. Mag.* **2**:351–354.

APPENDIX 3.1. STATISTICAL THEORY

A *random phenomenon* is one in which the fluctuations of the quantity under observation [in our present study, light field amplitude $V(r, t)$] as a function of time cannot be precisely predicted. No cause-and-effect relationship is available or known, due, for example, to the complexity of the situation.

The statistical approach toward random phenomena is quite different from the approach toward periodic or aperiodic phenomena. We are quite familiar with the mathematics of handling periodic or aperiodic functions: The response of a physical system to periodic or aperiodic functions is well determined. Whenever the same physical system (or replica) under the same conditions is used, the output is predictable. Response to a sudden excitation at the input is predictable with great precision. A delay in the time of occurrence of a transient merely translates the response. But with random functions, as we shall see, experiments carried out under essentially similar conditions do not produce similar results. With random functions we are almost always dealing with a *large class* of functions. It makes no sense to talk about one random function. The class of functions is referred to as an *ensemble*. It is made up of a large number of functions called *members*, all of which have one or several things in common. For example, all functions may have the same energy or the same central frequency.

We thus need a mathematical theory that is capable of dealing with a *class*, or ensemble, of functions rather than with a particular function. In an ensemble we have an infinite variety of wave forms. We therefore look for characteristics that are common to *all* members on the basis of average behavior.

An ordinary (periodic or aperiodic) function $f(t)$ can be specified by giving its precise value at each time t. Then we say that $f(t)$ is known. The specification of a random function is of a quite different style. It is specified in terms of probabilities. Let's suppose x is the random variable. We denote the ensemble by $\{x^{(k)}(t)\}$, where t is the time parameter and the superscript (k) labels the member functions. The curly brackets stand for the set of all functions belonging to the ensemble. Now let's also suppose for the moment that we have the necessary instruments to make precise measurements and we have all the time in the world to perform all the experiments to study the ensemble. The ensemble is made up of an extremely large number (approaching infinity) of members. We start with, say, a large number N of the members. At any given instant of time $t = t_1$, we count the number n_1 of members (among N) that have a value lying between x_1 and $x_1 + \Delta x_1$. The ratio n_1/N gives a probability of finding a member whose value lies between x_1 and $x_1 + \Delta x_1$ at $t = t_1$ when a random choice is made among the N members. Note that n_1/N depends on t_1 and is proportional to Δx_1 for small Δx_1. We now allow the number N to become larger to admit more and more members of the ensemble. For very large N (approaching infinity) the ratio just mentioned is by definition the probability of finding the values between x_1 and $x_1 + \Delta x_1$ at time $t = t_1$ for the ensemble as a whole. Such probabilities may be obtained for various values of x_1 and also for various values of time t_1. We define the probability of finding the values between x_1 and $x_1 + \Delta x_1$ of the random variable x as

$$\lim_{N \to \infty} \frac{n_1}{N} = W_1(x_1; t_1)\,\Delta x_1.$$

The function $W_1(x_1;t_1)$ is called the *first probability density function*. The next function, W_2, is the *joint probability density function*. It is defined as

$$W_2(x_1, x_2; t_1, t_2)\,\Delta x_1\,\Delta x_2$$

and is the probability of finding the values of the random variable x between x_1 and $x_1 + \Delta x_1$ at $t = t_1$ *and* between x_2 and $x_2 + \Delta x_2$ at $t = t_2$. Obviously a succession of such probability density functions $W_1, W_2, \ldots, W_n$ may be defined. Therefore it is also obvious that with a probabilistic description of an ensemble we cannot know anything about the individual member function $x^{(k)}(t)$. Our description necessarily deals with the ensemble as a whole. In a random process the dependence of the random variable x on t is only known probabilistically. We assume that all the density functions $W_1, W_2, \ldots, W_n$ for any physical process always exist.

A random process is said to be *statistically nonstationary* if the functions in the hierarchy $W_1, W_2, \ldots, W_n$ all depend on the absolute origin of time. That is, the underlying physical mechanism changes in the course of time, so the statistical properties become a function of time. As an analogy, we mention that an optical lens system is said to be nonstationary when the impulse response changes its shape from one region of the image plane to another.

A random process is said to be *statistically stationary* if the functions in the hierarchy $W_1, W_2, \ldots, W_n$ are *all invariant* under arbitrary (linear) time displacements. Thus the first function is independent of time, $W_1(x_1)$; the second depends on the time difference, $W_2(x_1, x_2; t_2 - t_1)$; the third depends on two time differences, $W_3(x_1, x_2, x_3; t_2 - t_1, t_3 - t_1)$; and so forth for all the functions.

The variable x is the random variable. The specification of the ensemble by means of the density functions $W_1, W_2, \ldots, W_n$ allows us to define the statistical averages over the entire ensemble for functions $f(x)$ of the random variable x. Suppose we wish to find the average value of x itself. The range of x is $-\infty$ to $+\infty$. Any one particular value of x, say between x_j and $x_j + \Delta x_j$, is obtained with probability $W_1(x_j; t)\,\Delta x_j$. The weighted mean—that is, the sum $\Sigma x_j W_1(x_j; t)\,\Delta x_j$—is the average of x. In the limit this sum approaches an integral. The statistical average of the random variable x over the entire ensemble is defined by

$$E\{x(t)\} = \int_{-\infty}^{\infty} x W_1(x; t)\, dx. \qquad \text{(A3-1)}$$

The notation $E\{x\}$ stands for the *expectation value*—sometimes called the ensemble average—of x. If $f(x)$ is a single-valued function of the random variable x, then the ensemble average is obviously

$$E\{f(x)\} = \int_{-\infty}^{\infty} f(x)\, W_1(x; t)\, dx. \qquad \text{(A3-2)}$$

In general, the expectation value of $f(x)$ will change with time. When the process is stationary, the function W_1 is independent of time, in which case the average of $f(x)$ is also time independent.

Another quantity of interest is the *covariance function*—or the correlation function—of the value x_1 of the random variable x at time t_1 and of the value x_2 at time t_2. It is denoted by $R_x(t_1, t_2)$ and is defined by

$$R_x(t_1, t_2) = E\{x_1 x_2\}. \tag{A3-3}$$

In terms of the joint probability density function W_2, we have

$$R_x(t_1, t_2) = \int\int x_1 x_2 W_2(x_1, x_2; t_1, t_2)\, dx_1\, dx_2. \tag{A3-4}$$

Since $R_x(t_1, t_2)$ deals with two values of the same random variable $x(t)$, we call it the *autocorrelation* function. This is distinguished from the *cross-correlation* function, $R_{xy}(t_1, t_2)$, defined for two different random variables $x(t)$ and $y(t)$:

$$R_{xy}(t_1, t_2) = E\{xy\}. \tag{A3-5}$$

Observe that when the process is statistically stationary the function $W_2(x_1, x_2; t_1, t_2)$ depends only on the time difference $t_2 - t_1$; as a result, the correlation functions also depend on $t_2 - t_1$, and not on t_1 and t_2 separately.

Since the strict mathematical definition of statistical stationarity is too severe to cope with in practice, the idea of *wide-sense stationarity* is introduced in the theory. A random process is said to be statistically *stationary in the wide sense if the mean is independent of time and the correlation function is a function of the time difference only.* This definition implies that W_1 is time independent and W_2 is a function of the time difference $t_2 - t_1$. The concept of wide-sense stationarity says nothing about the higher-order density functions W_3, W_4, and so on. In practice we will generally assume wide-sense stationarity and obtain W_1 and W_2 by considering finite (large) portions of the sample functions $x^{(k)}(t)$.

In Chapter 3, the field amplitude V itself was regarded as the random variable. The fields $V(P_1, t)$ and $V(P_2, t)$ from two different space points are to be regarded as two different random variables. Thus $\Gamma_{11}(t_1, t_2)$ and $\Gamma_{11}(\tau)$ are autocorrelation functions and $\Gamma_{12}(t_1, t_2)$ and $\Gamma_{12}(\tau)$ are cross correlations for the nonstationary and the stationary case, respectively.

Time Average and Statistical Average

We have up to now defined the average value and the correlation function for a random variable over the entire ensemble. In the formal discussion of Young's experiment, we defined a correlation function, denoted by $\Gamma_{12}(\tau)$, but we defined it in terms of the time-average definition. The question now arises whether the correlation functions and/or averages defined in these two ways are the same.

The ensemble average of a quantity is an average carried out from one member of the ensemble to another at a fixed time t:

$$E\{x(t)\} = \lim_{N\to\infty} \frac{1}{N} \sum_{k=1}^{N} x^{(k)}(t)$$
$$= \int_{-\infty}^{\infty} xW_1(x;t)\,dx, \qquad t \text{ fixed.} \tag{A3-6}$$

This may be called an average taken *across* the random process. The time average, on the other hand, is a number obtained by working with one typical member of the ensemble:

$$\langle x\rangle = \lim_{T\to\infty} \frac{1}{2T} \int_{-T}^{T} x(t)\,dt, \qquad k \text{ fixed.} \tag{A3-7}$$

This may be called an average taken *along* the random process. In general, these two definitions are obviously not the same. The ensemble average generally depends on time t, as to when the average was taken. The time average is a number independent of t. Note that the finite average defined in Eq. (A3-7) may not converge, and, in general, the average may not exist for any of the sample functions. Also, even if it did exist, the average in general may be different from one sample function to another. When the statistics are stationary, the ensemble average is independent of time. For the stationary case, we make the following argument regarding the equivalence of the two modes of averaging.

Consider a physical system specified by means of its initial and ambient conditions. Let a record of its output (some physical quantity) be made as a function of time. Imagine many such physical systems—replicas of the original. An ensemble may now be built by admitting a record from each such replica as one member. To discuss the equivalence of the ensemble and time averages, we need to compare the whole ensemble with one member. In practice, in many physical situations if we wait long enough it is a good approximation to assume that any one member function goes through all possible values (exhausts all possible states and almost equally often) that are available to all the various members of the ensemble. If this is the case, we may consider a single physical system and make a time record of its output. This very long time record may be cut up into many sections of length T such that T itself is very large compared to any of the time periods involved in the system. These various sections may themselves be regarded as possible members of the ensemble. What we are assuming is that any one member goes through all the "behavioral patterns" available for the various members of the ensemble. If this assumption is reasonable (that is, unless some *a priori* knowledge of the system prohibits us from making this assumption), then the average taken along any one member should be very nearly the same as the average taken (at *any* particular time) over many members of the ensemble. Thus when the process is stationary and when the above conditions are applicable, we shall assume that both methods of averaging give the same result:

$$
\begin{aligned}
E\{x\} &= \langle x \rangle \\
E\{x^n\} &= \langle x^n \rangle \\
E\{x_1 x_2\}_{t_{21}} &= \langle x_1 x_2 \rangle_{t_{21}}
\end{aligned} \tag{A3-8}
$$

where $t_{21} \equiv t_2 - t_1$.

The discussion of these assumptions is based on what we reasonably expect the physical system to do. The subject matter is a part of the more general *ergodic hypothesis* and the condition of *ergodicity*, which are beyond the scope of this appendix. We will for the most part assume the equivalence of the ensemble average and the time average. For further discussion on this point, see the book by Lee (1960, pp. 207–209) and the book by Papoulis (1965, pp. 327–330). To sum up, we may say that for a stationary random process that is ergodic, the limit $\langle x(t) \rangle$, Eq. (A3-7), exists for every sample function except for a set with probability of zero, and the time average $\langle x(t) \rangle$ is equal to the constant ensemble average, $E\{x(t)\}$, with probability of one.

Moments and Characteristic Function

The statistical average of the random variable x was defined in Eq. (A3-1). Omitting (for convenience) the explicit dependence of W_1 on time t, we have

$$E\{x\} = \int_{-\infty}^{\infty} x W_1(x)\, dx. \tag{A3-9}$$

The function W_1 is the first probability density function for the random variable x. The mean value of x is called the *first moment* of the probability density function W_1. The nth moment is similarly defined as the statistical average of the nth power of x,

$$E\{x^n\} = \int x^n W_1(x)\, dx. \tag{A3-10}$$

The zeroth moment is simply the total area under the probability density function, which by definition is unity:

$$E\{x^0\} = \int W_1(x)\,dx = 1. \qquad \text{(A3-11)}$$

This is so for all the density functions $W_1, W_2, \ldots, W_n$. That is, for example,

$$\int\int_{-\infty}^{\infty} W_2(x_1, x_2)\,dx_1\,dx_2 = 1. \qquad \text{(A3-12)}$$

Clearly, joint moments may be defined

$$E\{x_1^n x_2^m\} = \int\int_{-\infty}^{\infty} x_1^n x_2^m W_2(x_1, x_2)\,dx_1 dx_2. \qquad \text{(A3-13)}$$

For further properties of $W_1, W_2, \ldots, W_n$ we refer to Davenport and Root (1958) and Papoulis (1965).

The *characteristic function* of a probability density function W_1 is defined as the ensemble average of the Fourier kernel, and we will denote it by $c_1(z)$:

$$c_1(z) = E\{\exp(izx)\} = \int_{-\infty}^{\infty} \exp(izx)\,W_1(x)\,dx. \qquad \text{(A3-14)}$$

In other words, the characteristic function $c_1(z)$ is defined as the Fourier transform of the density function $W_1(x)$. If the characteristic function $c_1(z)$ has a Taylor series expansion in some interval about the origin, then

$$c_1(z) = \sum_{n=0}^{\infty} \left.\frac{d^n c_1(z)}{dz^n}\right|_{z=0} \frac{z^n}{n!}. \qquad \text{(A3-15)}$$

But from Eq. (A3-14), the nth derivative of $c_1(z)$ evaluated at $z = 0$ is seen to be proportional to the nth moment of the density function W_1. That is,

$$\left.\frac{d^n}{dz^n} c_1(z)\right|_{z=0} = i^n \int_{-\infty}^{\infty} x^n W_1(x)\,dx$$
$$= i^n E\{x^n\}. \qquad \text{(A3-16)}$$

Therefore, the power series expansion in Eq. (A3-15) may be rewritten as

$$c_1(z) = \sum_{n=0}^{\infty} E\{x^n\} \frac{(iz)^n}{n!}. \qquad \text{(A3-17)}$$

Observe that the moments of the random variable uniquely determine the characteristic function. The Fourier transform in Eq. (A3-14) is unique, and hence the moments uniquely determine the probability density function.

Probability Densities and Correlation Functions

We have seen up to now that in order to completely describe a random process we need to study the hierarchy of probability density functions $W_1, W_2, \ldots, W_n$. In practice it is impossible to determine all these functions. Even in finding any one of them, we can only deal with a large portion of the sample $x^{(k)}(t)$ instead of the entire sample $-\infty \leq t \leq \infty$, and we can study only a finite (large) number of such sample functions. We will restrict our attention to the stationary random processes that are ergodic; then it is possible to use the simpler procedure of time averaging instead of statistical averaging. In practice, if time averaging is allowed, the correlation functions of the random variable are more easily accessible to measurement than the probability density functions. Now if the process is ergodic (Papoulis, 1965, pp. 327–330), the correlation functions determined through the time averages are equated to their corresponding definitions in terms of the probability density functions. Two of the correlation functions of interest from the point of view of coherence theory are the second-order correlation function,

$$\langle x_1 x_2 \rangle = E\{x_1 x_2\} = \int\int x_1 x_2 W_2(x_1, x_2, \tau_2)\, dx_1\, dx_2, \quad \text{(A3-18)}$$

and the fourth-order correlation function,

$$\begin{aligned}\langle x_1 x_2 x_3 x_4 \rangle &= E\{x_1 x_2 x_3 x_4\} \\ &= \int\int\int\int x_1 x_2 x_3 x_4 W_4(x_1, x_2, x_3, x_4, \tau_2, \tau_3, \tau_4) \\ &\quad \times dx_1\, dx_2\, dx_3\, dx_4. \end{aligned} \quad \text{(A3-19)}$$

For optics in particular, the square of the random variable (which is also a random variable) is of interest. In this case, the contracted form of the fourth-order correlation function is useful, and it is related to a new second-order density function $W_2'(x_1^2, x_2^2, \tau)$. Thus,

$$\begin{aligned}\langle x_1^2 x_2^2 \rangle &= E\{x_1^2 x_2^2\} \\ &= \int\int x_1^2 x_2^2 W_2'(x_1^2, x_2^2, \tau)\, dx_1^2\, dx_2^2. \end{aligned}$$

Obviously a hierarchy of correlation functions may be generated to correspond to the various orders and types of the probability density functions of the random variable x. Observe that the various moments and joint moments of the probability density functions are simply the zero ordinates of the corresponding correlation functions.

For example, the second moment, $E\{x^2\}$, of the first probability density function, $W_1(x)$, is the zero ordinate of the second-order autocorrelation function, $\langle x_1 x_2 \rangle$, of the stationary process. Also, the zero ordinate of $\langle x_1 x_2 \rangle$ is the first moment of the density function $W_1'(x^2)$.

A random process can be described completely by means of the hierarchy of the probability density function $W_1, W_2, \ldots, W_n$. The specification of the hierarchy of correlation functions is also a complete and an equivalent description of the random process. In any case, the entire hierarchy of one type or another is necessary for a complete description of the process. For example, knowing the first probability density function W_1 gives us knowledge of all its moments, but they are not sufficient to determine the correlation function, $\langle x_1 x_2 \rangle$. On the other hand, the correlation function $\langle x_1 x_2 \rangle$ gives the second moment of W_1 but gives no information about the rest of the moments of W_1. Now the function $\langle x_1 x_2 \rangle$ is uniquely determined through the density function W_2. Knowledge of all its moments and joint moments is more information than is contained in $\langle x_1 x_2 \rangle$, but W_2 is not sufficient to determine the third-order correlation function $\langle x_1 x_2 x_3 \rangle$. Obviously, knowing the first few density functions and a first few correlation functions is not sufficient, in general, to describe the random process. There are, however, special exceptions to this statement. For example, if the random process is known to be Gaussian, then, as is well known, the probability density function is determined completely by knowing the mean value and the second moment. Furthermore, the correlation functions of all orders can be expressed in terms of the mean value and the second-order correlation function $\langle x_1 x_2 \rangle$.

We will not go into details of this special case here; we refer to Blackman (1966), Davenport and Root (1958), and Papoulis (1965). In general, however, the matter is not so simple, and we have to face the formidable task of finding the hierarchy of density functions or the various orders of correlation functions. Therefore, in practice we have to satisfy ourselves with much simpler characteristics of the random process. From the point of view of measurement we restrict ourselves to the mean value and the second- and perhaps fourth-order correlation functions. Together with these it is often possible to measure the density function W_1 and the joint function W_2. This, or course, is by no means a complete description, but it is the best we can generally do.

REFERENCES

Blackman, N. M. (1966). *Noise and Its Effect on Communication*, McGraw-Hill, New York, 212 pp.

Davenport, W. B., Jr., and W. L. Root (1958). *Random Signals and Noise*, McGraw-Hill, New York, 393 pp.

Lee, Y. W. (1960). *Statistical Theory of Communication*, Wiley, New York, 509 pp.

Papoulis, A. (1965). *Probability, Random Variables, and Stochastic Process*, McGraw-Hill, New York, 583 pp.

APPENDIX 3.2. AUTOCORRELATION FUNCTION AND ITS FOURIER TRANSFORM

We turn our attention to the discussion of the autocorrelation function and its Fourier transform for the case of the stationary random process that is ergodic. The following discussion is aimed to point out *the importance of the autocorrelation function and why we should avoid making statements about individual sample functions*. To begin this discussion, we will state certain facts about Fourier series and transforms. This will be followed by the study of the autocorrelation function of a periodic function. We will then extend the discussion to arbitrary functions as possible members of the stationary randòm process.

Fourier Theory

Let $x(t)$ be a real or complex, periodic function of a real variable t with period T. We define $\hat{x}(t)$ as the series

$$\hat{x}(t) = \sum_{n=-\infty}^{\infty} a_n \exp(i2\pi n f_0 t), \qquad f_0 = \frac{1}{T},$$

with

$$a_n = \frac{1}{T}\int_0^T x(t)\exp(-i2\pi n f_0 t)\,dt. \tag{A3-20}$$

Is $\hat{x}(t)$ as just defined the same as $x(t)$?

(i) If $x(t)$ is of bounded variation [at most finite discontinuities in $(0, T)$] *and* if it is *absolutely integrable* on $(0, T)$,

$$\int_0^T |x(t)|\,dt < \infty, \tag{A3-21}$$

then $\hat{x}(t) = x(t)$ wherever $x(t)$ is continuous.

(ii) If $x(t)$ is of bounded variation *and* if its modulus square is integrable on $(0, T)$,

$$\int_0^T |x(t)|^2\,dt < \infty, \tag{A3-22}$$

then $\hat{x}(t)$ is the same as $x(t)$ in the sense of the *limit in the mean*, denoted l.i.m. That is, if the limit,

$$\lim_{N\to\infty} \int_0^T \left| x(t) - \sum_{n=-N}^{N} a_n \exp(i2\pi n f_0 t) \right|^2 dt = 0,$$

then

$$x(t) = \underset{N\to\infty}{\text{l.i.m.}} \sum_{n=-N}^{N} a_n \exp(i2\pi n f_0 t). \tag{A3-23}$$

Here the mean square error goes to zero. This convergence is also called convergence-in-the-mean.

For abbreviation, we will drop l.i.m. and just assume $\hat{x}(t)$ is the same as $x(t)$, and use the following terminology:

$$\int_0^T |x(t)|^2\,dt = \text{total energy in } (0, T) \tag{A3-24}$$

and

$$\frac{1}{T}\int_0^T |x(t)|^2\,dt = \text{power or average energy in } (0, T). \tag{A3-25}$$

From Fourier theory, Parseval's theorem is

$$\frac{1}{T}\int_0^T |x(t)|^2\,dt = \sum_{n=-\infty}^{\infty} |a_n|^2 = \text{power}. \tag{A3-26}$$

When $x(t)$ is not periodic, we have the following: Consider a function $X(f)$ whose Fourier transform is $x(t)$; that is, if the integral exists,

$$x(t) = \int_{-\infty}^{\infty} X(f) \exp(i2\pi ft)\, df. \tag{A3-27}$$

(We temporarily leave our notation of $\hat{x}$ to denote Fourier transform.) Also, the *inverse* Fourier transform is defined by

$$\hat{X}(f) = \int_{-\infty}^{\infty} x(t) \exp(-i2\pi ft)\, dt \tag{A3-28}$$

if it exists. We have the following two cases.

(i) If

$$\int_{-\infty}^{\infty} |\hat{X}(f)|\, df < \infty, \tag{A3-29}$$

then

$$x(t) = \int_{-\infty}^{\infty} X(f) \exp(i2\pi ft)\, df,$$

and if

$$\int_{-\infty}^{\infty} |x(t)|\, dt < \infty,$$

then

$$X(f) = \hat{X}(f).$$

(ii) If the modulus square of $\hat{X}(f)$ is integrable,

$$\int_{-\infty}^{\infty} |\hat{X}(f)|^2\, df < \infty, \tag{A3-30}$$

then there exists a function $x(t)$ that is also such that

$$\int_{-\infty}^{\infty} |x(t)|^2\, dt < \infty, \tag{A3-31}$$

and it is related to $X(f)$ as follows:

$$x(t) = \underset{A\to\infty}{\text{l.i.m.}} \int_{-A}^{A} X(f) \exp(i2\pi ft)\, df \tag{A3-32}$$

and

$$X(f) = \underset{A\to\infty}{\text{l.i.m.}} \int_{-A}^{A} x(t)\exp(-i2\pi ft)\,dt. \tag{A3-33}$$

This is the important Plancherel's theorem. With the condition that the modulus square of $X(f)$ is integrable, we now have the analog of Parseval's theorem:

$$\int_{-\infty}^{\infty} |X(f)|^2\,df = \int_{-\infty}^{\infty} |x(t)|^2\,dt. \tag{A3-34}$$

We shall refer to $x(t)$ and $X(f)$ as a Fourier transform pair and omit l.i.m. With this summary of Fourier theory we now consider the decomposition of periodic functions.

Periodic Functions

Consider a periodic function $x(t)$ with *finite energy* in $(0, T)$. That is,

$$\int_0^T |x(t)|^2\,dt < \infty.$$

By Parseval's theorem, the power or its average energy is

$$\text{power} = \frac{1}{T}\int_0^T |x(t)|^2\,dt = \sum_{n=-\infty}^{\infty} |a_n|^2.$$

We define the *power spectral density function* $\mathcal{S}(f)$ as

$$\mathcal{S}(f) = \sum_{n=-\infty}^{\infty} |a_n|^2\,\delta(f - nf_0). \tag{A3-35}$$

$\mathcal{S}(f)$ is made up of a series of impulses at the component frequencies of $x(t)$. The strength $|a_n|^2$ of each impulse is equal to the power in that frequency. $\mathcal{S}(f)$ is the spectral power per unit frequency bandwidth. Hence,

$$\begin{aligned}\text{total power} &= \int_{-\infty}^{\infty} \mathcal{S}(f)\,df \\ &= \sum_{n=-\infty}^{\infty} |a_n|^2 \\ &= \frac{1}{T}\int_0^T |x(t)|^2\,dt. \end{aligned} \tag{A3-36}$$

The *time-average definition* of the autocorrelation function $\mathcal{R}(\tau)$ for a periodic function is

$$\mathcal{R}(\tau) = \lim_{T \to \infty} \frac{1}{2T} \int_{-T}^{T} x(t+\tau)\, x^*(t)\, dt. \tag{A3-37}$$

(For this subsection in the appendix we use $\mathcal{R}$ for the correlation function defined via the time average. When ensemble average definition is used, we will use R.)

For the periodic function $x(t)$ given by Eq. (A3-20), the autocorrelation function can be shown to be

$$\mathcal{R}(\tau) = \sum_{n=-\infty}^{\infty} |a_n|^2 \exp(+i2\pi n f_0 \tau). \tag{A3-38}$$

The Fourier transform of $\mathcal{R}(\tau)$ is then

$$\int_{-\infty}^{\infty} \mathcal{R}(\tau) \exp(-i2\pi f\tau)\, d\tau = \sum_{n=-\infty}^{\infty} |a_n|^2\, \delta(f - nf_0). \tag{A3-39}$$

Thus the power spectral density $\mathcal{S}(f)$ and the autocorrelation function $\mathcal{R}(\tau)$ form a Fourier pair

$$\mathcal{S}(f) = \int_{-\infty}^{\infty} \mathcal{R}(\tau) \exp(-i2\pi f\tau)\, d\tau \tag{A3-40}$$

and

$$\mathcal{R}(\tau) = \int_{-\infty}^{\infty} \mathcal{S}(f) \exp(+i2\pi f\tau)\, df. \tag{A3-41}$$

In particular,

$$\mathcal{R}(0) = \int_{-\infty}^{\infty} \mathcal{S}(f)\, df = \text{total power}. \tag{A3-42}$$

We summarize the properties as follows:

$\mathcal{S}(f)$ is nonnegative.
Phase information of complex numbers a_n is absent in $\mathcal{S}(f)$.
A whole class of functions $x(t)$ all having the same $|a_n|$ but differing in phase have the same $\mathcal{S}(f)$.

$\mathcal{S}(f)$ and $\mathcal{R}(\tau)$ are a Fourier pair.

Total power in $x(t)$ is $\int_{-\infty}^{\infty}\mathcal{S}(f)\,df = \mathcal{R}(0)$.

Random Functions

We now turn our attention to the study of random functions. Consider the random function $x(t)$ over the infinite interval $-\infty \leq t \leq \infty$. If $x(t)$ has *finite total energy*, that is, if

$$\int_{-\infty}^{\infty} |\, x(t)\,|^2\, dt < \infty,$$

then we have Parseval's theorem as a statement of conservation of energy,

$$\int_{-\infty}^{\infty} |\, x(t)\,|^2\, dt = \int_{-\infty}^{\infty} |\, X(f)\,|^2\, df < \infty,$$

and by Plancherel's theorem, $x(t)$ and $X(f)$ form a Fourier pair. However, the total power is

$$\text{total power} = \lim_{T\to\infty} \frac{1}{2T}\int_{-T}^{T} |\, x(t)\,|^2\, dt = 0. \tag{A3-43}$$

For functions that have finite total energy, the considerations of power spectral densities are *not* useful.

For the study of random processes we will adopt a different point of view. We will consider the functions to continue forever (as is indeed the case for stationary random processes) and assume they have *infinite energy over the infinite interval* but *finite energy over a finite interval.* In this way we may expect finite and nonzero total power; that is,

$$\lim_{T\to\infty} \frac{1}{2T}\int_{-T}^{+T} |\, x(t)\,|^2\, dt < \infty. \tag{A3-44}$$

We can then go ahead and define power spectral densities. This leaves us with another problem: now the modulus square of the function $x(t)$ is not integrable. Therefore, Plancherel's theorem does not apply. *The Fourier transform for the individual member functions of the ensemble may not exist*! Nevertheless, we will now define power spectral densities (which are measurable) and correlation functions for the random process and try to understand the implications of the ergodic hypothesis. We will proceed in two ways. We consider a single member of the ensemble and define the autocorrelation (and power spectral density) in the time-average sense. This will be followed by the definition of the autocorrelation function over the entire ensemble.

Consider a single member $x(t)$, whose total power is finite and nonzero, such that condition (A3-44) is satisfied. For such a function we will *assume the existence of the correlation function* $\mathcal{R}(\tau)$:

$$\mathcal{R}(\tau) = \lim_{T\to\infty} \frac{1}{2T} \int_{-T}^{T} x(t+\tau)\, x^*(t)\, dt < \infty. \tag{A3-45}$$

And, of course, by Eq. (A3-44),

$$\mathcal{R}(0) = \lim_{T\to\infty} \frac{1}{2T} \int_{-T}^{T} | x(t) |^2\, dt < \infty. \tag{A3-46}$$

By using the definition of $\mathcal{R}(\tau)$ given in Eq. (A3-37) we can show that

$$| \mathcal{R}(\tau) | \le \mathcal{R}(0). \tag{A3-47}$$

We further assume that $\mathcal{R}(\tau)$ *is absolutely integrable*:

$$\int_{-\infty}^{\infty} | \mathcal{R}(\tau) |\, d\tau < \infty. \tag{A3-48}$$

From the discussion of the Fourier theory, the Fourier transform of $\mathcal{R}(\tau)$ exists although that of $x(t)$ may not exist.

Thus the power spectral density $\mathcal{S}(f)$ defined as the Fourier transform of the autocorrelation function $\mathcal{R}(\tau)$ does exist. We define it as

$$\mathcal{R}(\tau) = \int_{-\infty}^{\infty} \mathcal{S}(f) \exp(+i2\pi f\tau)\, df. \tag{A3-49}$$

We also have that the total power in the random continuing function $x(t)$ is finite, due to Eq. (A3-46),

$$\mathcal{R}(0) = \int_{-\infty}^{\infty} \mathcal{S}(f)\, df = \text{total power}. \tag{A3-50}$$

Thus, although not much can be said about the member function $x(t)$, we may study its autocorrelation function and power spectral density.

Power Spectral Density Via Ensemble Average

We started with some basic facts about Fourier theory and dealt with a single member $x(t)$ of the ensemble at length. We had in mind a stationary random process in which the member functions continue forever. The study of single member functions is called the *generalized harmonic analysis*. This analysis was developed by Norbert Wiener. It is beyond the scope of this appendix; a particularly lucid account of it is given by Lee (1960). The autocorrelation function of a random function and the power density spectrum of the random function form a Fourier transform pair called the *Wiener–Khintchine theorem*. The fundamental assumption underlying this theorem is that the autocorrelation function exists *and* is common to all member functions of the ensemble. Because of the unique relationship between the function and its Fourier transform, the theorem implies that every member of the ensemble has the same power spectral density irrespective of its waveform. Thus, when dealing with stationary random processes, it is more appropriate to work with the autocorrelation function and the power spectral density than with the individual member functions of the ensemble.

Although it is true that a single member $x(t)$ of an ensemble does not have a Fourier transform, yet a function $\psi_T(f)$ defined by

$$\psi_T(f) = \int_0^T x(t) \exp(-i2\pi ft)\, dt \tag{A3-51}$$

can exist. Let us denote the autocorrelation of x over the interval $(0, T)$ by $R(\tau)$ and define it as in Eq. (A3-45). By using Eq. (A3-51) we can show that

$$R(\tau) = \int_{-\infty}^{\infty} df \left(\lim_{T\to\infty} \frac{|\psi_T(f)|^2}{2T} \right) \exp(i2\pi f\tau). \tag{A3-52}$$

Let us use $S(f)$ to denote the Fourier transform of $R(\tau)$; that is,

$$S(f) = \lim_{T\to\infty} \frac{|\psi_T(f)|^2}{2T}. \tag{A3-53}$$

But this limit as defined for a single member of the ensemble, in general, will not exist (see Beran and Parrent, 1964, p. 23, and Davenport and Root, 1958, p. 107). In spite of this we note that the limit may exist in the ensemble average sense. That is, we may write

$$S(f) = \lim_{T\to\infty} E\left\{ \frac{|\psi_T(f)|^2}{2T} \right\}. \tag{A3-54}$$

Whether this limit exists must be tested in every problem. To do so, we treat $|\psi_T(f)|^2$ as a random variable. It is necessary to show that the ensemble average of this new random variable approaches $S(f)$ *and* that its variance approaches zero as $T \to \infty$ (see, e.g., the discussion by Papoulis, 1965, p. 343). For our purposes we will assume the existence of the limit in Eq. (A3-54) and proceed to show that the function $S(f)$ defined as the Fourier transform of the autocorrelation function $R(\tau)$ is *nonnegative*. In this way $S(f)$ may be referred to as a true power spectral density. Since Eq. (A3-54) contains the ensemble average of a positive quantity, we must have

$$E\left\{\frac{1}{2T}|\psi_T(f)|^2\right\} \geq 0. \qquad \text{(A3-55)}$$

Upon using Eq. (A3-51) for the definition of $\psi_T(f)$, we have

$$\frac{1}{2T}\int_0^T\int_0^T R(t_1 - t_2)\exp[-i2\pi f(t_1 - t_2)]\,dt_1\,dt_2 \geq 0, \qquad \text{(A3-56)}$$

where $R(t_1 - t_2)$ is the correlation function of the stationary random process, defined through the *ensemble average*

$$R(t_1 - t_2) = E\{x(t_1)x(t_2)\}. \qquad \text{(A3-57)}$$

We make a change of variable in Eq. (A3-56) as follows:

$$t_1 + t_2 = t, \qquad t_1 - t_2 = \tau. \qquad \text{(A3-58)}$$

With this change, Eq. (A3-56) becomes

$$\left[\frac{1}{2T}\int_0^T d\tau\,R(\tau)\exp(-i2\pi f\tau) \times \int_\tau^{2T-\tau} dt \right.$$
$$\left. + \frac{1}{2T}\int_{-T}^0 d\tau\,R(\tau)\exp(i2\pi f\tau) \times \int_{-\tau}^{2T+\tau} dt\right] \geq 0. \qquad \text{(A3-59)}$$

Evaluation of the integrals on t finally produces

$$\int_{-T}^{T} d\tau\left(1 - \frac{|\tau|}{T}\right)R(\tau)\exp(-i2\pi f\tau) \geq 0.$$

The limit, $T \to \infty$, of this expression finally yields the desired result,

$$S(f) = \int_{-\infty}^{\infty} d\tau\,R(\tau)\exp(-i2\pi f\tau) \geq 0. \qquad \text{(A3-60)}$$

Thus $S(f)$, defined as the Fourier transform of $R(\tau)$, is a nonnegative function. Therefore, the function $S(f)$ is the true power spectral density in the ensemble-average sense for the stationary random process. We list the properties below:

$S(f)$ and $R(\tau)$ form a Fourier pair.
$\int_{-\infty}^{\infty} S(f)\,df = R(0) =$ total power.
$S(f) \geq 0$.

Summary

For periodic functions we found that the autocorrelation function $\mathcal{R}(\tau)$ defined in the time-average sense is such that its Fourier transform $\mathcal{S}(f)$ is nonnegative. In dealing with random functions we concluded that the Fourier transform of a single member function may not exist. It is more meaningful to talk about its autocorrelation function, whose Fourier transform exists. In Eq. (A3-60) we showed that the Fourier transform of the autocorrelation defined as an ensemble average is nonnegative. Thus the transform $S(f)$ is a true power spectral density. If ergodicity applies, then the ensemble-average autocorrelation is the same as the one defined via the time average. Then the Fourier transform of the autocorrelation defined with the time average is also a true power spectral density, since the Fourier relation is unique.

REFERENCES

Beran, M. J., and G. B. Parrent, Jr. (1964). *Theory of Partial Coherence*, Prentice-Hall, Englewood Cliffs, NJ, 189 pp.

Davenport, W. B., Jr., and W. L. Root (1958). *Random Signals and Noise*, McGraw-Hill, New York, 393 pp.

Lee, Y. W. (1960). *Statistical Theory of Communication*, Wiley, New York, 509 pp.

Papoulis, A. (1965). *Probability, Random Variables, and Stochastic Process*, McGraw-Hill, New York, 583 pp.

4

Mathematics of the Mutual Coherence Function

In this chapter we first study some mathematical properties that follow immediately from the definition of the mutual coherence function (MCF). With this preparation we then discuss the important quasimonochromatic approximation and the conditions of coherence and noncoherence.

We begin with the definition of the analytic signal,

$$V(t) = \frac{1}{\sqrt{2}} V^{(r)}(t) + \frac{i}{\sqrt{2}} V^{(i)}(t), \tag{4-1}$$

wherein the real and imaginary parts form a Hilbert-transform pair:

$$V^{(i)}(t) = \frac{1}{\pi} \fint_{-\infty}^{\infty} \frac{1}{t' - t} V^{(r)}(t')\, dt' \tag{4-2}$$

and

$$V^{(r)}(t) = -\frac{1}{\pi} \fint_{-\infty}^{\infty} \frac{1}{t' - t} V^{(i)}(t')\, dt'. \tag{4-3}$$

A simple change of the dummy variable under the integral and a suitable adjustment of the free variable on the left-hand side of Eq. (4-3) gives

$$V^{(r)}(t+\tau) = -\frac{1}{\pi}\int_{-\infty}^{\infty} \frac{1}{\tau'-\tau} V^{(i)}(t+\tau')\,d\tau'. \tag{4-4}$$

Such relationships with shifted arguments will be used frequently in what follows.

Now the MCF in terms of the time average may be defined in two equivalent ways:

$$\Gamma_{12}(\tau) = \langle V_1(t+\tau)V_2^*(t)\rangle$$

or

$$\Gamma_{12}(\tau) = \langle V_1(t)V_2^*(t-\tau)\rangle. \tag{4-5}$$

The subscripts 1 and 2 on V denote space coordinates, $V_1(t) = V(P_1, t)$. From this it follows that the MCF obeys the complex crossing symmetry relation, namely,

$$\Gamma_{12}(\tau) = \Gamma_{21}^*(-\tau). \tag{4-6}$$

This relocation of τ from the first function into the second will also be used often in dealing with the mathematical preliminaries. By using the real and imaginary parts of the analytic signal, we may express the MCF in terms of four cross correlations:

$$\Gamma_{12}(\tau) = \tfrac{1}{2}\left[\Gamma_{12}^{rr}(\tau) + \Gamma_{12}^{ii}(\tau) - i\Gamma_{12}^{ri}(\tau) + i\Gamma_{12}^{ir}(\tau)\right]. \tag{4-7}$$

The definitions of the individual cross correlations are no different from the definition of the MCF itself; for example,

$$\begin{aligned}\Gamma_{12}^{rr}(\tau) &= \langle V_1^{(r)}(t+\tau)V_2^{(r)}(t)\rangle \\ &= \langle V_1^{(r)}(t)V_2^{(r)}(t-\tau)\rangle.\end{aligned} \tag{4-8}$$

With this groundwork we shall verify three simple relationships,

(i) $\Gamma_{12}(\tau) = \Gamma_{12}^{rr}(\tau) - \mathrm{i}\Gamma_{12}^{ri}(\tau)$

(ii) $\Gamma_{11}^{ri}(0) = 0 = \Gamma_{11}^{ir}(0)$

(iii) $\Gamma_{11}(0) = \Gamma_{11}^{rr}(0),$

and the property that

(iv) $\Gamma_{12}(\tau)$ is an analytic signal.

The verifications are based on the assumption that the operation of the Hilbert transform may be interchanged with that of the time average.

To verify relationship (i), we must first show that $\Gamma_{12}^{rr}(\tau) = +\Gamma_{12}^{ii}(\tau)$. To do this, we first introduce subscript 1 on the field variable in Eq. (4-4) and multiply both sides by $V_2^{(r)}(t)$. A subsequent time-averaging operation over the variable t gives

$$\Gamma_{12}^{rr}(\tau) = -\frac{1}{\pi}\int_{-\infty}^{\infty} \frac{1}{\tau' - \tau} \langle V_1^{(i)}(t + \tau') V_2^{(r)}(t) \rangle \, d\tau'.$$

In this relation, the Hilbert operation is on the first function. We can make it work on the second by relocating τ' in the time average as mentioned before. A subsequent change of variable, $\tau' = -\tau''$, then gives

$$\Gamma_{12}^{rr}(\tau) = +\frac{1}{\pi}\int_{-\infty}^{\infty} \frac{1}{\tau'' - (-\tau)} \langle V_1^{(i)}(t)\, V_2^{(r)}(t + \tau'') \rangle \, d\tau''.$$

We are now ready to use the Hilbert relation of Eq. (4-2), which leads us to

$$\Gamma_{12}^{rr}(\tau) = +\Gamma_{12}^{ii}(\tau). \tag{4-9}$$

In a like manner we can show that

$$\Gamma_{12}^{ir}(\tau) = -\Gamma_{12}^{ri}(\tau). \tag{4-10}$$

These two relationships, when used in the expression of Eq. (4-7), immediately establish relationship (i), namely,

$$\Gamma_{12}(\tau) = \Gamma_{12}^{rr}(\tau) - i\Gamma_{12}^{ri}(\tau). \tag{4-11}$$

With all this work we are not far away from verifying relationship (ii). The coincidence of the two space points, $P_1 = P_2$, implies $V_1(t) = V_2(t)$; and the value $\tau = 0$ in Eq. (4-10) gives

$$\Gamma_{11}^{ir}(0) = -\Gamma_{11}^{ri}(0).$$

But since $\Gamma_{11}^{ri}(0)$ is obviously the same as $\Gamma_{11}^{ir}(0)$, we conclude that

$$\Gamma_{11}^{ir}(0) = 0 = \Gamma_{11}^{ri}(0), \tag{4-12}$$

which is relationship (ii).

Having verified relationship (ii) puts us squarely on the road to verifying relationship (iii), for Eq. (4-12), coupled with the expression of Eq. (4-11) for the case of coincident space points, yields

$$\Gamma_{11}(0) = \Gamma_{11}^{rr}(0). \tag{4-13}$$

This relationship assures us that the optical intensity calculated by using the analytic signal is the same as that found by using the real physical field.

The last item on the list is to show that the MCF $\Gamma_{12}(\tau)$ is an analytic signal like the field representation V, in terms of which it is defined. This entails showing that its real and imaginary parts form a Hilbert-transform pair; that is, we need to verify that

$$-\Gamma_{12}^{ri}(\tau) = +\frac{1}{\pi} ⨍_{-\infty}^{\infty} \frac{1}{\tau' - \tau} \Gamma_{12}^{rr}(\tau')\, d\tau'. \tag{4-14}$$

By looking at the right-hand side (RHS) we recognize that the Hilbert operation on the second function of $\Gamma_{12}^{rr}(\tau')$ will allow us to make contact with the left-hand side (LHS). To this end, we begin with the definition of $\Gamma_{12}^{rr}(\tau)$ in Eq. (4-8) and then change the variable τ' to $-\tau''$, to get

$$\begin{aligned} \text{RHS} &= +\frac{1}{\pi} ⨍_{-\infty}^{\infty} \frac{1}{\tau' - \tau} \langle V_1^{(r)}(t)\, V_2^{(r)}(t - \tau')\rangle \, d\tau' \\ &= -\frac{1}{\pi} ⨍_{-\infty}^{\infty} \frac{1}{\tau'' + \tau} \langle V_1^{(r)}(t)\, V_2^{(r)}(t + \tau'')\rangle \, d\tau'' \\ &= -\langle V_1^{(r)}(t)\, V_2^{(i)}(t - \tau)\rangle \\ &= -\Gamma_{12}^{ri}(\tau) = \text{LHS}. \end{aligned}$$

Thus the MCF $\Gamma_{12}(\tau)$ is also an analytic signal.

This particular property of the MCF is also verifiable for the case of nonstaionary statistics, $\Gamma_{12}(t_1, t_2)$, in which case it depends on t_1 and t_2 separately (see Appendix 4.1). As shown in the appendix, the temporal

Fourier transform is one-sided on both frequency variables,

$$\Gamma_{12}(t_1, t_2) = \iint_0^{\infty} \hat{\Gamma}_{12}(\nu_1, \nu_2) \exp[+i2\pi(\nu_1 t_1 - \nu_2 t_2)]\, d\nu_1\, d\nu_2. \quad (4\text{-}15)$$

Likewise, the Fourier transform of the MCF is one-sided; for it we shall use the notation

$$\hat{\Gamma}_{12}(\nu) = \begin{cases} \int_{-\infty}^{\infty} \Gamma_{12}(\tau) \exp(+i2\pi\nu\tau)\, d\tau, & \nu > 0 \\ 0, & \nu < 0 \end{cases} \quad (4\text{-}16)$$

with the inverse relationship

$$\Gamma_{12}(\tau) = \int_0^{\infty} \hat{\Gamma}_{12}(\nu) \exp(-i2\pi\nu\tau)\, d\nu. \quad (4\text{-}17)$$

The function $\hat{\Gamma}_{12}(\nu)$ is called the *mutual spectral density function* (MSDF). Since $\Gamma_{12}(\tau)$ obeys the complex crossing symmetry, we have

$$\hat{\Gamma}_{12}(\nu) = \hat{\Gamma}_{21}^{*}(\nu). \quad (4\text{-}18)$$

For the autocorrelation function, we have

$$\Gamma_{11}(\tau) = \int_0^{\infty} \hat{\Gamma}_{11}(\nu) \exp(-i2\pi\nu\tau)\, d\nu, \quad (4\text{-}19)$$

where $\hat{\Gamma}_{11}(\nu)$ is the optical spectral intensity (spectral density function). It is real and positive. We discussed this quantity in Chapter 3.

When the statistics are stationary, the function $\Gamma_{12}(t_1, t_2)$ assumes the form

$$\Gamma_{12}(t_1, t_2) \rightarrow \Gamma_{12}(\tau), \qquad \tau = t_2 - t_1. \quad (4\text{-}20)$$

To see what this special form implies in the frequency domain, we use the identity

$$\nu_1 t_1 - \nu_2 t_2 = -\tfrac{1}{2}(\nu_1 + \nu_2)(t_2 - t_1) + \tfrac{1}{2}(\nu_1 - \nu_2)(t_2 + t_1)$$

in the exponent of the double Fourier kernel in Eq. (4-15). By inspection, we conclude that the condition in Eq. (4-20) implies and is implied by

$$\hat{\Gamma}_{12}(\nu_1, \nu_2) \to \hat{\Gamma}_{12}(\nu_1)\,\delta(\nu_1 - \nu_2). \tag{4-21}$$

Either form may be used to indicate that the statistics are stationary.

WAVE EQUATION

The real physical field obeys the wave equation,

$$\left(\nabla_1^2 - \frac{1}{c^2}\frac{\partial^2}{\partial t_1^2}\right) V^{(r)}(\mathbf{r}_1, t_1) = 0,$$

where the subscript 1 on the Laplacian ∇^2 refers to the operation on the space coordinate $\mathbf{r}_1$. Now the imaginary part $V^{(i)}$ is defined as a linear operation (Hilbert transform) over $V^{(r)}$. The reader may easily discover that $V^{(i)}$ so defined is also a solution of the wave equation, either by going back to its Fourier decomposition or by making use of shifted arguments in $V^{(r)}(t)$ obtained after a change of variable in the Hilbert transform.

The analytic signal, being a linear combination of these two solutions, is itself a solution,

$$\left(\nabla_1^2 - \frac{1}{c^2}\frac{\partial^2}{\partial t_1^2}\right) V(\mathbf{r}_1, t_1) = 0. \tag{4-22}$$

We regard V as the random variable and $V(\mathbf{r}_1, t_1)$ as a member of the associated ensemble. Multiplication of the above equation from the right by another member, $V^*(\mathbf{r}_2, t_2)$, of the ensemble gives

$$\left(\nabla_1^2 - \frac{1}{c^2}\frac{\partial^2}{\partial t_1^2}\right) V(\mathbf{r}_1, t_1)\, V^*(\mathbf{r}_2, t_2) = 0$$

because the differential operator is ineffective on $V^*(\mathbf{r}_2, t_2)$. Since the equation holds for all the members of the ensemble, it holds for the ensemble average. The averaging procedure, however, leaves the deterministic (differential) operator unaffected and gives

$$\left(\nabla_1^2 - \frac{1}{c^2}\frac{\partial^2}{\partial t_1^2}\right) \Gamma(\mathbf{r}_1, t_1; \mathbf{r}_2, t_2) = 0. \tag{4-23}$$

Furthermore, since the field and its complex conjugate both satisfy the wave equation, a procedure similar to the above also yields

$$\left(\nabla_2^2 - \frac{1}{c^2} \frac{\partial^2}{\partial t_2^2} \right) \Gamma(\mathbf{r}_1, t_1; \mathbf{r}_2, t_2) = 0. \tag{4-24}$$

Thus the MCF obeys the wave equation on both sets of variables.

If the statistics are stationary, then the MCF is a function of the time difference $\tau = t_2 - t_1$. With this constraint the second partial derivative on time fulfills

$$\frac{\partial^2}{\partial t_1^2} = \frac{\partial^2}{\partial \tau^2} = \frac{\partial^2}{\partial t_2^2}.$$

For this case, we find that

$$\left(\nabla_s^2 - \frac{1}{c^2} \frac{\partial^2}{\partial \tau^2} \right) \Gamma(\mathbf{r}_1, \mathbf{r}_2, \tau) = 0, \qquad s = 1, 2. \tag{4-25}$$

Thus the MCF obeys the wave equation with respect to both space variables, $\mathbf{r}_1$ and $\mathbf{r}_2$. The two wave equations are *coupled* through the MCF itself and have to be solved together.

The equation satisfied by the mutual spectral density function (MSDF) $\hat{\Gamma}(\mathbf{r}_1, \mathbf{r}_2, \nu)$ is found by letting the above differential operator work on the Fourier decomposition,

$$\Gamma(\mathbf{r}_1, \mathbf{r}_2, \tau) = \int_0^\infty \hat{\Gamma}(\mathbf{r}_1, \mathbf{r}_2, \nu) \exp(-i2\pi\nu\tau)\, d\nu.$$

The operation results in

$$\int_0^\infty (\nabla_s^2 + k^2)\, \hat{\Gamma}(\mathbf{r}_1, \mathbf{r}_2, \nu) \exp(-i2\pi\nu\tau)\, d\nu = 0,$$

where $k = 2\pi\nu/c = \omega/c$ is the propagation constant. The above relationship is true for all τ, and since the Fourier frequencies are linearly independent it may be satisfied by setting the integrand equal to zero. Therefore, the MSDF obeys the Helmholtz equation,

$$(\nabla_s^2 + k^2)\, \hat{\Gamma}(\mathbf{r}_1, \mathbf{r}_2, \nu) = 0, \qquad s = 1, 2, \tag{4-26}$$

with respect to both space variables.

The wave equation and the Helmholtz equation apply to a region of space devoid of sources. In the next section we relate the MSDF within the source-free volume to that on the surface enclosing the volume.

INTEGRAL FORMULATION

We seek the mutual spectral density function $\hat{\Gamma}(\mathbf{r}_1, \mathbf{r}_2, \nu)$ in the source-free volume $\mathcal{V}$ in terms of the MSDF $\hat{\Gamma}(\mathbf{s}_1, \mathbf{s}_2, \nu)$ specified on the surface $\mathcal{S}$ enclosing the volume. The variable $\mathbf{r}$ denotes points in the volume, and $\mathbf{s}$ stands for points on the surface. The integral formulation expresses the desired solution in terms of a Green's function. In this section we present a formal treatment of the problem (see Beran and Parrent, 1964, Art. 3.2). In Chapter 5 we shall study an application to diffraction from an aperture in a plane.

A Green's function of the problem is a solution of

$$\left(\nabla_1^2 + k^2\right) \hat{G}_0(\mathbf{r}_1, \mathbf{r}_3, \nu) = -\delta(\mathbf{r}_1 - \mathbf{r}_3)$$

and

$$\left(\nabla_3^2 + k^2\right) \hat{G}_0(\mathbf{r}_1, \mathbf{r}_3, \nu) = -\delta(\mathbf{r}_1 - \mathbf{r}_3). \tag{4-27}$$

The function $\hat{G}_0$ is a known function; it exemplifies the simple physical situation of an expanding spherical wave emanating from a point source represented by the delta function. For our choice of the time dependence, $\exp(-i\omega t)$, we may choose $\hat{G}_0 = [\exp(ik\rho)]/4\pi\rho$, where $\rho = |\mathbf{r}_1 - \mathbf{r}_3|$. If the prospect of including the point source in the volume is found troublesome, remember that the infinitesimal volume enclosing that source may be omitted from the volume $\mathcal{V}$ by an appropriate distortion of the surface $\mathcal{S}$. The particular choice of the Green's function given above is referred to as the *primitive Green's function*. It is not unique. Its linear combination with any arbitrary solution ψ of the homogeneous equation

$$(\nabla^2 + k^2)\psi = 0$$

yields the "modified function"

$$G = G_0 + \psi,$$

which is also an acceptable Green's function that fulfills Eqs. (4-27). This freedom of choice is particularly useful in constructing a Green's function that obeys the desired boundary conditions of the problem. (Additional discussion of this procedure may be found in Chapter 5 and Appendix 5.2.) We shall assume that such a modified Green's function has been constructed and use it in what is to follow.

In the formal procedure we consider the pair of equations

$$\nabla_3^2 \hat{\Gamma}(\mathbf{r}_3, \mathbf{r}_2, \nu) = -k^2 \hat{\Gamma}(\mathbf{r}_3, \mathbf{r}_2, \nu)$$

and

$$\nabla_3^2 \hat{G}(\mathbf{r}_1, \mathbf{r}_3, \nu) = -k^2 \hat{G}(\mathbf{r}_1, \mathbf{r}_3, \nu) - \delta(\mathbf{r}_1 - \mathbf{r}_3).$$

Multiply, from the left, the first equation by $\hat{G}$ and the second by $\hat{\Gamma}$, subtract, and integrate over $\mathbf{r}_3$ throughout the volume $\mathcal{V}$. Due to the delta function, one side clearly gives $\hat{\Gamma}(\mathbf{r}_1, \mathbf{r}_2, \nu)$; thus,

$$\hat{\Gamma}(\mathbf{r}_1, \mathbf{r}_2, \nu) = \iiint_{\mathcal{V}} \big[\hat{G}(\mathbf{r}_1, \mathbf{r}_3, \nu)\, \nabla_3^2 \hat{\Gamma}(\mathbf{r}_3, \mathbf{r}_2, \nu) - \hat{\Gamma}(\mathbf{r}_3, \mathbf{r}_2, \nu)\, \nabla_3^2 \hat{G}(\mathbf{r}_1, \mathbf{r}_3, \nu)\big]\, d^{(3)}\mathbf{r}_3,$$

where $d^{(3)}\mathbf{r}_3$ signifies the volume element. The volume integration may be converted to a surface integral by use of Green's identity (Arfken, 1970, p. 49; Goodman, 1968, p. 34), which is essentially integration by parts in three dimensions. This procedure results in

$$\hat{\Gamma}(\mathbf{r}_1, \mathbf{r}_2, \nu) = \iint_{\mathcal{S}} \left[\hat{G}(\mathbf{r}_1, \mathbf{s}_1, \nu) \frac{\partial \hat{\Gamma}(\mathbf{s}_1, \mathbf{r}_2, \nu)}{\partial n(\mathbf{s}_1)} - \hat{\Gamma}(\mathbf{s}_1, \mathbf{r}_2, \nu) \frac{\partial \hat{G}(\mathbf{r}_1, \mathbf{s}_1, \nu)}{\partial n(\mathbf{s}_1)}\right] d\mathcal{S}_1.$$

In this notation $\mathbf{s}_1$ is a point on the surface $\mathcal{S}$ whose area element is $d\mathcal{S}_1$. The symbol $\partial/\partial n$ calls for a normal derivative, and the label $\mathbf{s}_1$ specifies that it be taken at the site of $\mathbf{s}_1$ in a direction pointing out of the volume.

A modified Green's function $\hat{G}$ that fulfills Eq. (4-27) may be chosen so that it vanishes when either of its space arguments is on $\mathcal{S}$. Or it may be possible to choose a $\hat{G}$ that vanishes when its arguments are on a portion of $\mathcal{S}$ and subsequently show that the contribution of the remaining portion of the surface integral tends to zero as that portion recedes to infinity (see, e.g., Marathay, 1975). In any case, use of the modified Green's function $\hat{G}$

formally reduces the surface integral to

$$\hat{\Gamma}(\mathbf{r}_1, \mathbf{r}_2, \nu) = -\iint_{\mathbb{S}} \hat{\Gamma}(\mathbf{s}_1, \mathbf{r}_2, \nu) \frac{\partial \hat{G}(\mathbf{r}_1, \mathbf{s}_1, \nu)}{\partial n(\mathbf{s}_1)} d\mathbb{S}_1.$$

The treatment administered to the first argument of $\hat{\Gamma}$ is also effective in a like manner on the second argument. The end result is

$$\hat{\Gamma}(\mathbf{r}_1, \mathbf{r}_2, \nu) = +\iint_{\mathbb{S}} \iint_{\mathbb{S}} \hat{\Gamma}(\mathbf{s}_1, \mathbf{s}_2, \nu) \frac{\partial \hat{G}(\mathbf{r}_1, \mathbf{s}_1, \nu)}{\partial n(\mathbf{s}_1)} \times \frac{\partial \hat{G}^*(\mathbf{r}_2, \mathbf{s}_2, \nu)}{\partial n(\mathbf{s}_2)} d\mathbb{S}_1 \, d\mathbb{S}_2, \tag{4-28}$$

which gives $\hat{\Gamma}$ at any pair of points $\mathbf{r}_1, \mathbf{r}_2$ in the volume in terms of $\hat{\Gamma}$ at *all* pairs of points $\mathbf{s}_1, \mathbf{s}_2$ on the surface. The second Green's function appears as a complex conjugate, since the field variable corresponding to $\mathbf{s}_2$ is conjugated. The MCF $\Gamma(\mathbf{r}_1, \mathbf{r}_2, \tau)$ in the volume is obtained by use of the Fourier decomposition integral. It is interesting to observe that the optical spectral intensity $\hat{\Gamma}(\mathbf{r}, \mathbf{r}, \nu)$ at any point $\mathbf{r}$ in the volume may be found only through the *mutual* spectral density function at all pairs of points on the surface, a feature characteristic of calculations in coherence theory.

COHERENT LIMIT

If by coherence we mean good fringes, then we might put $|\gamma_{12}(\tau)|$ identically equal to 1 for *all* τ. It would imply an unattainable situation of fringes covering the entire plane of observation. Nevertheless, its mathematical implication is of interest in studying the coherent limit of our calculations.

Before plunging into the mathematics, we observe from the very definition of the normalized MCF that the expression

$$|\gamma_{12}(\tau)| \equiv 1 \tag{4-29}$$

indicates the factored form

$$|\Gamma_{12}(\tau)| = [\Gamma_{11}(0)]^{1/2} [\Gamma_{22}(0)]^{1/2}.$$

The full implications were studied in an elegant paper by Mehta et al. (1966). Their results are in terms of three theorems, which we summarize here.

The theorems are about $\gamma(\mathbf{r}_1, \mathbf{r}_2, \tau)$ defined for all pairs of points $\mathbf{r}_1$ and $\mathbf{r}_2$ in a domain D. The first theorem deals with a single point $\mathbf{R} = \mathbf{r}_1 = \mathbf{r}_2$ in the domain D. It specifies the most general form that the degree of "self-coherence" $\gamma(\mathbf{R}, \mathbf{R}, \tau)$ can have if we demand that it be unimodular (i.e., its absolute value is unity) for all delays τ.

Theorem 1. If $|\gamma(\mathbf{R}, \mathbf{R}, \tau)| \equiv 1$ for all real τ, that is, $-\infty \leq \tau \leq \infty$, with $\mathbf{R}$ fixed, then

$$\gamma(\mathbf{R}, \mathbf{R}, \tau) = \exp(-i2\pi\nu_0\tau), \qquad \mathbf{R} \in D,$$

where ν_0 is a real parameter and $\nu_0 > 0$.

Furthermore, if the conditions of Theorem 1 are satisfied and if $\mathbf{r}$ is any point in D, then

$$\gamma(\mathbf{r}, \mathbf{R}, \tau) = \gamma(\mathbf{r}, \mathbf{R}, 0) \exp(-i2\pi\nu_0\tau)$$

and

$$\gamma(\mathbf{R}, \mathbf{r}, \tau) = \gamma(\mathbf{R}, \mathbf{r}, 0) \exp(-i2\pi\nu_0\tau).$$

The second theorem deals with a pair of fixed points $\mathbf{R}_1$ and $\mathbf{R}_2$ in D. It gives the most general form that the degree of coherence $\gamma(\mathbf{R}_1, \mathbf{R}_2, \tau)$ can have, if we demand that it be unimodular for all delays τ.

Theorem 2. If $|\gamma(\mathbf{R}_1, \mathbf{R}_2, \tau)| \equiv 1$ for all real τ, that is, $-\infty \leq \tau \leq \infty$, with $\mathbf{R}_1$ and $\mathbf{R}_2$ any two fixed points in D, then

$$\gamma(\mathbf{R}_1, \mathbf{R}_2, \tau) = \exp[i(\beta - 2\pi\nu_0\tau)],$$

where ν_0 and β are real constants and $\nu_0 > 0$.

Furthermore, if the conditions of Theorem 2 are satisfied and if $\mathbf{r}$ is any point in D, then as a corollary we find

$$\gamma(\mathbf{R}_1, \mathbf{R}_1, \tau) = \gamma(\mathbf{R}_2, \mathbf{R}_2, \tau) = \exp(-i2\pi\nu_0\tau)$$

and

$$\gamma(\mathbf{r}, \mathbf{R}_2, \tau) = \gamma(\mathbf{R}_1, \mathbf{R}_2, 0)\, \gamma(\mathbf{r}, \mathbf{R}_1, \tau).$$

The third theorem deals with arbitrary points $\mathbf{r}_1$ and $\mathbf{r}_2$ belonging to the domain D. It gives the most general form that the degree of coherence $\gamma(\mathbf{r}_1, \mathbf{r}_2, \tau)$ can have, if we demand that it be unimodular for all delays τ.

Theorem 3. If $|\gamma(\mathbf{r}_1, \mathbf{r}_2, \tau)| \equiv 1$ for *all* real τ, that is, $-\infty \leq \tau \leq \infty$, and for *all* pairs $\mathbf{r}_1$ and $\mathbf{r}_2$ belonging to D, then

$$\gamma(\mathbf{r}_1, \mathbf{r}_2, \tau) = \exp\{i[\beta(\mathbf{r}_1, \mathbf{r}_2) - 2\pi\nu_0\tau]\},$$

where

$$\beta(\mathbf{r}_1, \mathbf{r}_2) = \alpha(\mathbf{r}_1) - \alpha(\mathbf{r}_2),$$

and where $\alpha(\mathbf{r})$ is an arbitrary real function of a single point $\mathbf{r}$ belonging to D.

This same form for the complex degree of coherence was proved by Barakat (1966) by use of the Fourier–Stieltjes representation.

From these theorems, we observe that if the complex degree of coherence is unimodular, $|\gamma_{12}(\tau)| = 1$, in a domain D for all τ, then the MCF assumes the factored form

$$\Gamma(\mathbf{r}_1, \mathbf{r}_2, \tau) = U(\mathbf{r}_1)\, U^*(\mathbf{r}_2) \exp(-i2\pi\nu_0\tau), \tag{4-30}$$

where the optical field is monochromatic, of frequency $\nu_0 = ck_0/2\pi$, and $U(\mathbf{r})$ is any solution of the Helmholtz equation, $(\nabla^2 + k_0^2)U = 0$.

In this sense, only monochromatic fields are coherent. In practice, however, we find that fields that are not strictly monochromatic can also exhibit "good" coherence. Thus the requirement of Eq. (4-29) is too severe to be useful as a condition of coherence in practice. We shall relax the requirement and yet maintain its general features as shown in the next section.

QUASIMONOCHROMATIC APPROXIMATION

Optical fields are found in practice for which the spectral spread $\Delta\nu$ is much smaller than the mean frequency $\bar{\nu}$. These are the so-called quasimonochromatic fields (QM fields). The full implication of this condition constitutes the quasimonochromatic approximation.

For a QM field we shall assume that the MSDF $\hat{\Gamma}_{12}(\nu)$ attains appreciable values only in a small region $\Delta\nu$ around $\bar{\nu}$ and that it falls to zero or to negligible values outside this region:

$$\hat{\Gamma}_{12}(\nu) \simeq 0, \qquad |\nu - \bar{\nu}| > \Delta\nu, \tag{4-31}$$

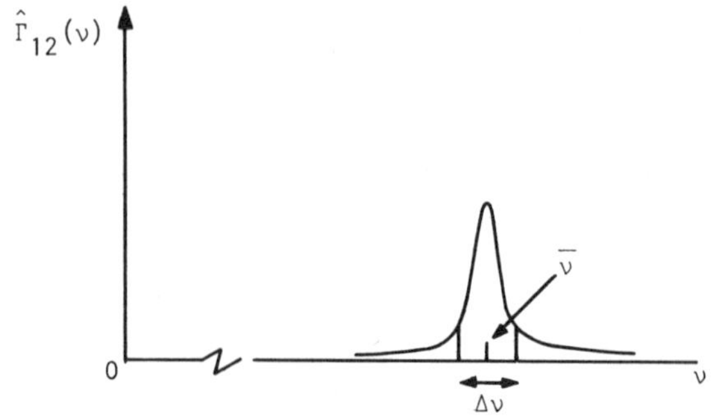

Fig. 4-1. Mutual spectral density $\hat{\Gamma}_{12}(\nu)$ for a quasimonochromatic field. $\hat{\Gamma}_{12}(\nu)$ is significant when $\bar{\nu} - \Delta\nu/2 \leq \nu \leq \bar{\nu} + \Delta\nu/2$ and $\hat{\Gamma}_{12}(\nu) \simeq 0$ for $|\nu - \bar{\nu}| > \Delta\nu$. There is no symmetry restriction about $\bar{\nu}$ on this narrow band function.

where $\Delta\nu \ll \bar{\nu}$. The situation is sketched in Fig. 4-1. The Fourier decomposition of the MCF may be rewritten in the form

$$\Gamma_{12}(\tau) = \exp(-i2\pi\bar{\nu}\tau)\int_0^\infty \hat{\Gamma}_{12}(\nu)\exp[-i2\pi(\nu - \bar{\nu})\tau]\,d\nu. \quad (4\text{-}32)$$

A significant contribution to this integral comes from within the range $|\nu - \bar{\nu}| \leq \frac{1}{2}\Delta\nu$. (An asymmetric spectrum calls for some minor modification of this condition, but the general argument is still valid.) In this case, if we should limit our attention to small enough time delays,

$$\tau \lesssim \frac{1}{\Delta\nu}, \quad (4\text{-}33)$$

then the exponential under the integral may be approximated as unity for those values of ν for which $\hat{\Gamma}_{12}(\nu)$ is significant. Hence,

$$\Gamma_{12}(\tau) \simeq \exp(-i2\pi\bar{\nu}\tau)\int_0^\infty \hat{\Gamma}_{12}(\nu)\,d\nu.$$

We have arrived at the result

$$\Gamma_{12}(\tau) = \Gamma_{12}(0)\exp(-i2\pi\bar{\nu}\tau), \qquad \tau \lesssim \frac{1}{\Delta\nu}, \quad (4\text{-}34)$$

which is the cornerstone of quasimonochromatic approximation. It describes the behavior of $\Gamma_{12}(\tau)$ for a *limited range of* τ *values* for QM fields. In this range the QM field behaves almost like a monochromatic field of frequency $\bar{\nu}$. There is an important difference, however. Due to the factor

$\Gamma_{12}(0)$, the QM field may be coherent, partially coherent, or even noncoherent! The attractive feature is that in most optical problems where it is possible to satisfy condition (4-33), we may work with the simpler function $\Gamma_{12}(0)$ instead of the full function $\Gamma_{12}(\tau)$—a considerable work saver. However, when condition (4-33) is violated, then Eq. (4-34) does not apply and the QM field may *not* be treated as almost monochromatic even if $\Delta\nu \ll \bar{\nu}$. In fact, a full polychromatic treatment via $\Gamma_{12}(\tau)$ is then called for.

The situation is now ripe for relaxing the condition of coherence. A QM field may be called *coherent* if for *all* pairs of points 1 and 2 in domain D

$$|\gamma_{12}(\tau)| = 1, \tag{4-35}$$

whenever τ satisfies the inequality of Eq. (4-33):

$$\tau \lesssim \frac{1}{\Delta\nu}.$$

Theorems 1 through 3 of Mehta et al. and the special form of Eq. (4-34) indicate that the MCF for a coherent QM field has the factored form

$$\Gamma(\mathbf{r}_1, \mathbf{r}_2, \tau) = B(\mathbf{r}_1)\, B^*(\mathbf{r}_2) \exp(-i2\pi\bar{\nu}\tau), \tag{4-36}$$

where $B(\mathbf{r})$ is a function of the single point $\mathbf{r}$. Observe that the factorization has been achieved in the presence of $\Delta\nu$.

The MCF, in general, obeys two coupled wave equations,

$$\left(\nabla_s^2 - \frac{1}{c^2}\frac{\partial^2}{\partial\tau^2}\right)\Gamma_{12}(\tau) = 0, \qquad s = 1, 2.$$

The factorization decouples them; the function $B(\mathbf{r})$ satisfies a single Helmholtz equation,

$$\left(\nabla_s^2 + \bar{k}^2\right) B(\mathbf{r}) = 0, \tag{4-37}$$

where $\bar{k} = 2\pi\bar{\nu}/c$. This is generally the procedure adopted in practice for treating the wave amplitude, $B(\mathbf{r})$, but the QM approximation tells us under what conditions it is justified. Thus, simply requiring $\Delta\nu \ll \bar{\nu}$ is not sufficient; we must also fulfill conditions (4-33) and (4-35) so that the field is coherent.

Finally, for a QM field, the familiar form of the wave amplitude,

$$V(\mathbf{r}, t) = A(\mathbf{r}, t) \exp[+i\phi(\mathbf{r}, t) - i2\pi\bar{\nu}t], \tag{4-38}$$

is physically meaningful because, for $\Delta\nu \ll \bar{\nu}$, the envelope A and the phase ϕ are slowly varying compared to the carrier, $\cos 2\pi\bar{\nu}t$. But the condition $\Delta\nu \ll \bar{\nu}$ in no way demands that the same variation of A or ϕ exist for all points in the field. Hence, $|\gamma_{12}(\tau)|$ need not be unity. In fact, it may even be zero for a particular choice of points 1 and 2.

NONCOHERENT LIMIT

If the absence of fringes in the plane of observation is to be construed as the state of noncoherence of the light from the two points P_1 and P_2, then we may set $|\gamma_{12}(\tau)| = 0$ as the condition of noncoherence for *all* pairs of points in the pinhole plane and all time delays. The implications of this condition were studied by Parrent in his Ph.D. dissertation (see Beran and Parrent, 1964, Art. 4.2).

Requirement for Noncoherence

Let there exist a noncoherent and a nonnull field in a domain D of space. Its description in terms of the mutual spectral density function is

$$|\hat{\Gamma}(\mathbf{s}_1, \mathbf{s}_2, \nu)| \equiv 0, \qquad \text{for } \mathbf{s}_1 \neq \mathbf{s}_2 \text{ in } D$$

and

$$|\hat{\Gamma}(\mathbf{s}_1, \mathbf{s}_2, \nu)| \not\equiv 0, \qquad \text{for } \mathbf{s}_1 = \mathbf{s}_2. \tag{4-39}$$

Consider a surface $\mathbb{S}$ embedded in D over which we may perform the surface integrals

$$\hat{\Gamma}(\mathbf{r}_1, \mathbf{r}_2, \nu) = \iint_{\mathbb{S}} \iint_{\mathbb{S}} \hat{\Gamma}(\mathbf{s}_1, \mathbf{s}_2, \nu) \frac{\partial \hat{G}(\mathbf{r}_1, \mathbf{s}_1, \nu)}{\partial n(\mathbf{s}_1)} \times \frac{\partial \hat{G}^*(\mathbf{r}_2, \mathbf{s}_2, \nu)}{\partial n(\mathbf{s}_2)} d\mathbb{S}_1 \, d\mathbb{S}_2.$$

The conclusions of the present discussion remain unaffected and yet gain simplification if we regard the surface $\mathbb{S}$ to be planar and suppose that the MSDF has a factored form. To describe this factorization we shall replace the point description $\mathbf{s}_1$ and $\mathbf{s}_2$ in favor of the "average" and "difference" vectors,

$$\mathbf{s} = \tfrac{1}{2}(\mathbf{s}_1 + \mathbf{s}_2)$$

and

$$\mathbf{s}_{12} = \mathbf{s}_1 - \mathbf{s}_2,$$

respectively. In terms of these variables, the factored form is expressed as

$$\hat{\Gamma}(\mathbf{s}_1, \mathbf{s}_2, \nu) = \hat{f}(\mathbf{s}, \nu)\, \hat{g}(\mathbf{s}_{12}, \nu). \tag{4-40}$$

Consistent with the conditions of Eq. (4-39) we ask that the function $\hat{g}(\mathbf{s}_{12}, \nu)$ be zero over the entire $\mathbf{s}_{12}$ axis except at the origin, $\mathbf{s}_{12} = 0$, where it should be unity. In addition, we ask that the function $\hat{f}(\mathbf{s}, \nu)$ be real and nonnegative because at the origin of the $\mathbf{s}_{12}$ axis it represents the optical spectral intensity,

$$\hat{f}(\mathbf{s}, \nu) = \hat{\Gamma}(\mathbf{s}_1, \mathbf{s}_1, \nu).$$

This particular choice of the MSDF is much like the Carter and Wolf source (see Carter and Wolf, 1977), which we shall study in detail in Chapter 5.

Due to the factored form, the area integrals become

$$\hat{\Gamma}(\mathbf{r}_1, \mathbf{r}_2, \nu) = \iint_{\mathcal{S}} \hat{f}(\mathbf{s}, \nu) \left[\iint_{\mathcal{S}_{12}} \hat{g}(\mathbf{s}_{12}, \nu) \frac{\partial \hat{G}(\mathbf{r}_1, \mathbf{s} + \frac{1}{2}\mathbf{s}_{12}, \nu)}{\partial n_+} \times \frac{\partial \hat{G}^*(\mathbf{r}_2, \mathbf{s} - \frac{1}{2}\mathbf{s}_{12}, \nu)}{\partial n_-} \, d\mathcal{S}_{12} \right] d\mathcal{S}.$$

The subscripted symbol $n_{\pm}$ is meant to indicate that the respective normal derivatives are taken at the surface points $\mathbf{s} \pm \frac{1}{2}\mathbf{s}_{12}$. The area integral over the difference variable shown in square brackets is of interest. It is zero because the integrand containing $\hat{g}$ is zero everywhere, except at an isolated point where it is nonzero and *finite*. Hence the MSDF in the field $\hat{\Gamma}(\mathbf{r}_1, \mathbf{r}_2, \nu)$ is zero everywhere, even for $\mathbf{r}_1 = \mathbf{r}_2$. The implication is that the field is null, contrary to the initial assumption.

What is amiss is that in a continuum the reduction of a fourfold integral to a twofold type cannot be accomplished with a finite integrand. If we should allow the integrand to grow indefinitely as the area element about $\mathbf{s}_{12} = 0$ goes to zero, suggesting a delta-function behavior, the integral may result in a nonzero value.

Despite the nature of our derivation, the conclusion undoubtedly survives for an arbitrary surface $\mathcal{S}$. Thus we have Parrent's theorem 1:

> A completely noncoherent field [in the sense of condition (4-39)] cannot exist in free space.

As a corollary we also have:

A completely noncoherent source surface cannot radiate.

As the discussion up to now suggests, if the MSDF $\hat{\Gamma}(\mathbf{s}_1, \mathbf{s}_2, \nu)$ is endowed with a delta-function singularity,

$$\hat{\Gamma}(\mathbf{s}_1, \mathbf{s}_2, \nu) = \beta \hat{\Gamma}(\mathbf{s}_1, \mathbf{s}_1, \nu)\, \delta^{(2)}(\mathbf{s}_1 - \mathbf{s}_2), \tag{4-41}$$

then indeed the MSDF in the field is nonzero and has the expression

$$\hat{\Gamma}(\mathbf{r}_1, \mathbf{r}_2, \nu) = \beta \iint_{\mathbb{S}} \hat{\Gamma}(\mathbf{s}_1, \mathbf{s}_1, \nu) \frac{\partial \hat{G}(\mathbf{r}_1, \mathbf{s}_1, \nu)}{\partial n(\mathbf{s}_1)} \times \frac{\partial \hat{G}^*(\mathbf{r}_2, \mathbf{s}_1, \nu)}{\partial n(\mathbf{s}_1)} d\mathbb{S}_1. \tag{4-42}$$

The symbol β is a constant and has the units of area. The delta function is two-dimensional, and because of it we now have the prospect of infinite optical intensity value at every point on $\mathbb{S}$. We should therefore look for a weaker requirement for noncoherence and in particular offer a meaningful representation of a noncoherent source.

Weaker Requirement for Noncoherence

For the following discussion we use the two-dimensional Fourier transform over the space coordinates. The notation used is displayed in the list of notation and symbols at the beginning of this book. For the basic understanding of the topic of the angular spectrum or the spatial frequency description, the reader is referred to Goodman (1968) and Gaskill (1978).

We consider a planar source and model it by use of the factored form of Eq. (4-40), much like the Carter and Wolf source. Let us suppose the source is in the plane $z = 0$, as shown in Fig. 4-2. We are interested in the radiation emanating from it into the right half-space, $z > 0$. To describe the source, we adapt for our use the symbolism and nomenclature from the radiometric literature (see Nicodemus, 1976, and Klein, 1970).

For a radiating source, one uses the term *radiant exitance*, literally describing its ability to radiate, and denote it by M. For work in the temporal frequency domain, one uses $\hat{\mathsf{M}}$ and defines it by

$$\hat{\mathsf{M}}(x_1, y_1, 0, \nu) = \mathsf{C}\hat{\Gamma}(x_1, y_1, x_1, y_1, 0, \nu). \tag{4-43}$$

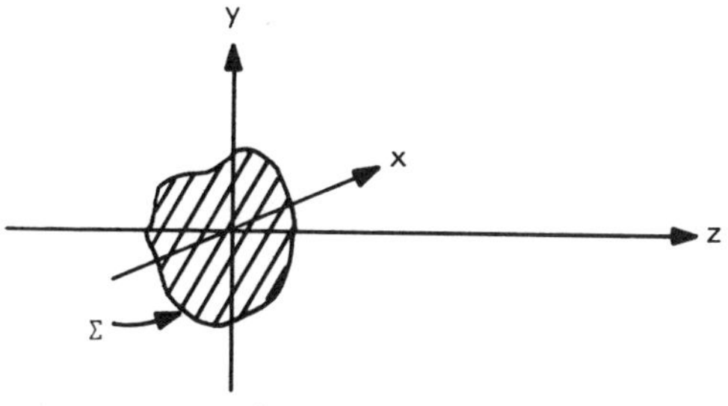

Fig. 4-2. Plane source geometry and coordinate system. Σ is the planar source of arbitrary shape in the x-y plane $z = 0$. The physical extent and the shape description are included in the functional details of $\hat{\Gamma}(x_1, y_1, x_1, y_1, 0, \nu)$. We are interested in studying the radiation in the right half-space, $z > 0$, from such a source.

The constant **C** is the same as that used before in relation to irradiance. The value zero in the argument stands for the source plane, $z = 0$. The display of the arguments x_1, y_1 of $\hat{\Gamma}$, denoting the points in the source plane, is particularly convenient in dealing with a pair of two-dimensional spatial transforms. In the temporal frequency domain, $\hat{\mathbf{M}}$ is called the *spectral radiant exitance*, with units W m^{-2} Hz^{-1} to describe the source.

Let us choose the MSDF of the source in the form

$$\hat{\Gamma}(x_1, y_1, x_2, y_2, 0, \nu) = \hat{\Gamma}(x_1, y_1, x_1, y_1, 0, \nu)\, \hat{g}(x_1 - x_2, y_1 - y_2, \nu). \tag{4-44}$$

The coherence factor $\hat{g}$ may be a real or a complex autocorrelation function, such that its zero ordinate is real and unity, $\hat{g}(0, 0, \nu) = 1$.

To mimic the delta function that was used with some success to describe noncoherence, we may put

$$|\hat{g}(x_{12}, y_{12}, \nu)| \begin{cases} \leq 1, & \text{for } \left(x_{12}^2 + y_{12}^2\right)^{1/2} \leq \varepsilon \\ = 0, & \text{for } \left(x_{12}^2 + y_{12}^2\right)^{1/2} > \varepsilon, \end{cases} \tag{4-45}$$

where $x_{12} = x_1 - x_2$ and $y_{12} = y_1 - y_2$ are the difference variables and ε is a small number. Light from some point x_1, y_1 may be regarded as coherent with the light from all points inside an ε neighborhood of it and mutually noncoherent with light from points outside it. The interval of spatial coherence is ε; in any particular application it is chosen smaller than or on the order of some characteristic dimension involved. As such we may choose

ε on the order of a few mean wavelengths,

$$\varepsilon \simeq \mathcal{O}(\bar{\lambda}). \tag{4-46}$$

Then for all intents and purposes the source will function as a noncoherent source.

However, more insight may be gained by an analysis in the spatial frequency domain. The delta-function limit of $\hat{g}$ is characterized by a *constant* spatial frequency spectrum (SFS) over both the real and evanescent waves. These latter exhibit an exponential decay as detailed in the angular spectrum formulation of diffraction theory (Appendix 5.1 and Goodman, 1968, p. 49). As such, they do not play an essential role in the radiation zone although the representation of plane waves is not complete without them. It can be established easily that, on the source, distances on the order of or smaller than the wavelength correspond to evanescent waves. Sources that exhibit field oscillations within such distances do not radiate, consistent with the corollary to Parrent's theorem.

Let us therefore focus our attention on the real waves. We describe *noncoherence* via *sharply peaked functions whose SFS is constant or very nearly constant over all the real plane waves*. Thus we may use the following types† of functions to describe noncoherence:

$$\hat{g}(x_{12}, y_{12}, \nu) = \mathrm{Sinc}(kx_{12})\,\mathrm{Sinc}(ky_{12})$$

or

$$= \mathrm{Besinc}\left(k\left(x_{12}^2 + y_{12}^2\right)^{1/2}\right)$$

or

$$= \exp\left[-k^2\left(x_{12}^2 + y_{12}^2\right)\right], \tag{4-47}$$

and other similar functions. These functions and their spectra are sketched in Fig. 4-3. Beran and Parrent (1963 and 1964, Art. 4.4) expressed the delta function as a limit of a Gaussian function and showed that the Besinc form and the delta function both yield the same result in the radiation zone provided the constant β is chosen to be (see Problem 4-1)

$$\beta = \frac{\lambda^2}{\pi}. \tag{4-48}$$

† For the definition of Besinc see List of Symbols, Notation, and Abbreviations.

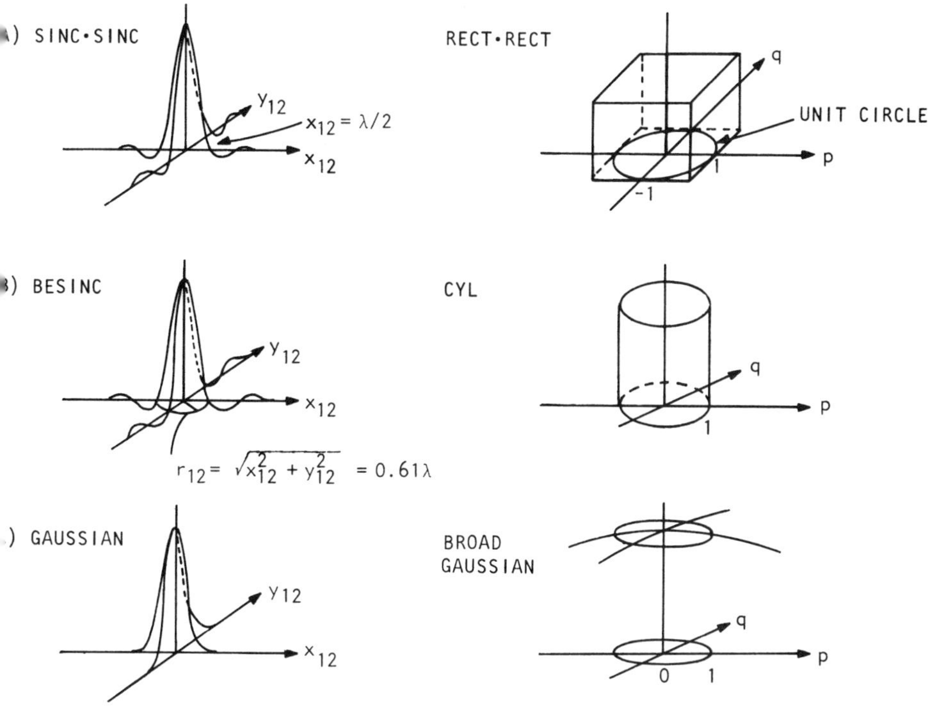

Fig. 4-3. Noncoherence: sharply peaked functions $\hat{g}(x_{12}, y_{12}, \nu)$ and their spatial frequency spectra. The rows A, B, and C picture three possibly sharply peaked functions [see Eqs. (4-45) and (4-47) of the text] in the x_{12}, y_{12} plane. In the right-hand column is given the pictorial representation of the spatial frequency spectrum. The real waves are represented by points within the unit circle in the *p-q* plane. Cases A and B have a constant spectrum over the unit circle. In case C the spectrum is very nearly a constant.

For most theoretical calculations, the delta function is a convenient shortcut and we shall use it whenever appropriate.

The essence of the discussion is that the details of the mathematical forms of the above class of sharply peaked functions are unimportant. They are all alike as far as the description of noncoherence is concerned. The most that may be said about them is that they have significant values only over short distances and attain a small fraction of their maximum value or go to zero over distances on the order of a few wavelengths (see Problem 4-2).

Such noncoherent sources are akin to the "white noise" sources of electrical engineering; see, for example, Davenport and Root (1958). These latter are delta correlated and are called "white" in analogy to white light. Their temporal frequency spectrum is a constant and is sometimes also referred to as an "equal energy" spectrum. Our noncoherent source is *white in the spatial frequency domain*. It, too, has an equal energy spectrum in the sense that it delivers the same amount of (average) power in each direction dictated by each spatial frequency in the right half-space.

Of course, for the consideration of such large angles from the normal the validity of scalar theory must be taken into account. By itself scalar theory is not a complete theory of light. Nevertheless, since scalar theory has been found remarkably useful, its full implications should be examined and its consistency should be demonstrated within its own framework.

FILTERING OF LIGHT

The discussion in this last section is aimed toward the statistical aspect of the MCF $\Gamma_{12}(\tau)$. Light is often filtered through narrow-band color filters or very-narrow-band interference filters. These are categorized as deterministic, in the sense that they have a known, fixed, spectral transmittance function, $f(\nu)$, over the entire aperture.

Consider Young's two-pinhole experiment, Fig. 4-4. Suppose the conditions of illumination (due to Σ) of the holes P_1 and P_2 are such that they are mutually noncoherent as demonstrated by the absence of the fringes even on axis. Obviously, by introducing a spectral filter $f(\nu)$ in front of the source Σ, one cannot change the mutual coherence of the light from P_1 and P_2. That is, if they were noncoherent to begin with, they will remain so after filtering. This fact is equally evident mathematically.

A pair of points for which $\Gamma_{12}(\tau) = 0$ will also have $\hat{\Gamma}_{12}(\nu) = 0$ for all ν. Now we may define $\hat{\Gamma}_{12}(\nu)$ as the ensemble average (see Chapter 3 and

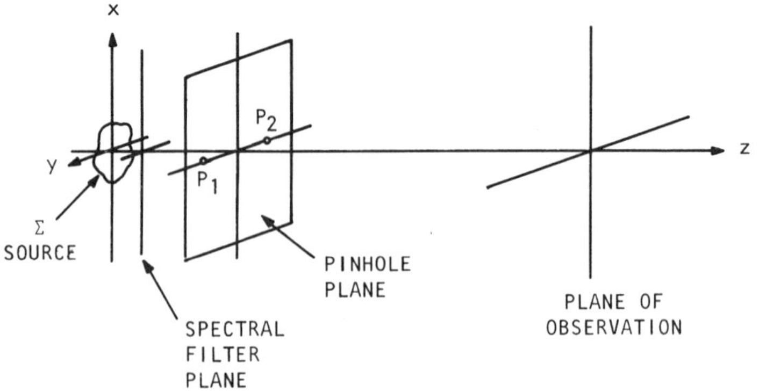

Fig. 4-4. Young's two-pinhole experiment with a spectral filter over the source Σ.

Appendix 3.2),

$$\hat{\Gamma}_{12}(\nu) = \lim_{T\to\infty} E\left\{ \frac{\hat{V}_T(\mathbf{r}_1, \nu)\hat{V}_T^*(\mathbf{r}_2, \nu)}{2T} \right\},$$

where $\hat{V}_T$ is the spectrum of a single member of the ensemble over the time interval $-T$ to T. By spectral filtering we have

$$\hat{V}_T'(\mathbf{r}, \nu) = f(\nu)\,\hat{V}_T(\mathbf{r}, \nu).$$

But since $f(\nu)$ is common to *all* members (it is deterministic with respect to this ensemble), the MSDF after filtering is

$$\hat{\Gamma}_{12}'(\nu) = |f(\nu)|^2\,\hat{\Gamma}_{12}(\nu), \qquad \text{for all } \nu.$$

Thus, if $\hat{\Gamma}_{12}(\nu) = 0$ before filtering, the MSDF after filtering will also be zero for the two points P_1 and P_2. Parrent studied this situation (see Beran and Parrent, 1964, Art. 4.2). We may state Parrent's theorem 2:

> Mutually noncoherent light from two points will continue to remain non-coherent even after spectral filtering.

The introduction of the spectral filter does change something, but that's a different story. Thus, if the thermal illumination with spectral band $\Delta\nu$ is such that the light from P_1 and P_2 is mutually coherent, then a fairly good set of fringes is seen over a region of dimensions $c\tau \lesssim c/\Delta\nu$, in the

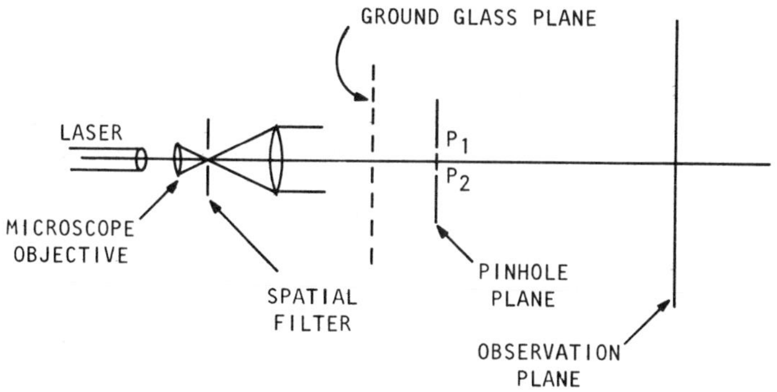

Fig. 4-5. Interference experiment with a ground glass in front of the pinhole plane. The scheme is the same as in Fig. 4-4 except that the source Σ is replaced by a beam-broadened collimated laser light.

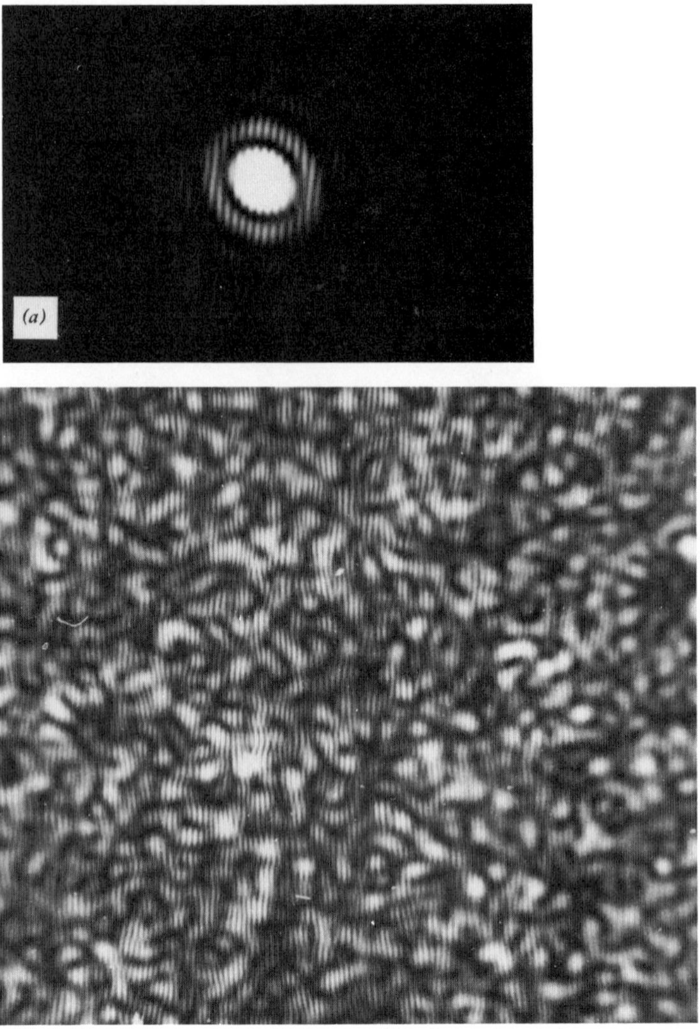

Fig. 4-6. Interference of light from two pinholes (experimental results of Considine, 1963, reproduced from Beran and Parrent, 1964). (*a*) Pattern produced before introduction of ground glass. (*b*) Fragmented pattern produced by introduction of single, stationary ground glass.

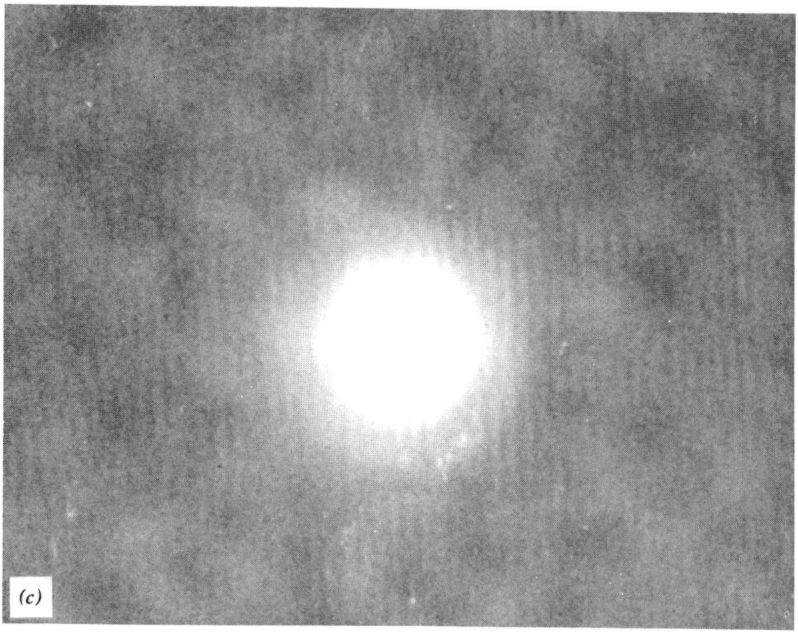

Fig. 4-6. (c) Ensemble-averaged pattern with the ground glass in slow motion.

observation plane (see the section on the quasimonochromatic approximation). Introduction of a much narrower band filter than $\Delta\nu$ merely increases the size of this region.

The approach taken up to now also tells us that the introduction of a fine-grained ground glass in front of the pinholes (see Fig. 4-5) will not suddenly change the state of coherence of the light from points P_1 and P_2 to noncoherence. A ground glass in an aperture basically spreads coherent radiation, due to its fine structure. With diligence we may plot in detail the rapid and irregular phase variations contributed by it on the transmitted beam, and it will be a fixed deterministic function. The coherence of the laser-illuminated points P_1 and P_2 is not changed by the introduction of the ground glass. That is, $\Gamma_{12}(\tau)$ has not changed in the sense that the visibility of the fringes has remained essentially the same except that the fringe pattern is now irregularly fragmented (see Figs. 4-6a and b).

However, if this same experiment is repeated using different, similarly prepared ground glasses, the story is different. First, the different ground glasses may be regarded as members of the ground-glass ensemble. Second, introducing each member in turn in front of the pinholes yields a different fragmentation of fringes. If the resulting fringe patterns are averaged, as, for

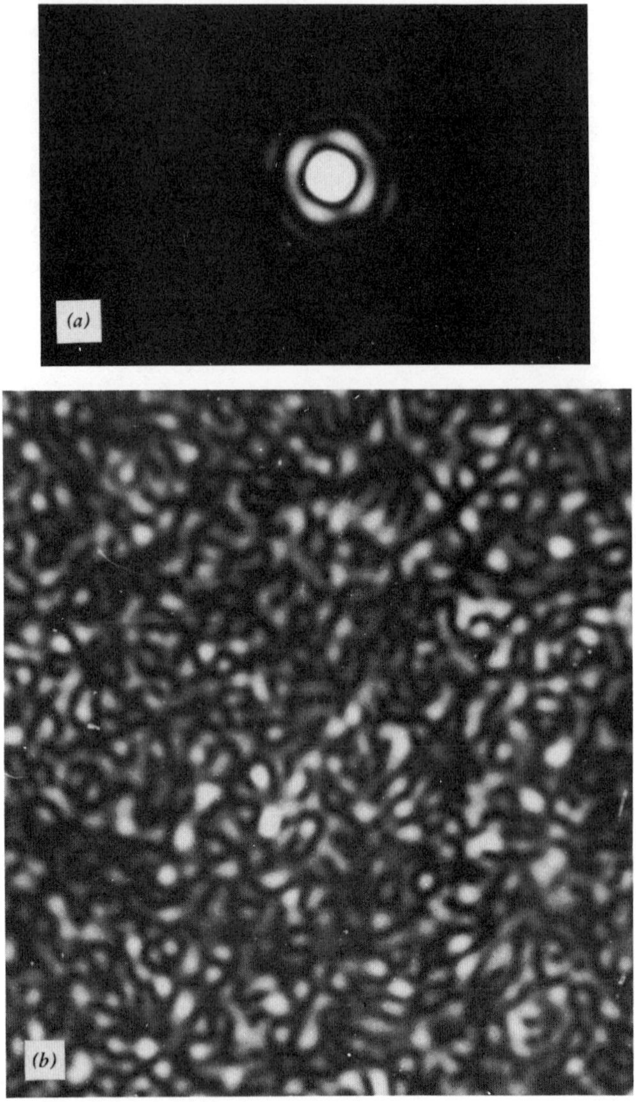

Fig. 4-7. Diffraction pattern from a single (circular) aperture (experimental results of Considine, 1963, reproduced from Beran and Parrent, 1964). (*a*) Pattern produced before introduction of ground glass. (*b*) Fragmented pattern produced by introduction of single, stationary ground glass.

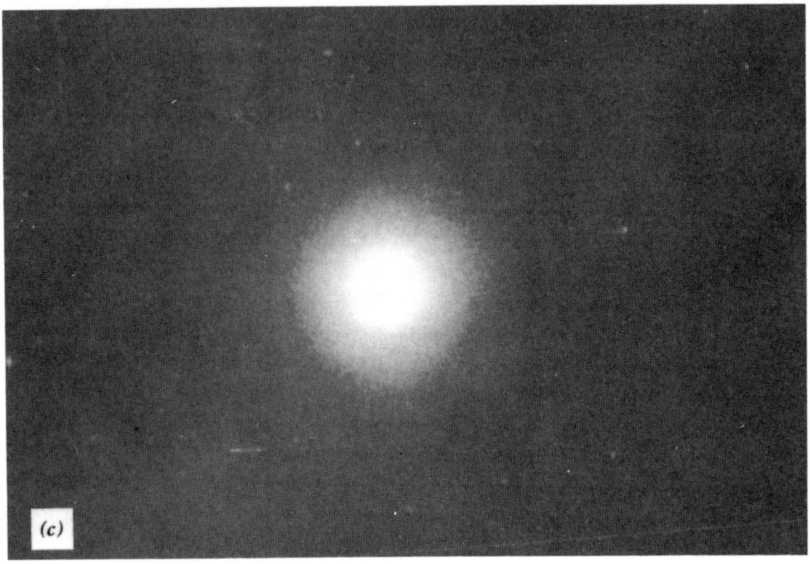

Fig. 4-7. (*c*) Ensemble-averaged pattern with the ground glass in slow motion.

example, by multiple exposure of a single photographic plate, an ensemble-averaged fringe pattern results. In this way the statistics of the resulting interfering beams are changed. In practice, one does not have to use a large number of ground glasses to bring about this ensemble average. The same effect is realized if a single member is slowly moved in front of the pinhole plane. The situation is much like the equivalence of statistical (ensemble) averaging and time averaging of Chapter 3. The evidence of the change in $\Gamma_{12}(\tau)$ via the reduction of visibility is shown in Fig. 4-6*c*. Thus, light statistics are changed, but only after the effect of the ground-glass ensemble is brought in.

The same theme is found in another experiment, wherein the diffraction pattern of a single aperture is looked at (see Fig. 4-7*a*). The introduction of a single, stationary ground glass fragments the pattern (Fig. 4-7*b*). When the glass is moved, a series of different members of the ensemble are effective during the time of measurement (exposure time). The average diffraction pattern so obtained is displayed in Fig. 4-7*c*. The reader may refer to Considine (1963, 1966) for the experiments related to the photographs of Figs. 4-6 and 4-7.

Thus, in statistical considerations, it is important to clearly define what the quantity of interest is and how it is measured, for on it depends what constitutes an ensemble and its members.

PROBLEMS

4-1. Consider a uniform source whose radiant exitance is constant and use the Besinc form of Eq. (4-47) to describe its noncoherence. Use the condition that the SFS of the Besinc function is the same as that of the delta function over the domain of real waves (i.e., for $p^2 + q^2 \leq 1$), and show that the constant β used in Eq. (4-41) has the value $\beta = \lambda^2/\pi$.

4-2. Consider a Gaussian form, $\exp[-k^2\alpha^2(x_{12}^2 + y_{12}^2)]$, for the coherence factor of Eq. (4-47). The parameter α is to be determined by imposing the condition that the SFS of the coherence factor be very nearly constant over the real waves. This is done by asking, for example, that the SFS attain values no less than 90% of its maximum value over the unit circle $p^2 + q^2 \leq 1$. For this value of α determine the coherence distance, $(x_{12}^2 + y_{12}^2)^{1/2}$, over which the coherence factor has significant values, that is, values no less than 0.8.

REFERENCES

Arfken, G. (1970). *Mathematical Methods for Physicists*, 2nd ed., Academic Press, New York, 815 pp.

Barakat, R. (1966). Theorem in coherence theory, *J. Opt. Soc. Am.* **56** (6):739–740.

Beran, M. J., and G. B. Parrent, Jr. (1963). The mutual coherence of incoherent radiation, *Nuovo Cimento* **27**:1049–1065.

Beran, M. J., and G. B. Parrent, Jr. (1964). *Theory of Partial Coherence*, Prentice-Hall, Englewood Cliffs, NJ, 189 pp.

Carter, W. H., and E. Wolf (1977). Coherence and radiometry with quasihomogeneous planar sources, *J. Opt. Soc. Am.* **67** (6):785–796.

Considine, P. S. (1963). Experimental study of coherent imaging (Abstr.), *J. Opt. Soc. Am.* **53** (11):1351.

Considine, P. S. (1966). Effects of coherence on imaging systems, *J. Opt. Soc. Am.* **56** (8):1001–1009.

Davenport, W. B., Jr., and W. L. Root (1958). *Random Signals and Noise*, McGraw-Hill, New York, Art. 6-6, pp. 104–108.

Gaskill, J. D. (1978). *Linear Systems, Fourier Transforms and Optics*, Wiley, New York, 554 pp.

Goodman, J. W. (1968). *Introduction to Fourier Optics*, McGraw-Hill, New York, 287 pp.

Klein, M. V. (1970). *Optics*, Wiley, New York, 647 pp.

Marathay, A. S. (1975). Diffraction of light from an aperture on a spherical surface, *J. Opt. Soc. Am.* **65** (8):909–913.

Mehta, C. L., E. Wolf, and A. P. Balachandran (1966). Some theorems on the unimodular complex degree of optical coherence, *J. Math. Phys.* **7** (1):133–138.

Nicodemus, F. W., Ed. (1976). *Self-Study Manual on Optical Radiation Measurements*, Part 1, Concepts, Chapters 1–3, National Bureau of Standards Technical Note 910-1.

APPENDIX 4.1. NONSTATIONARY MUTUAL COHERENCE FUNCTION

When the statistics are nonstationary, the mutual coherence function may be properly defined through the ensemble average

$$\Gamma(\mathbf{r}_1, t_1; \mathbf{r}_2, t_2) = E\{V(\mathbf{r}_1, t_1)\, V^*(\mathbf{r}_2, t_2)\}, \tag{A4-1}$$

where $\mathbf{r}_1$ and $\mathbf{r}_2$ are two space points at which the field variable V is specified; these points are described in a suitably chosen three-dimensional coordinate system. We shall use the notation of Eq. (A4-1) or abbreviate it as $\Gamma_{12}(t_1, t_2)$. This function depends on the two times t_1 and t_2 separately. We shall now undertake to define the appropriate complex function to be associated with a real function of two time variables.

Let us first define a Hilbert-transform operator [e.g., Eq. (2-18)]:

$$\hat{\mathcal{H}}_j = +\frac{1}{\pi} \int_{-\infty}^{\infty} \frac{dt_j'}{t_j' - t_j}, \qquad j = 1, 2. \tag{A4-2}$$

Now consider the real function $\Gamma_{12}^{rr}(t_1, t_2)$ defined through the ensemble average

$$\Gamma_{12}^{rr}(t_1, t_2) = E\{V^{(r)}(\mathbf{r}_1, t_1)\, V^{(r)}(\mathbf{r}_2, t_2)\}. \tag{A4-3}$$

The Hilbert transform of $\Gamma_{12}^{rr}(t_1, t_2)$ on the first time variable ($j = 1$) gives

$$\Gamma_{12}^{ir}(t_1, t_2) = +\frac{1}{\pi} \int_{-\infty}^{\infty} \frac{dt_1'}{t_1' - t_1} \Gamma_{12}^{rr}(t_1', t_2). \tag{A4-4}$$

This follows immediately from the Hilbert-transform relation between $V^{(r)}$ and $V^{(i)}$. The definition of the Hilbert-transform operator in Eq. (A4-2) allows us to write Eq. (A4-4) in the abbreviated form

$$\Gamma_{12}^{ir} = \hat{\mathcal{H}}_1 \Gamma_{12}^{rr}. \tag{A4-5}$$

Similarly we also have

$$\Gamma_{12}^{ri} = \hat{\mathcal{H}}_2 \Gamma_{12}^{rr}. \tag{A4-6}$$

Furthermore, with a double application of the Hilbert-transform operator, we find

$$\Gamma_{12}^{ii} = \hat{\mathcal{H}}_1 \hat{\mathcal{H}}_2 \Gamma_{12}^{rr}. \tag{A4-7}$$

Now since Γ_{12}^{rr} depends on two time variables, its Fourier transform $\hat{\Gamma}_{12}^{rr}$ is a function of two frequency variables,

$$\hat{\Gamma}_{12}^{rr}(\nu_1, \nu_2) = \iint_{-\infty}^{\infty} \Gamma_{12}^{rr}(t_1, t_2)\exp[-i2\pi(\nu_1 t_1 - \nu_2 t_2)]\, dt_1\, dt_2. \quad \text{(A4-8)}$$

This is the mutual spectral density function for the mutual coherence function defined through the real field variable $V^{(r)}$. Mutual spectral densities for the functions Γ_{12}^{ii}, Γ_{12}^{ri}, and Γ_{12}^{ir} may be similarly defined. These latter densities may be expressed in terms of $\hat{\Gamma}_{12}^{rr}(\nu_1, \nu_2)$ by using Eqs. (A4-5)–(A4-7). It can be shown that

$$\begin{aligned} \hat{\Gamma}_{12}^{ir}(\nu_1, \nu_2) &= -i\,\mathrm{sgn}(\nu_1)\,\hat{\Gamma}_{12}^{rr}(\nu_1, \nu_2), \\ \hat{\Gamma}_{12}^{ri}(\nu_1, \nu_2) &= +i\,\mathrm{sgn}(\nu_2)\,\hat{\Gamma}_{12}^{rr}(\nu_1, \nu_2), \\ \hat{\Gamma}_{12}^{ii}(\nu_1, \nu_2) &= +\mathrm{sgn}(\nu_1)\,\mathrm{sgn}(\nu_2)\,\hat{\Gamma}_{12}^{rr}(\nu_1, \nu_2), \end{aligned} \quad \text{(A4-9)}$$

where the sgn function is defined as

$$\mathrm{sgn}(\nu) = \begin{cases} +1, & \nu > 0 \\ -1, & \nu < 0. \end{cases}$$

Now following Eq. (4-7) for the stationary case, the mutual coherence function for the nonstationary case may be expressed as

$$\Gamma_{12}(t_1, t_2) = \tfrac{1}{2}\left[\Gamma_{12}^{rr}(t_1, t_2) + \Gamma_{12}^{ii}(t_1, t_2) - i\Gamma_{12}^{ri}(t_1, t_2) + i\Gamma_{12}^{ir}(t_1, t_2)\right]. \quad \text{(A4-10)}$$

The mutual spectral density $\hat{\Gamma}_{12}(\nu_1, \nu_2)$, defined as the Fourier transform of $\Gamma_{12}(t_1, t_2)$, may be expressed in terms of $\hat{\Gamma}_{12}^{rr}$ alone by using Eq. (A4-9). We find that

$$\hat{\Gamma}_{12}(\nu_1, \nu_2) = [1 + \mathrm{sgn}(\nu_1)\,\mathrm{sgn}(\nu_2) + \mathrm{sgn}(\nu_2) + \mathrm{sgn}(\nu_1)]\,\hat{\Gamma}_{12}^{rr}(\nu_1, \nu_2). \quad \text{(A4-11)}$$

The reason for doing all the work starting from Eq. (A4-2) to Eq. (A4-11) was to establish that the mutual spectral density function $\hat{\Gamma}_{12}(\nu_1, \nu_2)$ is identically zero for negative arguments. From Eq. (A4-11) it follows that

$$\hat{\Gamma}_{12}(\nu_1, \nu_2) = \begin{cases} 2\,\hat{\Gamma}_{12}^{rr}(\nu_1, \nu_2), & \text{for } \nu_1 > 0 \text{ and } \nu_2 > 0 \\ 0, & \text{otherwise.} \end{cases} \quad \text{(A4-12)}$$

In this sense $\Gamma_{12}(t_1, t_2)$ is also an analytic signal of *two* time variables. Due to Eq. (A4-12), the Fourier transform relation for $\Gamma_{12}(t_1, t_2)$ is

$$\Gamma_{12}(t_1, t_2) = \int\int_0^{\infty} \hat{\Gamma}_{12}(\nu_1, \nu_2) \exp[+i2\pi(\nu_1 t_1 - \nu_2 t_2)] \, d\nu_1 \, d\nu_2, \tag{A4-13}$$

which was quoted in Eq. (4-15).

5

Propagation of the Mutual Coherence Function

In the framework of the theory of partial coherence, optical problems are formulated in the language of the mutual coherence function (MCF). The fast-oscillating optical field amplitude need never be mentioned. It is replaced by the relatively slow-varying MCF. In applications, either the MCF itself or the more familiar irradiance is measured. The basic problem is to find the MCF in the observation plane illuminated by light from an aperture in a parallel plane.

APERTURE IN A PLANE

Analysis of the Problem

The geometry of the problem is shown in Fig. 5-1. The Rayleigh–Sommerfeld diffraction theory is basic to our discussion, and it is described briefly in Appendix 5.1. In addition, the reader may refer to one of several

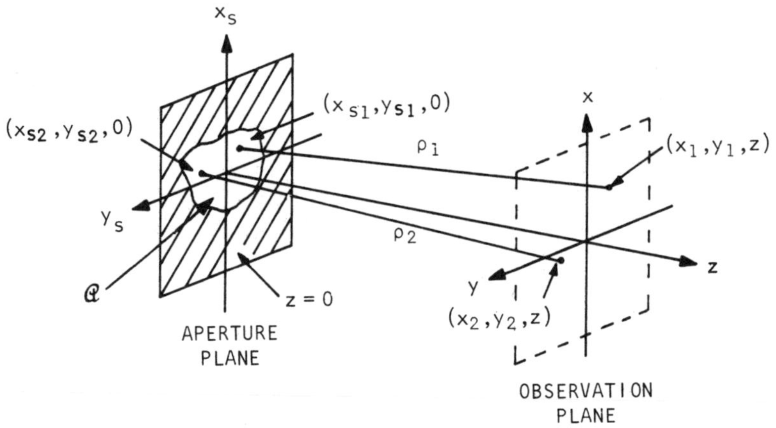

Fig. 5-1. Aperture geometry and the plane of observation for the propagation of the mutual coherence function. The aperture $\mathcal{A}$ is in the plane $z = 0$. The maximum extent d of the aperture is small compared to the distance z (i.e., $d/z \ll 1$). The symbol ρ_1 equals $|\mathbf{r}_1 - \mathbf{s}_1|$ and r_1 equals $|\mathbf{r}_1|$. Components of $\mathbf{s}_1$ are $(x_{s1}, y_{s1}, 0)$ and a similar scheme for subscript 2 is used.

excellent books on the topic, for example, Goodman (1968) and Sommerfeld (1972, Art. 34.C).

The MCF satisfies two coupled wave equations, and the mutual spectral density function (MSDF) obeys the corresponding Helmholtz equations, Eq. (4-26). The Green's function for the Helmholtz equation allows us to relate the MSDF at any pair of points in a volume to the MSDF on the surface enclosing that volume. For the problem at hand, the aperture plane is taken as a part of that surface, which in turn is enclosed by a large sphere intersecting that plane, extending into the right half-space, $z > 0$, and centered at one of the space arguments of $\hat{\Gamma}(\mathbf{r}_1, \mathbf{r}_2, \nu)$.

It is possible to construct a Green's function that vanishes when either one or the other of its space arguments assumes a position anywhere in the plane of the aperture. With this Green's function, let us discuss the surface integrals. The contribution of the sphere to the surface integral may be shown to approach zero as its radius tends to infinity, if we invoke Sommerfeld's radiation condition. This condition in the context of the MSDF may be formulated by starting with the spectrum $\hat{V}_T(\mathbf{r}, \nu)$ of the field $V(\mathbf{r}, t)$ over the time interval $-T$ to T. The ensemble average definition, Eq. (3-22), of the MSDF suggests that the radiation condition may be expressed in the form

$$\hat{\Gamma}(\mathbf{r}_1, \mathbf{r}_2, \nu) \simeq f(\theta_1, \phi_1, \theta_2, \phi_2, \nu) \frac{\exp[+ik(r_1 - r_2)]}{r_1 r_2}$$

for large distances from the aperture, that is, for $r_1, r_2 \to \infty$ (see Problem 5-1). The above form is analogous to Eq. (A5-5).

We are thus left with the area integration on the aperture plane only. In this plane, it is customary to make the following assumptions regarding the MSDF—these are called its boundary values. At all pairs of points within the aperture the MSDF is taken to have the same values (unperturbed) that it would have had in the absence of the obstructing aperture plane. In the portion of the plane outside the aperture, the MSDF is taken to be zero even for coincident space arguments. The surface integral is thus limited to the open areas $\mathscr{A}$ and contains only the normal derivative of the Green's functions,

$$\hat{\Gamma}(\mathbf{r}_1, \mathbf{r}_2, \nu) = \iint_{\mathscr{A}} \iint_{\mathscr{A}} \hat{\Gamma}(\mathbf{s}_1, \mathbf{s}_2, \nu) \frac{\partial \hat{G}(\mathbf{r}_1, \mathbf{s}_1, \nu)}{\partial n(\mathbf{s}_1)} \frac{\partial \hat{G}^*(\mathbf{r}_2, \mathbf{s}_2, \nu)}{\partial n(\mathbf{s}_2)} \times dx_{s1}\, dy_{s1}\, dx_{s2}\, dy_{s2}.$$

In relation to the aperture, the outward normal points along the $-z$ axis.

The derivative taken with respect to this variable and evaluated at $z = 0$ for the aperture plane leads to the expression

$$\hat{\Gamma}(x_1, y_1, x_2, y_2, z, \nu) = \iint_{\mathcal{A}} \iint_{\mathcal{A}} \hat{\Gamma}(x_{s1}, y_{s1}, x_{s2}, y_{s2}, 0, \nu)$$
$$\times \left[\frac{1}{2\pi} \cos\theta_1 (1 - ik\rho_1) \frac{\exp(ik\rho_1)}{\rho_1^2} \right]$$
$$\times \left[\frac{1}{2\pi} \cos\theta_2 (1 + ik\rho_2) \frac{\exp(-ik\rho_2)}{\rho_2^2} \right]$$
$$\times dx_{s1}\, dy_{s1}\, dx_{s2}\, dy_{s2}. \tag{5-1}$$

The display of the arguments of $\hat{\Gamma}$ shows the pairs of points in a given plane, $z =$ constant. The distance between an aperture point and a point in the observation plane is denoted by

$$\rho_j = |\mathbf{r}_j - \mathbf{s}_j|$$
$$= \left[(x_j - x_{sj})^2 + (y_j - y_{sj})^2 + z^2 \right]^{1/2}, \qquad j = 1, 2, \tag{5-2}$$

and the direction cosine with respect to the z axis is

$$\cos\theta_j = \frac{z}{\rho_j}, \qquad j = 1, 2. \tag{5-3}$$

The second bracketed expression in Eq. (5-1) with subscript 2 replaced by 1 is the complex conjugate of the first one. As an alternative derivation of the MSDF in the observation plane, we might begin with the expression of the spectrum $\hat{V}_T(\mathbf{r}, \nu)$ of the field $V(\mathbf{r}, t)$ in the time interval $-T$ to T. The derivation of the expression in Appendix 5.1 is in no way affected by the presence of the interval $-T$ to T. A product of the spectrum $\hat{V}_T(\mathbf{r}_1, \nu)$ with its complex conjugate spectrum $\hat{V}_T^*(\mathbf{r}_2, \nu)$ at another field point $\mathbf{r}_2$ is formed. The use of the ensemble average definition of $\hat{\Gamma}_{12}(\nu)$ of Eq. (3-22) on both sides of the equation will lead us to Eq. (5-1).

The remark in Appendix 5.1 about the boundary values of the field also applies in relation to the MSDF. Thus, the MSDF in the portion of the plane outside the aperture is never identically zero, and its values inside the aperture are, in fact, perturbed by the presence of the aperture. However, the assumed boundary conditions may be conveniently used if the aperture dimensions are very large compared to the wavelength λ, as is usually the case in practice. Conversely, the boundary conditions may turn out to be wholly inappropriate if the primary interest is in the field fluctuations over distances on the order of λ in $\mathcal{Q}$ or if the aperture dimensions are comparable to λ. In the latter case, the field outside the aperture may not be neglected.

In order to complete the parallelism with the development of the Rayleigh–Sommerfeld theory, we seek the spatial frequency domain version of Eq. (5-1). The spatial Fourier transform over a pair of points in a plane is

$$\mathring{\Gamma}(\kappa p_1, \kappa q_1, \kappa p_2, \kappa q_2, z, \nu) = \iiiint_{-\infty}^{\infty} \hat{\Gamma}(x_1, y_1, x_2, y_2, z, \nu) \times \exp[-i2\pi\kappa(p_1 x_1 + q_1 y_1 - p_2 x_2 - q_2 y_2)] \times dx_1\, dy_1\, dx_2\, dy_2. \tag{5-4}$$

The desired product relationship in the domain of spatial frequencies is

$$\mathring{\Gamma}(\kappa p_1, \kappa q_1, \kappa p_2, \kappa q_2, z, \nu) = \mathring{\Gamma}(\kappa p_1, \kappa q_1, \kappa p_2, \kappa q_2, 0, \nu) \exp[+ik(m_1 - m_2^*)z], \tag{5-5}$$

which is analogous to Eq. (A5-8) for the wave amplitude. For real waves the third direction cosine m is real,

$$m = +(1 - p^2 - q^2)^{1/2}, \qquad p^2 + q^2 \leq 1.$$

For evanescent waves it is pure imaginary,

$$m = +i(p^2 + q^2 - 1)^{1/2}, \qquad p^2 + q^2 > 1.$$

The choice of this sign assures us that the factor containing $m_1 - m_2^*$ will exponentially decay away ($z > 0$) from the aperture plane when $p^2 + q^2 > 1$ with appropriate subscripts.

Let us return to Eq. (5-1), which connects the MSDF in the aperture plane to that in the plane of observation. It represents the propagation of MSDF from one plane to another downstream. The MCF itself is recovered through the Fourier integral,

$$\Gamma(\mathbf{r}_1, \mathbf{r}_2, \tau) = \int_0^{\infty} \hat{\Gamma}(\mathbf{r}_1, \mathbf{r}_2, \nu) \exp(-i2\pi\nu\tau)\, d\nu. \tag{5-6}$$

The theory is applicable to partially coherent and polychromatic fields. This law of propagation is called the *generalized* van Cittert–Zernike theorem (see Beran and Parrent, 1964). Historically, van Cittert and Zernike derived their theorem for a noncoherent source, which we shall discuss later in this chapter. We observe from the generalized law:

1. It is linear, as is the basic wave equation.
2. The MSDF in the observation plane is found through the MSDF over the aperture $\mathcal{Q}$ with the pair of points exploring it independently.
3. The optical spectral intensity $\hat{\Gamma}(\mathbf{r}, \mathbf{r}, \nu)$ in the observation plane is found by knowing the full MSDF over $\mathcal{Q}$. The knowledge of just the optical spectral intensity over $\mathcal{Q}$ is insufficient.

Example of a Partially Coherent Source

We wish to calculate the spatial coherence in a plane $z =$ constant, by way of an example of a partially coherent source (Marathay, 1981). Let us suppose that the plane $z = 0$ contains a Carter–Wolf type source (see Carter and Wolf, 1977). Its MSDF is most easily described by means of average and difference variables,

$$\begin{aligned} x_s &= \tfrac{1}{2}(x_{s1} + x_{s2}) \quad \text{and} \quad y_s = \tfrac{1}{2}(y_{s1} + y_{s2}) \\ x_{s12} &= x_{s1} - x_{s2} \quad \text{and} \quad y_{s12} = y_{s1} - y_{s2}. \end{aligned} \tag{5-7}$$

Similarly, the direction cosines may be expressed by

$$\begin{aligned} p &= \tfrac{1}{2}(p_1 + p_2) \quad \text{and} \quad q = \tfrac{1}{2}(q_1 + q_2) \\ p_{12} &= p_1 - p_2 \quad \text{and} \quad q_{12} = q_1 - q_2. \end{aligned} \tag{5-8}$$

In the Fourier kernel we may use

$$p_1 x_{s1} + q_1 y_{s1} - p_2 x_{s2} - q_2 y_{s2} = p x_{s12} + q y_{s12} + p_{12} x_s + q_{12} y_s, \tag{5-9}$$

and for the area elements note that

$$dx_{s1}\, dy_{s1}\, dx_{s2}\, dy_{s2} = dx_s\, dy_s\, dx_{s12}\, dy_{s12}, \tag{5-10}$$

since the Jacobian of transformation is unity. For this topic the reader may refer to any one of several books on mathematical physics, for example, Sokolnikoff and Redheffer (1958).

The MSDF of the Carter–Wolf source is

$$\begin{aligned}\hat{\Gamma}_s(x_{s1}, y_{s1}, x_{s2}, y_{s2}, 0, \nu) &= \hat{\Gamma}_s(x_s + \tfrac{1}{2}x_{s12}, y_s + \tfrac{1}{2}y_{s12}, x_s - \tfrac{1}{2}x_{s12}, y_s - \tfrac{1}{2}y_{s12}, 0, \nu) \\ &= \hat{I}_s(x_s, y_s, \nu)\, \hat{g}_s(x_{s12}, y_{s12}, \nu). \qquad (5\text{-}11)\end{aligned}$$

The function $\hat{I}_s$ describes the optical spectral intensity (OSI) across the source, and $\hat{g}_s$ dictates the spatial coherence. The function $\hat{I}_s$ is real and nonnegative and is very broad compared to $\hat{g}_s$; for this reason the source is often called quasihomogeneous. The unitless function $\hat{g}_s$ is a suitably chosen autocorrelation function with the subsidiary conditions

$$\begin{aligned}\hat{g}_s(0, 0, \nu) &= 1, \\ \hat{g}_s(x_{s12}, y_{s12}, \nu) &= \hat{g}_s^*(x_{s21}, y_{s21}, \nu).\end{aligned} \qquad (5\text{-}12)$$

With the average and difference variables it is not difficult to show that the total Fourier transform is

$$\begin{aligned}\mathring{\Gamma}_s(\kappa p + \tfrac{1}{2}\kappa p_{12}, \kappa q + \tfrac{1}{2}\kappa q_{12}, \kappa p - \tfrac{1}{2}\kappa p_{12}, \kappa q - \tfrac{1}{2}\kappa q_{12}, 0, \nu) \\ = \mathring{I}_s(\kappa p_{12}, \kappa q_{12}, \nu)\, \mathring{g}_s(\kappa p, \kappa q, \nu).\end{aligned} \qquad (5\text{-}13)$$

For a partially coherent source, the spatial coherence in a plane $z > 0$ is found by using Eq. (5-1), or, equivalently, by taking the spatial Fourier transform of the product relationship of Eq. (5-5). For the present example the second method is simpler. It gives

$$\begin{aligned}\hat{\Gamma}(x + \tfrac{1}{2}x_{12}, y + \tfrac{1}{2}y_{12}, x - \tfrac{1}{2}x_{12}, y - \tfrac{1}{2}y_{12}, z, \nu) \\ = \iiiint_{-\infty}^{\infty} \mathring{I}_s(\kappa p_{12}, \kappa q_{12}, \nu)\, \mathring{g}_s(\kappa p, \kappa q, \nu) \exp[ik(m_1 - m_2^*)z] \\ \times \exp[i2\pi\kappa(p x_{12} + q y_{12} + p_{12} x + q_{12} y)] \\ \times d(\kappa p)\, d(\kappa q)\, d(\kappa p_{12})\, d(\kappa q_{12}).\end{aligned}$$

As before, we used the average and difference variables also in the plane $z =$ constant, and the differentials are fashioned after Eq. (5-10). For the forward direction (small angles), the combination $m_1 - m_2^*$ may be approximated up to quadratic terms to give

$$m_1 - m_2^* \simeq -(p p_{12} + q q_{12}). \qquad (5\text{-}14)$$

Thus, to a good approximation, the MSDF may be expressed in one of two equivalent ways:

$$\begin{aligned}
&\hat{\Gamma}\left(x+\tfrac{1}{2}x_{12},\, y+\tfrac{1}{2}y_{12},\, x-\tfrac{1}{2}x_{12},\, y-\tfrac{1}{2}y_{12},\, z,\, \nu\right)\\
&\quad=\iint_{-\infty}^{\infty} \mathring{I}_s(\kappa p_{12}, \kappa q_{12}, \nu)\, \hat{g}_s(x_{12}-p_{12}z,\, y_{12}-q_{12}z,\, \nu)\\
&\qquad\times \exp\left[i2\pi\kappa(p_{12}x+q_{12}y)\right] d(\kappa p_{12})\, d(\kappa q_{12})\\
&\quad=\iint_{-\infty}^{\infty} \hat{I}_s(x-pz,\, y-qz,\, \nu)\, \mathring{g}_s(\kappa p, \kappa q, \nu)\\
&\qquad\times \exp\left[i2\pi\kappa(px_{12}+qy_{12})\right] d(\kappa p)\, d(\kappa q).
\end{aligned} \tag{5-15}$$

It contains information about the OSI as well as about the spatial coherence of light in the plane $z=$ constant. In the first expression, the x, y dependence of the MSDF is found by Fourier transforming a product of two narrow functions. In the second expression, the x_{12}, y_{12} dependence is found by transforming a product of two broad functions. Therefore, in the illuminated plane the OSI, which is a function of x and y, is a broad function compared with the spatial coherence, which is a function of x_{12} and y_{12}.

By setting $x_{12}=0=y_{12}$, we may express the OSI in the plane $z=$ constant by the convolution

$$\begin{aligned}
&\hat{\Gamma}(x, y, x, y, z, \nu)\\
&\quad=\iint_{-\infty}^{\infty} \hat{I}_s(x-pz,\, y-qz,\, \nu)\, \mathring{g}_s(\kappa p, \kappa q, \nu)\, d(\kappa p)\, d(\kappa q).
\end{aligned} \tag{5-16}$$

This expression is nonnegative because, like $\hat{I}_s$, the function $\mathring{g}_s$ is real and nonnegative, since it is a spatial Fourier transform of an autocorrelation function $\hat{g}_s$.

To further our understanding, let us take a specific example wherein $\hat{I}_s$ and $\hat{g}_s$ are both Gaussian functions:

$$\begin{aligned}
\hat{I}_s(x_s, y_s, \nu) &= I_Q \exp\left(-\frac{x_s^2+y_s^2}{2\sigma_Q^2}\right),\\
\hat{g}_s(x_{s12}, y_{s12}, \nu) &= \exp\left(-\frac{x_{s12}^2+y_{s12}^2}{2\sigma_g^2}\right),
\end{aligned} \tag{5-17}$$

where we ask that $\sigma_Q \gg \sigma_g$. This choice of functions is consistent with the conditions discussed in Eqs. (5-11) and (5-12). Since the Fourier transform of a Gaussian is also a Gaussian, the calculation is straightforward (see Problem 5-2). For the MSDF in the observation plane $z > 0$ we get

$$\hat{\Gamma}(x + \tfrac{1}{2}x_{12}, y + \tfrac{1}{2}y_{12}, x - \tfrac{1}{2}x_{12}, y - \tfrac{1}{2}y_{12}, z, \nu)$$
$$= I_Q\left(\frac{k^2\sigma_Q^2\sigma_g^2}{z^2 + k^2\sigma_Q^2\sigma_g^2}\right)\exp\left[-\frac{x^2 + y^2}{2\sigma_Q^2}\left(\frac{k^2\sigma_Q^2\sigma_g^2}{z^2 + k^2\sigma_Q^2\sigma_g^2}\right)\right]$$
$$\times\exp\left[\frac{ik(xx_{12} + yy_{12})z}{z^2 + k^2\sigma_Q^2\sigma_g^2}\right]\exp\left[-\frac{x_{12}^2 + y_{12}^2}{2\sigma_g^2}\left(\frac{k^2\sigma_Q^2\sigma_g^2}{z^2 + k^2\sigma_Q^2\sigma_g^2}\right)\right]. \tag{5-18}$$

Observe that when $z = 0$ the quadratic phase factor goes to unity and the rest of the expression reduces to the MSDF of the source. Furthermore, the spatial coherence in the observation plane is also Gaussian, much like the source function $\hat{g}_s$ with a modified standard deviation,

$$\sigma_g'(z) = \left(\sigma_g^2 + \frac{z^2}{k^2\sigma_Q^2}\right)^{1/2}, \tag{5-19}$$

which is such that $\sigma_g'(z) > \sigma_g$. The distance over which the spatial coherence is essentially constant, namely, the coherence interval, increases with z. The light gets more coherent as it propagates away from the source.

The OSI on the observation plane is found by setting $x_{12} = 0 = y_{12}$:

$$\hat{\Gamma}(x, y, x, y, z, \nu) = I_Q\left(\frac{k^2\sigma_Q^2\sigma_g^2}{z^2 + k^2\sigma_Q^2\sigma_g^2}\right)\exp\left[-\frac{x^2 + y^2}{2\sigma_Q^2}\left(\frac{k^2\sigma_Q^2\sigma_g^2}{z^2 + k^2\sigma_Q^2\sigma_g^2}\right)\right]. \tag{5-20}$$

This expression may be calculated directly by convolving two Gaussians as prescribed by Eq. (5-16). The OSI in the observation plane is much like the source function $\hat{I}_s$ with a modified standard deviation:

$$\sigma_Q'(z) = \left(\sigma_Q^2 + \frac{z^2}{k^2\sigma_g^2}\right)^{1/2}. \tag{5-21}$$

The area over which the OSI remains nearly constant increases with z and, just as in the source plane, $\sigma_Q'(z) \gg \sigma_g'(z)$. In this example, the coherence properties of the illuminated region are just the scaled version of those of the source.

TOTAL RADIANT POWER

In the problem of the propagation of the MSDF, light is incident on an aperture $\mathcal{A}$ in an otherwise opaque plane. The aperture $\mathcal{A}$ admits light into the right half-space, $z > 0$. We expect that the amount of light intersecting *any* plane $z =$ constant, parallel to the aperture plane, will be the same as the amount that the aperture admitted into the right half-space.

In Eq. (4-43) we talked about the spectral radiant exitance $\hat{\mathsf{M}}$ from the source aperture,

$$\hat{\mathsf{M}}(x_1, y_1, 0, \nu) = \mathsf{C}\hat{\Gamma}(x_1, y_1, x_1, y_1, 0, \nu),$$

with units of W m^{-2} Hz^{-1}. If this quantity is integrated over the whole area of the aperture, we obtain the total spectral radiant power, $\hat{\mathbf{\Phi}}_{in}$, with units W Hz^{-1}, contained in it:

$$\hat{\mathbf{\Phi}}_{in}(0, \nu) = \int\int_{-\infty}^{\infty} \hat{\mathsf{M}}(x_1, y_1, 0, \nu)\, dx_1\, dy_1. \tag{5-22}$$

The limits of integration are only formally infinite, since the assumed boundary condition dictates the integration over the open area of the aperture $\mathcal{A}$. The spatial Fourier transform of the integrand is given in Eq. (5-4). The indicated area integration over x_1, y_1 yields the spatial frequency version,

$$\hat{\mathbf{\Phi}}_{in}(0, \nu) = \int\int_{-\infty}^{\infty} \mathring{\mathsf{M}}(\kappa p_1, \kappa q_1, 0, \nu)\, \kappa^2\, dp_1\, dq_1, \tag{5-23}$$

which is simply a restatement of Parseval's theorem, where $\mathring{\mathsf{M}}$ is identified with $\mathsf{C}\mathring{\Gamma}$.

A similar description is possible over a plane of observation at z in the radiated field. With respect to this plane, the spectral radiant power $\hat{\mathbf{\Phi}}_{out}$ is calculated in terms of the spectral irradiance $\hat{\mathsf{E}}$. The area integral over the plane $z =$ constant gives

$$\hat{\mathbf{\Phi}}_{out}(z, \nu) = \int\int_{-\infty}^{\infty} \hat{\mathsf{E}}(x, y, z, \nu)\, dx\, dy, \tag{5-24}$$

which in the spatial frequency domain has the form

$$\hat{\Phi}_{\text{out}}(z,\nu)=\iint_{-\infty}^{\infty}\mathring{\mathsf{E}}(\kappa p,\kappa q,z,\nu)\,\kappa^2\,dp\,dq. \tag{5-25}$$

The spectral radiant power in the plane at z can be connected to that admitted by the aperture only via a diffraction calculation. With the results of the previous section, $\hat{\Phi}_{\text{out}}$ is first expressed in terms of the boundary values over the aperture,

$$\hat{\Phi}_{\text{out}}(z,\nu)=\mathsf{C}\iint_{-\infty}^{\infty}dx_{s1}\,dy_{s1}\iint_{-\infty}^{\infty}dx_{s2}\,dy_{s2}\hat{\Gamma}(\mathbf{s}_1,\mathbf{s}_2,\nu)$$
$$\times\left[\iint_{-\infty}^{\infty}dx\,dy\frac{\partial\hat{G}(\mathbf{r},\mathbf{s}_1,\nu)}{\partial n(\mathbf{s}_1)}\,\frac{\partial\hat{G}^*(\mathbf{r},\mathbf{s}_2,\nu)}{\partial n(\mathbf{s}_2)}\right]. \tag{5-26}$$

The integrals over the aperture variables x_s, y_s are restricted only over the open areas of $\mathcal{Q}$, due to the assumed boundary conditions. The vector **s** stands for the aperture variables in the plane at $z=0$, while **r** has components x, y in the plane at z. The area integral over x, y, to be denoted by D, is more manageable in the spatial frequency domain since

$$\frac{-\partial\hat{G}(\mathbf{r},\mathbf{s}_1,\nu)}{\partial n(\mathbf{s}_1)}=\iint_{-\infty}^{\infty}\kappa^2\,dp_1\,dq_1\exp(ikmz)$$
$$\times\exp\{ik[p_1(x-x_{s1})+q_1(y-y_{s1})]\}, \tag{5-27}$$

as discussed in Appendixes 5.1 and 5.2. The desired integral assumes the form

$$D=\iint_{-\infty}^{\infty}\exp[ik(m_1-m_1^*)z]$$
$$\times\exp\{-ik[p_1(x_{s1}-x_{s2})+q_1(y_{s1}-y_{s2})]\}\,\kappa^2\,dp_1\,dq_1.$$

For the real waves, the exponential containing m_1 is unity, indicating a constant spatial frequency spectrum in this domain. The evanescent wave spectrum decays with z away from the aperture plane. Evaluation of the integral of the real waves gives us

$$D = \frac{k^2}{4\pi} \frac{2J_1\left\{k\left[(x_{s1} - x_{s2})^2 + (y_{s1} - y_{s2})^2\right]^{1/2}\right\}}{k\left[(x_{s1} - x_{s2})^2 + (y_{s1} - y_{s2})^2\right]^{1/2}}$$
$$+ \iint_{p_1^2+q_1^2>1} \exp(-2k\,|m_1|\,z)$$
$$\times \exp\{-ik[p_1(x_{s1} - x_{s2}) + q_1(y_{s1} - y_{s2})]\}\kappa^2\,dp_1\,dq_1. \quad (5\text{-}28)$$

Thus, since $D \neq \delta(x_{s1} - x_{s2})\,\delta(y_{s1} - y_{s2})$, the implied message is that $\hat{\mathbf{\Phi}}_{\text{out}} \neq \hat{\mathbf{\Phi}}_{\text{in}}$! The same answer is found via Eq. (5-5).

The most that may be said is that far enough away from the aperture (as $kz \to \infty$) the evanescent wave contribution tends to zero and the real waves arrive unchanged. Also from Eq. (5-5), if we set $p_1 = p_2$ and $q_1 = q_2$, the real-wave spectrum in the radiated field is the same as that in the aperture plane. Thus

$$\hat{\mathbf{\Phi}}_{\text{out}}(z, \nu) = \hat{\mathbf{\Phi}}_{\text{in}}(0, \nu) \quad (5\text{-}29)$$

will hold to any degree of verification far enough away from the aperture, where the evanescent waves may be neglected.

In optics the quantity $\hat{\mathsf{M}}$ is directly accessible to measurement. We shall therefore continue to use it and accept the equality in Eq. (5-29).

PROPAGATION OF QUASIMONOCHROMATIC FIELDS

The basic problem of finding the MCF $\Gamma_{12}(\tau)$ in the observation plane illuminated by light from an aperture in a parallel plane was dealt with in the beginning of this chapter. It was accomplished by connecting the mutual spectral densities from the aperture to the observation plane via Eq. (5-1). The MCF itself is found through the temporal Fourier transform of the MSDF in the respective planes. The procedure accounts for all the spectral components of the polychromatic field.

Considerable simplification is obtained for quasimonochromatic (QM) fields characterized by the spectral spread $\Delta\nu$ much smaller than the mean frequency $\bar{\nu}$, $\Delta\nu \ll \bar{\nu}$. The "small-path-difference" condition discussed before in Chapter 4 imposes certain conditions on the propagation problem. These are frequently fulfilled in practice, in which case we may simply study the mutual optical intensity function $\Gamma_{12}(0)$,

$$\Gamma_{12}(0) = \int_0^{\infty} \hat{\Gamma}_{12}(\nu)\, d\nu, \tag{5-30}$$

instead of the full MCF $\Gamma_{12}(\tau)$. We therefore seek a relationship between the $\Gamma_{12}(0)$ in the aperture plane and the $\Gamma_{12}(0)$ in the observation plane.

We begin with the relationship

$$\Gamma(\mathbf{r}_1, \mathbf{r}_2, \tau) = \int_0^{\infty} \exp(-i2\pi\nu\tau) \iint_{\mathscr{A}} \iint_{\mathscr{A}} \hat{\Gamma}(\mathbf{s}_1, \mathbf{s}_2, \nu) \times \left[\frac{\partial \hat{G}(\mathbf{r}_1, \mathbf{s}_1, \nu)}{\partial n(\mathbf{s}_1)} \frac{\partial \hat{G}^*(\mathbf{r}_2, \mathbf{s}_2, \nu)}{\partial n(\mathbf{s}_2)} \right] dx_{s1}\, dy_{s1}\, dx_{s2}\, dy_{s2}\, d\nu.$$

For a QM field, limiting our attention to a small range of τ values, $|\tau| \ll 1/\Delta\nu$, it gives

$$\begin{aligned} \Gamma(\mathbf{r}_1, \mathbf{r}_2, \tau) &= \Gamma(\mathbf{r}_1, \mathbf{r}_2, 0) \exp(-i2\pi\bar{\nu}\tau) \\ &= \exp(-i2\pi\bar{\nu}\tau) \int_0^{\infty} \iint_{\mathscr{A}} \iint_{\mathscr{A}} \hat{\Gamma}(\mathbf{s}_1, \mathbf{s}_2, \nu) \\ &\quad \times \left[\frac{\partial \hat{G}(\mathbf{r}_1, \mathbf{s}_1, \nu)}{\partial n(\mathbf{s}_1)} \frac{\partial \hat{G}^*(\mathbf{r}_2, \mathbf{s}_2, \nu)}{\partial n(\mathbf{s}_2)} \right] \\ &\quad \times dx_{s1}\, dy_{s1}\, dx_{s2}\, dy_{s2}\, d\nu. \end{aligned}$$

The MSDF $\hat{\Gamma}(\mathbf{s}_1, \mathbf{s}_2, \nu)$ is some sharply peaked function of the frequency ν. It has significant values only over the band of frequencies $\Delta\nu$ at $\bar{\nu}$. If the normal derivatives of the Green's functions enclosed within the square brackets are sufficiently slowly varying functions of ν over the band $\Delta\nu$ on which $\hat{\Gamma}$ is significant, they may be removed from under the integral on ν and evaluated at the mean frequency $\bar{\nu}$. This approximation coupled with Eq. (5-30) leads us to the desired relationship

$$\Gamma(\mathbf{r}_1, \mathbf{r}_2, 0) = \iint_{\mathscr{A}} \iint_{\mathscr{A}} \Gamma(\mathbf{s}_1, \mathbf{s}_2, 0) \left[\frac{\partial \hat{G}(\mathbf{r}_1, \mathbf{s}_1, \bar{\nu})}{\partial n(\mathbf{s}_1)} \frac{\partial \hat{G}^*(\mathbf{r}_2, \mathbf{s}_2, \bar{\nu})}{\partial n(\mathbf{s}_2)} \right] \times dx_{s1}\, dy_{s1}\, dx_{s2}\, dy_{s2}. \tag{5-31}$$

This is what makes the QM-field approximation attractive: it permits us to study the mutual optical intensity function instead of the full coherence function. It must be clearly understood that the condition $\Delta\nu \ll \bar{\nu}$ alone is *not* sufficient to bring about the simplification. We need to ask for small time delays $|\tau| \ll 1/\Delta\nu$ *and* be able to show that the appropriate functions describing the optics are slowly varying compared to $\hat{\Gamma}(\mathbf{s}_1, \mathbf{s}_2, \nu)$. The optical implication of these conditions is discussed at length in the statement of Problem 5-3. The calculation asked for in the problem will delineate the region of validity of the above and hence show why a full polychromatic treatment is called for if the said conditions are not fulfilled, even if $\Delta\nu \ll \bar{\nu}$. Equation (5-31) will be referred to as the basic propagation equation for QM fields.

When the spectral spread $\Delta\nu$ equals zero, the field is monochromatic and the states of partial coherence are out of the question. The problem is deterministic. As long as $\Delta\nu$ is nonzero—howsoever small—an ensemble of fields is possible, allowing for partial coherence and even noncoherence. Thus the mutual spectral density, $\hat{\Gamma}_{12}(\nu)$, presupposes an infinitesimal spread $\delta\nu$ about ν. The discussion of this section indicates that the optics studied for ν are valid over as large a region as desired in the plane of observation if $\delta\nu$ is kept as small as is required.

We now turn our attention to the coherent and noncoherent limits of the propagation of QM fields. The study of the polychromatic fields in these limits will be omitted, since in this case the theoretical development is essentially similar to the theory discussed in what is to follow (see Beran and Parrent, 1964).

PROPAGATION IN THE COHERENT LIMIT

In the coherent limit, the MCF $\Gamma_{12}(\tau)$ assumes the factored form applicable in the quasimonochromatic approximation,

$$\Gamma(\mathbf{r}_1, \mathbf{r}_2, \tau) = U(\mathbf{r}_1)\, U^*(\mathbf{r}_2) \exp(-i2\pi\bar{\nu}\tau), \tag{5-32}$$

where

$$\tau < \frac{1}{\Delta\nu} \quad \text{and} \quad \Delta\nu \ll \bar{\nu}.$$

Under these conditions the mutual optical intensity function of the aperture $\mathscr{A}$ illuminated with this light becomes

$$\Gamma(\mathbf{s}_1, \mathbf{s}_2, 0) = U(\mathbf{s}_1)\, U^*(\mathbf{s}_2). \tag{5-33}$$

Use of this in the basic propagation equation, Eq. (5-31), splits the pair of area integrals:

$$\Gamma(\mathbf{r}_1, \mathbf{r}_2, 0) = \left[\iint_{\mathcal{A}} U(\mathbf{s}_1) \frac{\partial \hat{G}(\mathbf{r}_1, \mathbf{s}_1, \bar{\nu})}{\partial n(\mathbf{s}_1)} dx_{s1}\, dy_{s1} \right] \times \left[\iint_{\mathcal{A}} U^*(\mathbf{s}_2) \frac{\partial \hat{G}^*(\mathbf{r}_2, \mathbf{s}_2, \bar{\nu})}{\partial n(\mathbf{s}_2)} dx_{s2}\, dy_{s2} \right].$$

The factored form on the right-hand side indicates that a coherent field remains coherent in propagation. For convenience, we may denote the field in the right half-space with a prime,

$$\Gamma(\mathbf{r}_1, \mathbf{r}_2, 0) = U'(\mathbf{r}_1)\, U'^*(\mathbf{r}_2),$$

and meaningfully talk about the wave amplitude at $\mathbf{r}$ with the property

$$U'(\mathbf{r}) = \iint_{\mathcal{A}} U(\mathbf{s}_1) \frac{\partial \hat{G}(\mathbf{r}, \mathbf{s}_1, \nu)}{\partial n(\mathbf{s}_1)} dx_{s1}\, dy_{s1}. \tag{5-34}$$

Furthermore, in a similar manner, the pair of coupled Helmholtz equations decouples, each one satisfied by the wave amplitude,

$$\left[\nabla_j^2 + \bar{k}^2\right] U(\mathbf{r}_j) = 0, \tag{5-35}$$

for the subscript j equal to either 1 or 2. Thus, in the coherent limit we recover the entire Rayleigh–Sommerfeld diffraction theory applicable for the propagation of the wave amplitude.

PROPAGATION IN THE NONCOHERENT LIMIT

Recall the discussion headed Noncoherent Limit in Chapter 4. For convenience we shall use the δ-function representation of noncoherence, Eqs. (4-41) and (4-48), and apply it to the QM approximation. Thus, if the aperture in the propagation problem is assumed to be the aperture of the noncoherent QM source, its mutual optical intensity is

$$\Gamma(\mathbf{s}_1, \mathbf{s}_2, 0) = \frac{\bar{\lambda}^2}{\pi} \Gamma(\mathbf{s}_1, \mathbf{s}_1, 0)\, \delta^{(2)}(\mathbf{s}_1 - \mathbf{s}_2), \tag{5-36}$$

where $\bar{\lambda}$ denotes the mean wavelength. In this limit we do not distinguish between $\mathbf{s}_1$ and $\mathbf{s}_2$, used in the coherence function. The effect of this on the

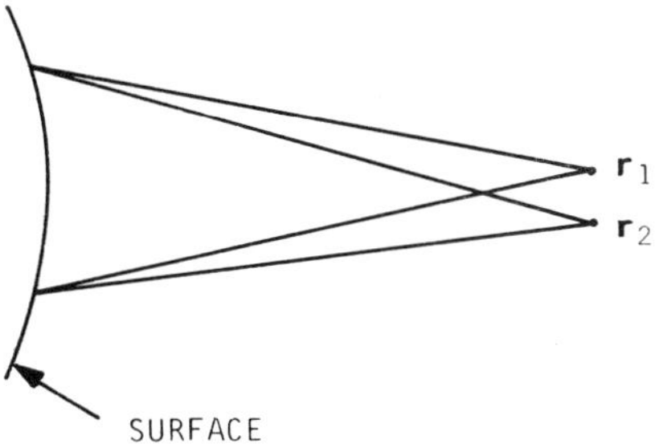

Fig. 5-2. Coherence by propagation.

basic propagation equation, Eq. (5-31), is to reduce it to a single area integral over the aperture, $\mathcal{A}$, namely,

$$\Gamma(\mathbf{r}_1, \mathbf{r}_2, 0) = \frac{\bar{\lambda}^2}{\pi} \iint_{\mathcal{A}} \Gamma(\mathbf{s}, \mathbf{s}, 0) \left[\frac{\partial \hat{G}(\mathbf{r}_1, \mathbf{s}, \bar{\nu})}{\partial n(\mathbf{s})} \frac{\partial \hat{G}^*(\mathbf{r}_2, \mathbf{s}, \bar{\nu})}{\partial n(\mathbf{s})} \right] dx_s\, dy_s. \tag{5-37}$$

The term $\Gamma(\mathbf{s}, \mathbf{s}, 0)$ is real and positive. The source radiant exitance is $\mathsf{M}(\mathbf{s},0) = \mathsf{C}\Gamma(\mathbf{s}, \mathbf{s}, 0)$. For a noncoherent source the radiant exitance (with the aperture boundary $\mathcal{A}$) function determines the MOI $\Gamma(\mathbf{r}_1, \mathbf{r}_2, 0)$ of the field. Thus the field from a noncoherent source *acquires coherence by the very process of propagation*.

In the aperture of a noncoherent source the radiation from two points has no correlation. During propagation each point in the field receives light from all points of the source. Thus at some distance from the source (see Fig. 5-2), it is reasonable to expect that the fields at a pair of points close enough will be similar in many respects since both receive light from the same set of points of a previous surface, subtending very nearly the same solid angle at $\mathbf{r}_1$ and $\mathbf{r}_2$. The acquisition of coherence by propagation is the essence of the van Cittert–Zernike theorem, which we study in more detail in the next section.

VAN CITTERT–ZERNIKE THEOREM: RADIATION FROM A PLANE, FINITE, NONCOHERENT SOURCE

Historically the van Cittert–Zernike theorem was formulated in the framework of diffraction theory applied to a plane, finite, noncoherent source aperture. We refer to Fig. 5-1 and make the customary approximations, namely, that the maximum extent, d, of the aperture is small compared to the distance, z, of the aperture from the plane of observation ($d/z \ll 1$), and that at optical wavelengths $k\rho$ is much greater than unity. In this context, the variation of the direction cosine, $\cos\gamma = z/\rho$, may be neglected as the point x_s, y_s explores the aperture, $\mathcal{a}$, while the point x, y remains fixed. The expression for the normal derivative of the Green's function used in Eq. (5-1) may thus be approximated to

$$\frac{\partial \hat{G}(\mathbf{r}_1, \mathbf{s}, \bar{\nu})}{\partial n(\mathbf{s})} = \frac{1}{2\pi}\cos\theta_1\left(1 - i\bar{k}\rho_1\right)\frac{\exp\left(i\bar{k}\rho_1\right)}{\rho_1^2}$$
$$\simeq \frac{-i}{\bar{\lambda} z}\exp\left(i\bar{k}\rho_1\right), \tag{5-38}$$

where $\bar{k} = 2\pi/\bar{\lambda} = 2\pi\bar{\nu}/c$ in the QM approximation. The propagation equation in the noncoherent limit, Eq. (5-37), reduces to

$$\Gamma(\mathbf{r}_1, \mathbf{r}_2, 0) = \frac{1}{\pi z^2}\iint_{\mathcal{a}} \Gamma(\mathbf{s}, \mathbf{s}, 0)\exp\left[i\bar{k}\left(\rho_1 - \rho_2\right)\right] dx_s\, dy_s, \tag{5-39}$$

where $\Gamma(\mathbf{s}, \mathbf{s}, 0)$ is the optical intensity at $\mathbf{s}$ in the source plane. Up to and including terms quadratic in the aperture variables, the difference $\rho_1 - \rho_2$ approximates to

$$\rho_1 - \rho_2 \simeq \left(r_1 - \frac{x_1 x_s + y_1 y_s}{z} + \frac{x_s^2 + y_s^2}{2z}\right)$$
$$- \left(r_2 - \frac{x_2 x_s + y_2 y_s}{z} + \frac{x_s^2 + y_s^2}{2z}\right). \tag{5-40}$$

(For a variation of this approximation, see footnote.†)

† The approximation for ρ_1 may be written in the form $\rho_1 \simeq r_1 - (x_1 x_s + y_1 y_s)/r_1 + (x_s^2 + y_s^2)/2z$. In this way, the actual direction cosines, x_1/r_1, y_1/r_1, and z/r_1 of the vector $\mathbf{r}_1$ appear in the theory. It is sometimes useful in diffraction theory to study the diffraction pattern on a sphere of radius r_1, centered at the origin in the aperture. It in no way increases the domain of validity of the theory because the angle restrictions ensuing from the approximation have to be obeyed. See, for example, the discussion in Born and Wolf (1970, Chapter VIII, Sec. 8.3.3). We have avoided this refinement since, from a practical standpoint, it does not seem important in coherence considerations.

The obvious cancellation of the quadratic terms leads us to the van Cittert–Zernike result,

$$\Gamma(\mathbf{r}_1, \mathbf{r}_2, 0) = \frac{\exp[+i\bar{k}(r_1 - r_2)]}{\pi z^2} \times \iint_{\mathscr{A}} \Gamma(\mathbf{s}, \mathbf{s}, 0) \exp\left(-i2\pi \frac{x_{12}x_s + y_{12}y_s}{\bar{\lambda} z}\right) dx_s\, dy_s, \tag{5-41}$$

where $x_{12} \equiv x_1 - x_2$ and $y_{12} \equiv y_1 - y_2$. The phase factor $\exp[+i\bar{k}(r_1 - r_2)]$ depends on the position of the pair of observation points. Apart from this factor, the statement of the van Cittert–Zernike theorem is: *The spatial coherence* $\Gamma(\mathbf{r}_1, \mathbf{r}_2, 0)$ *in the observation plane is the two-dimensional spatial Fourier transform of the optical intensity distribution in the noncoherent source aperture* $\mathscr{A}$.

The consequences of this theorem are:

1. The absolute value of the spatial coherence is a function of the coordinate difference,

$$|\Gamma(\mathbf{r}_1, \mathbf{r}_2, 0)| \rightarrow \Gamma(x_{12}, y_{12}, z, 0). \tag{5-42}$$

2. The irradiance distribution in the observation plane is a constant,

$$\begin{aligned} \mathsf{E}(\mathbf{r}, 0) &= \mathsf{C}\Gamma(\mathbf{r}, \mathbf{r}, 0) \\ &= \frac{\mathsf{C}}{\pi z^2} \iint_{\mathscr{A}} \Gamma(\mathbf{s}, \mathbf{s}, 0)\, dx_s\, dy_s, \end{aligned} \tag{5-43}$$

and it displays the characteristic $1/z^2$ falloff, where $\mathbf{r}$ is a point in the plane at z.

It must be clearly understood that the spatial Fourier transform relationship appeared *without* our invoking the "far-field" condition that is necessary in transit from the Fresnel to the Fraunhofer approximation in diffraction theory. Furthermore, the transformation is over the optical intensity in the aperture, unlike the Fraunhofer case wherein the field variable itself is transformed. In the van Cittert–Zernike theorem, the Fourier transform came about purely as a consequence of the property of *noncoherence* of the source aperture. The δ-function representation was used, but the reader may verify that the other representations discussed in Chapter 4 will result in the same answer (see Problem 5-4).

The fact that the irradiance distribution is a constant is a property of a spatially stationary coherence function. The van Cittert–Zernike theorem describes the state of spatial coherence of the radiation illuminating the observation plane. Unlike the Fraunhofer pattern, it is not something that is "visible" in that plane. It is manifested only in the quality of fringes in an interference experiment. For example, in a Young's two-pinhole experiment, light from two points $\mathbf{r}_1$ and $\mathbf{r}_2$ is isolated by use of two pinholes at these locations in an otherwise opaque mask. The light from them interferes in the overlapping region in the space behind the mask. The visibility of the fringes observed is directly related to the absolute value $|\Gamma(\mathbf{r}_1, \mathbf{r}_2, 0)|$, and the fringe shift is related to the phase of Γ.

Example 5-1

Consider a uniform, noncoherent source aperture in the shape of a rectangle, as shown in Fig. 5-3. The pinhole mask plane is at a distance z_0 from the source; it is in this plane that one calculates the spatial coherence function. For simplicity the pinholes, P_1 at $\mathbf{r}_1$ and P_2 at $\mathbf{r}_2$, are shown only along the x axis. By a uniform source we mean one of constant radiant exitance, which is proportional to

$$\Gamma(\mathbf{s}, \mathbf{s}, 0) = I_0 \,\mathrm{Rect}\left(\frac{x_s}{a_0}\right) \mathrm{Rect}\left(\frac{y_s}{b_0}\right). \tag{5-44}$$

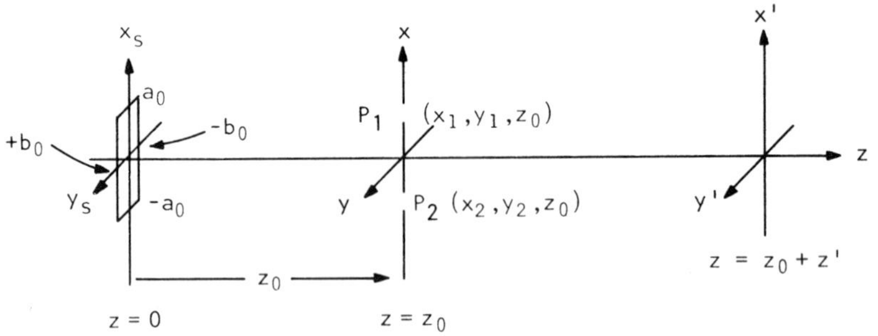

Fig. 5-3. Rectangular source aperture in the $z = 0$ plane. The z axis is perpendicular to the source aperture. The mask with pinholes P_1 and P_2 is in the $z = z_0$ plane. Pinholes P_1 and P_2 are along the x axis and are placed symmetrically about z. The interference plane has coordinates x' and y' and is at $z = z_0 + z'$.

The spatial coherence function in the plane $z = z_0$ is

$$\Gamma(\mathbf{r}_1, \mathbf{r}_2, 0) = \frac{4a_0 b_0}{\pi z_0^2} I_0 \exp[+i\bar{k}(r_1 - r_2)] \times \mathrm{Sinc}\left(\frac{2\pi a_0 x_{12}}{\bar{\lambda} z_0}\right) \mathrm{Sinc}\left(\frac{2\pi b_0 y_{12}}{\bar{\lambda} z_0}\right). \tag{5-45}$$

The irradiance distribution at $\mathbf{r} = \hat{\mathbf{i}}x + \hat{\mathbf{j}}y + \hat{\mathbf{k}}z_0$ in the pinhole plane is

$$\begin{aligned} \mathsf{E} &= \mathsf{C}\Gamma(\mathbf{r}, \mathbf{r}, 0) \\ &= \mathsf{C} I_0 \frac{4a_0 b_0}{\pi z_0^2}. \end{aligned} \tag{5-46}$$

Example 5-2

As another example, we consider a uniform noncoherent source with a circular aperture of radius a_0. In the source plane we have

$$\Gamma(\mathbf{s}, \mathbf{s}, 0) = I_0 \,\mathrm{cyl}\left(\frac{r_s}{a_0}\right), \tag{5-47}$$

where the radial coordinate $r_s = (x_s^2 + y_s^2)^{1/2}$. The spatial coherence function at $z = z_0$ is

$$\Gamma(\mathbf{r}_1, \mathbf{r}_2, 0) = \frac{\pi a_0^2}{\pi z_0^2} I_0 \exp[i\bar{k}(r_1 - r_2)] \mathrm{Besinc}\left(\frac{2\pi a_0 r_{12}}{\bar{\lambda} z_0}\right), \tag{5-48}$$

where

$$r_{12} = \left(x_{12}^2 + y_{12}^2\right)^{1/2}.$$

The irradiance at $\mathbf{r}$ in the plane $z = z_0$ is

$$\begin{aligned} \mathsf{E} &= \mathsf{C}\Gamma(\mathbf{r}, \mathbf{r}, 0) \\ &= \mathsf{C} I_0 \left(\frac{a_0^2}{z_0^2}\right). \end{aligned} \tag{5-49}$$

Example 5-3

In this example we start with a partially coherent source, namely, the Carter–Wolf source of Eq. (5-11), and proceed to the limit of noncoherence.

The spatial coherence in the plane $z =$ constant is given in Eq. (5-15). The state of noncoherence is characterized by a uniform spatial frequency spectrum (see Chapter 4), that is,

$$\mathring{g}_s(\kappa p, \kappa q, \nu) = \text{constant}, \qquad p^2 + q^2 \le 1.$$

For this case, with a simple change of variables $x' = x - pz$ and $y' = y - qz$, Eq. (5-15) converts to

$$\begin{aligned}\hat{\Gamma}&(x + \tfrac{1}{2}x_{12}, y + \tfrac{1}{2}y_{12}, x - \tfrac{1}{2}x_{12}, y - \tfrac{1}{2}y_{12}, z, \nu)\\ &= \frac{1}{\lambda^2 z^2}\mathring{g}_s(0,0,\nu)\int\!\!\int_{-\infty}^{\infty} \hat{I}_s(x', y', \nu)\\ &\quad\times \exp\left[i2\pi\frac{(x - x')x_{12} + (y - y')y_{12}}{\lambda z}\right] dx'\, dy'.\end{aligned}$$

With a minor rearrangement and use of the old variables x_1, y_1 and x_2, y_2 the spatial coherence in the plane $z > 0$ is given by

$$\begin{aligned}\hat{\Gamma}(x_1, y_1, x_2, y_2, z, \nu) &= \left[\frac{\pi}{\lambda^2}\mathring{g}_s(0,0,\nu)\right]\frac{\exp\left[i2\pi\left(x_1^2 + y_1^2 - x_2^2 - y_2^2\right)/\lambda z\right]}{\pi z^2}\\ &\quad\times \mathring{I}_s\left(\frac{x_1 - x_2}{\lambda z}, \frac{y_1 - y_2}{\lambda z}, \nu\right).\end{aligned} \tag{5-50}$$

Apart from the constant, unitless factor up front, the rest of the expression is the van Cittert–Zernike result; its quasimonochromatic version appears in Eq. (5-41). This example shows that the condition of uniform SFS is important for characterizing noncoherence and that the quadratic approximation of Eq. (5-14) in the angular spectrum domain and the quadratic approximation in the coordinate space give equivalent results.

For the case of a rectangular aperture, the nature of the spatial coherence function is sketched in Fig. 5-4. The function describes the mutual coherence of the light from the pinholes P_1 and P_2. It depends on the separation distance x_{12} and/or y_{12}. The Sinc function may be centered at P_1 or P_2; its value at one pinhole (P_1) *relative* to the other pinhole (P_2) is what is important. The broken lines in Fig. 5-4*a* and *b* mark the zeros of one or the other Sinc function. The value of the Sinc function at P_1 as shown in the figure is the visibility of the fringes in any interference plane behind the mask. The change in the location of the interference plane changes only

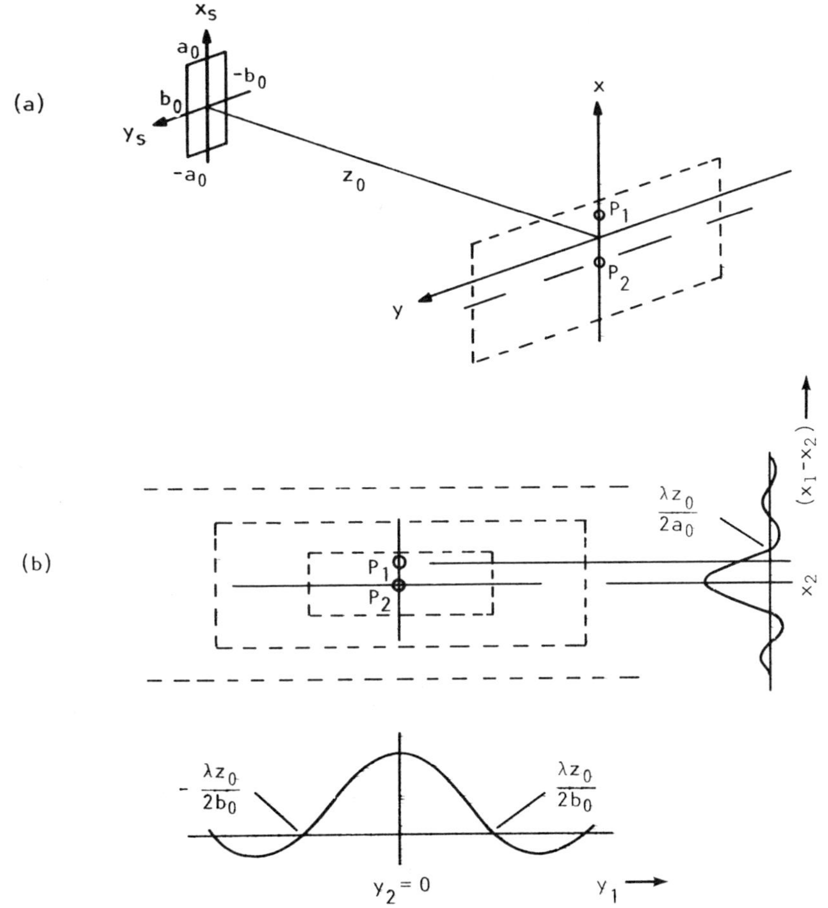

Fig. 5-4. Nature of the spatial coherence function for a rectangular source aperture. (*a*) Source aperture is of height $2a_0$ and width $2b_0$. Pinholes P_1 and P_2 are along the x axis in the x-y plane at a distance z_0 from the source. The broken lines indicate the zeros of the Sinc functions. (*b*) The Sinc function along the x axis is narrow compared to the one along the y axis. The broken lines mark off the zeros of one or the other Sinc function. The spatial coherence of the light from P_1 relative to the light from P_2 is studied. The Sinc functions are centered at P_2. As the pinhole P_1 is moved away from P_2, the coherence is reduced until it reaches zero. Upon further displacement of P_1 the coherence becomes negative. The visibility of fringes is given by the absolute value, and the negative sign indicates reversal of contrast.

the scale of the fringes but not their visibility. For P_1 and P_2 disposed symmetrically about the origin along the x axis, $r_1 = r_2$, and since the Sinc functions are real, the phase function is unity. That is, the center fringe (interference maximum) is at the origin along the x' axis in the observation plane. However, if P_1 and P_2 are relocated to change the distance r_{12} between them and yet keep them symmetric, there comes a point when P_1 is in the first sidelobe of the Sinc function centered at P_2; that is, $\bar{\lambda} z_0/2a_0 < r_{12} < 2(\bar{\lambda} z_0/2a_0)$. The visibility is still given by the absolute value of the Sinc function, while the negative sign indicates a π phase shift. That is, the center fringe, instead of being an interference maximum, is found to be a minimum (see Problem 5-5).

For coherence considerations in practice, it is necessary to define in some convenient way what we mean by fringes of good visibility. In the instance of the Sinc function, we may choose the criterion $\mathscr{v} \geq 2/\pi$ for fringes of good visibility. This is the value of the Sinc when its argument equals $\pi/2$, a point halfway to its first zero. If we use $\mathscr{w}$ to denote the *coherence width* of the beam in the direction perpendicular to its propagation, then for the x and y directions in the mask plane we have, respectively,

$$\begin{aligned} \mathscr{w}_x &= \frac{\bar{\lambda} z_0}{2(2a_0)} = \frac{\bar{\lambda}}{2\theta_a}, \\ \mathscr{w}_y &= \frac{\bar{\lambda} z_0}{2(2b_0)} = \frac{\bar{\lambda}}{2\theta_b}. \end{aligned} \tag{5-51}$$

We observe that the coherence widths are inversely proportional to the respective angular subtense, θ_a and θ_b, of the source to mask plane. In this way, the magnitude of the *coherence area*, $\mathscr{a}_c$, in the mask plane is

$$\mathscr{a}_c = \mathscr{w}_x \mathscr{w}_y = \frac{\bar{\lambda}^2}{4\theta_a\theta_b}. \tag{5-52}$$

We refer to Problems 5-6 and 5-7 for typical numerical size of the coherence area and width.

THOMPSON AND WOLF EXPERIMENT

The Thompson and Wolf experiment is of fundamental importance in coherence theory. It verifies the various aspects of the states of partial coherence through a modification of the basic Young's two-pinhole experiment. It makes use of an instrument called the *diffractometer*, which is described in Born and Wolf (1970, Sec. 10.4.3); see also Thompson and Wolf (1957).

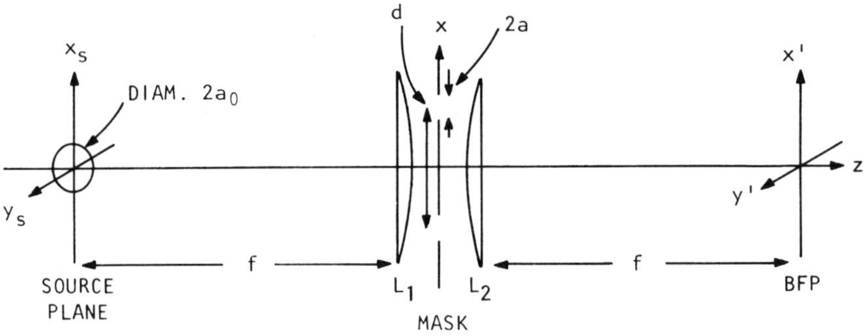

Fig. 5-5. Thompson and Wolf experimental setup. *Source plane:* Coordinates, x_s, y_s, and $r_s = (x_s^2 + y_s^2)^{1/2}$. Uniform noncoherent circular source of diameter $2a_0$. Lenses L_1 and L_2 have focal length f with common optical axis along the z axis. *Mask plane:* Coordinates, x, y, and $r = (x^2 + y^2)^{1/2}$. Two identical circular apertures of diameter $2a$, symmetrically placed about the z axis with center-to-center separation d. *BFP:* Back focal plane of L_2, with coordinates x', y', and $r' = (x'^2 + y'^2)^{1/2}$.

The basic experimental setup is sketched in Fig. 5-5. It consists of a noncoherent circular source of radius a_0 and a plane mask midway between two identical lenses L_1 and L_2 of focal length f. The mask consists of two identical circular apertures of radius a and center-to-center separation d, which are situated symmetrically about the optical axis. The source is in the front focal plane of L_1, and the interference fringes appear in the back focal plane (BFP) of L_2, where they are examined by means of a microscope. In the experiment, the aperture separation d was varied in order to study the spatial coherence function in the mask plane. Due to this procedure, the fringe frequency in the BFP of L_2 changes, yet allows for the investigation of the visibility and phase. In a subsequent experiment, Thompson (1958) examined the coherence function without varying d, but by changing the angular subtense of the source to the mask plane.

The procedure for calculating the light distribution in the BFP of L_2 is as follows. The application of the van Cittert–Zernike theorem, Eq. (5-41), allows us to calculate the spatial coherence function in the aperture of L_1. Since L_1 and L_2 are close together and the light between them is essentially parallel, the spatial coherence at the mask is taken to be the same as that on the aperture of L_1. Next, the coherence function is multiplied by two mask functions, one for each of the pair of independent arguments, to give the coherence behind the mask. The presence of the lens L_2 brings about a spatial Fourier transformation of the light behind the mask. This Fourier operation is manifested in the BFP of L_2 and contains the desired information. The details of the action of the lenses will be studied in Chapter 6. We shall now proceed to do this calculation.

The coherence in the mask plane is given by Eq. (5-48). Its specialization for the aperture symmetry may be denoted by

$$\Gamma(r_{12}) = I_0\left(\frac{a_0^2}{f^2}\right)\text{Besinc}\left(\frac{2\pi a_0 r_{12}}{\bar{\lambda} f}\right), \tag{5-53}$$

where for the center-to-center distance r_{12} equals d. For each circular aperture of radius a, the amplitude transmittance is a cylinder function. For a displacement of $\pm d/2$ along x, we shall denote it by

$$c_{\pm}(x, y) \equiv \text{cyl}\left\{\frac{\left[(x \mp d/2)^2 + y^2\right]^{1/2}}{a}\right\}, \tag{5-54}$$

respectively. The mask function is

$$M(x, y) = [c_+(x, y) + c_-(x, y)]. \tag{5-55}$$

The coherence function behind the mask is the product of the function in front and the mask function for both points,

$$\begin{aligned}\Gamma(x_1, y_1, x_2, y_2, f, 0) &= \Gamma(r_{12})\, M(x_1, y_1)\, M^*(x_2, y_2)\\ &= \Gamma(r_{12})[c_+(x_1, y_1)\, c_+(x_2, y_2)\\ &\quad + c_+(x_1, y_1)\, c_-(x_2, y_2)\\ &\quad + c_-(x_1, y_1)\, c_+(x_2, y_2)\\ &\quad + c_-(x_1, y_1)\, c_-(x_2, y_2)].\end{aligned} \tag{5-56}$$

The spatial Fourier transform of this appears in the BFP of L_2 (see Problem 5-8). In the present case, the coherence function $\Gamma(r_{12})$ is rather broad compared to the diameter, $2a$, of each aperture (see Problem 5-9). Thus we may make the optically reasonable and simplifying assumption that the coherence is essentially constant over any one aperture. Its value at the aperture center is $\Gamma(0) = I_0 a_0^2/f^2$. The coherence of one aperture relative to the other is approximated to $\Gamma(d)$ and is what is to be studied. The transform of $c_{\pm}$ is

$$\tilde{c}_{\pm}\left(\frac{r'}{\bar{\lambda} f}\right) = \pi a^2\,\text{Besinc}\left(\frac{2\pi a r'}{\bar{\lambda} f}\right)\exp\left(\mp i\pi\frac{x'd}{\bar{\lambda} f}\right). \tag{5-57}$$

The irradiance distribution in the BFP of L_2 is thus

$$\mathsf{E}(x', y') = \mathsf{C}I_0\left(\frac{2\pi^2 a_0^2 a^4}{\bar{\lambda}^2 f^4}\right)\left[\text{Besinc}\left(\frac{2\pi a r'}{\bar{\lambda} f}\right)\right]^2 \times\left[1 + \text{Besinc}\left(\frac{2\pi a_0 d}{\bar{\lambda} f}\right)\cos\left(\frac{2\pi x' d}{\bar{\lambda} f}\right)\right]. \tag{5-58}$$

It shows the familiar cosine fringe pattern with visibility equal to the absolute value of the complex degree of coherence,

$$|\gamma(d)| = \frac{|\Gamma(d)|}{\Gamma(0)} = \left|\text{Besinc}\left(\frac{2\pi a_0 d}{\bar{\lambda} f}\right)\right|. \tag{5-59}$$

The negative sidelobes of the Besinc indicate a π phase shift, and they may be included as an argument, $\beta(d)$, in the cosine, where $\beta(d) = 0$ or π, according to whether $\gamma(d)$ is greater or less than 0. The maximum and minimum values of E are

$$\mathsf{E}_{\max} = \mathsf{E}_0\left[\text{Besinc}\left(\frac{2\pi a r'}{\bar{\lambda} f}\right)\right]^2 [1 + |\gamma(d)|]$$

and

$$\mathsf{E}_{\min} = \mathsf{E}_0\left[\text{Besinc}\left(\frac{2\pi a r'}{\bar{\lambda} f}\right)\right]^2 [1 - |\gamma(d)|], \tag{5-60}$$

where E_0 is the value of the irradiance at $x' = 0 = y'$.

The results of the Thompson and Wolf experiment are shown in Fig. 5-6. Under each photomicrograph of the fringe pattern is shown the theoretical curve dictated by Eq. (5-58). The broken line envelopes represent the $\mathsf{E}_{\max}$ and $\mathsf{E}_{\min}$ of Eq. (5-60). The microdensitometer trace of the photographic (positive) transparency will come close to the theoretical curves. The numerical values of the experimental parameters are listed below each curve. From one case to the next the circular aperture separation d is varied; all other parameters are held fixed. Each value of d corresponds to a certain point marked by a dot (•) on the curve of the complex degree of coherence, $|\gamma(d)|$, displayed in Fig. 5-7. Observe the π phase shifts for cases D and E. They are in agreement with our theoretical discussion, indicating a relative minimum at $x' = 0$ rather than a relative maximum.

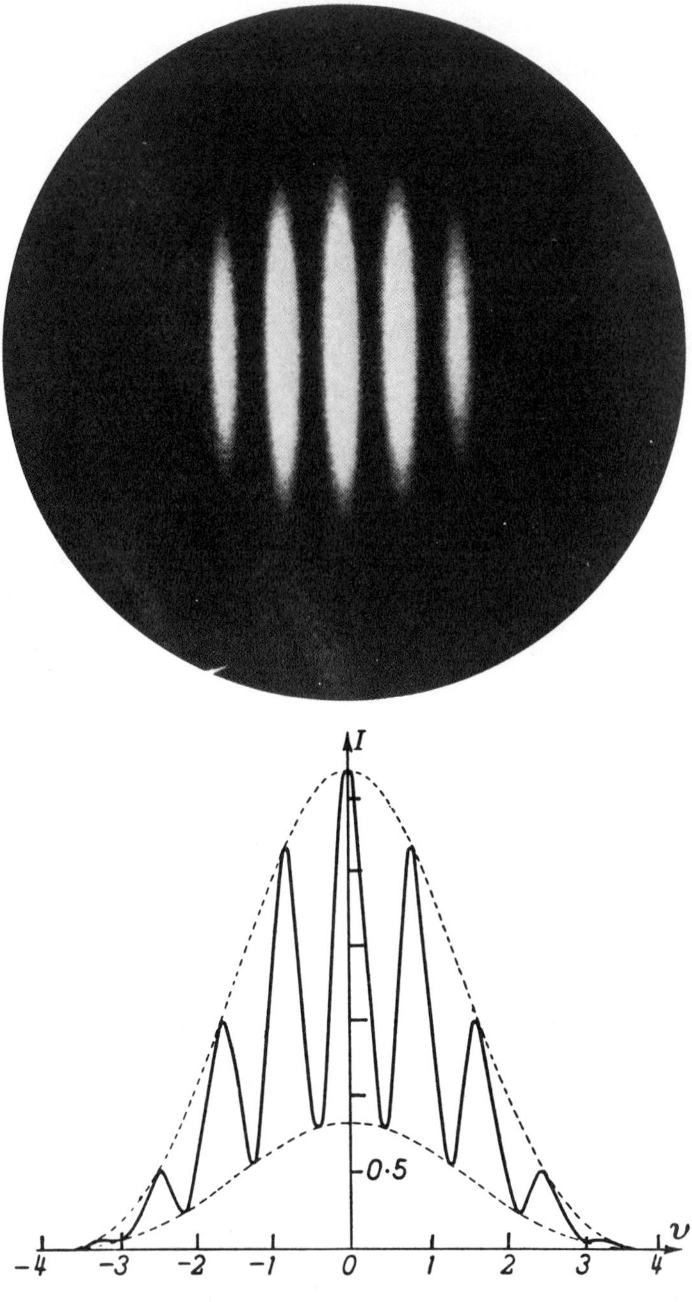

d = 0.6 cm,
$|\gamma(d)| = 0.593$, $\beta(d) = 0$

Fig. 5-6A. Two-beam interference with partially coherent light. The associated theoretical curve is shown below each observed pattern, along with the values of the experimental parameters. For the theoretical curves shown, the label I stands for the irradiance E of Eq. (5-58) and the label v stands for the argument $2\pi x'd/\bar{\lambda}f$. [Reproduced by permission from B. J. Thompson and E. Wolf, *J. Opt. Soc. Am.* **47**(10), 895–902 (1957).]

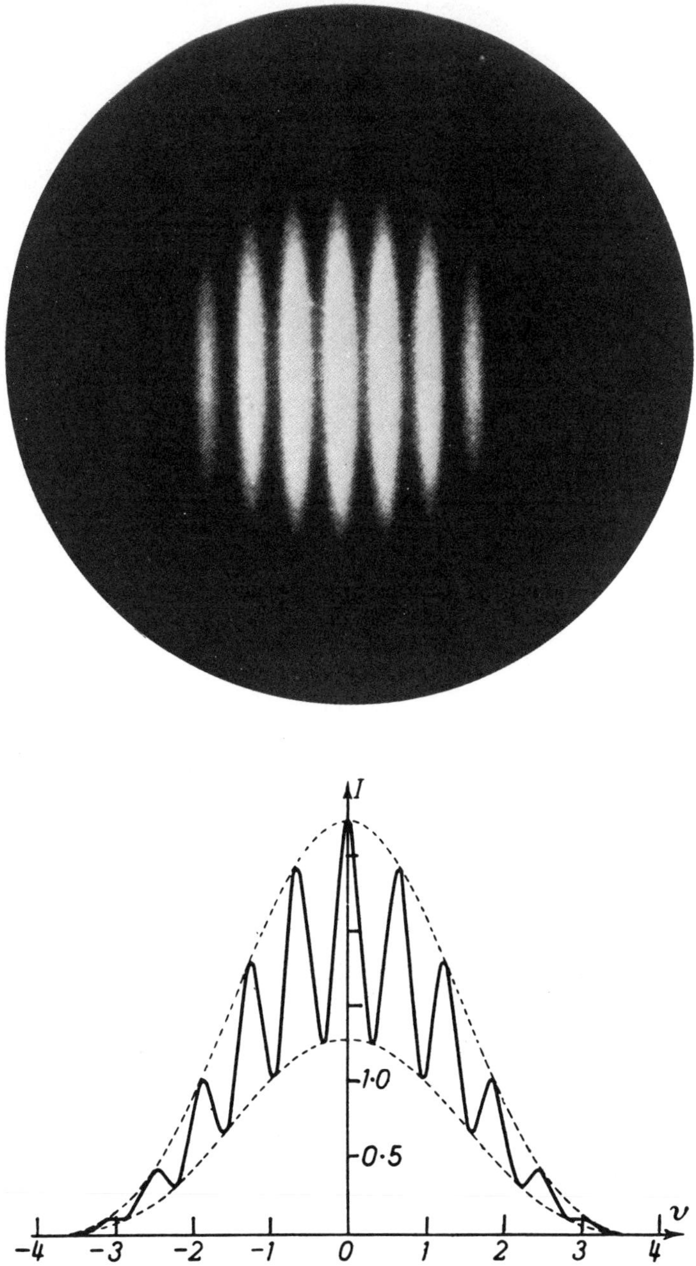

d = 0.6 cm,
$|\gamma(d)| = 0.361$, $\beta(d) = 0$

Fig. 5-6B

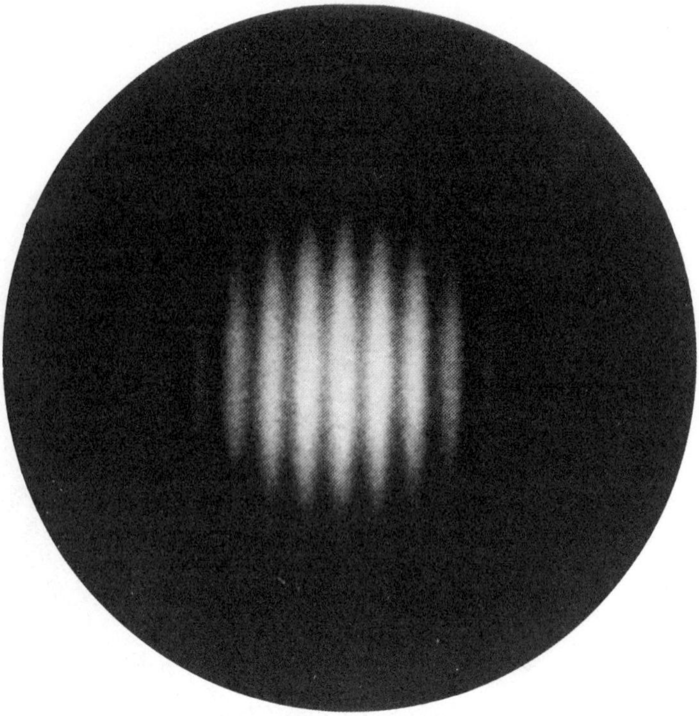

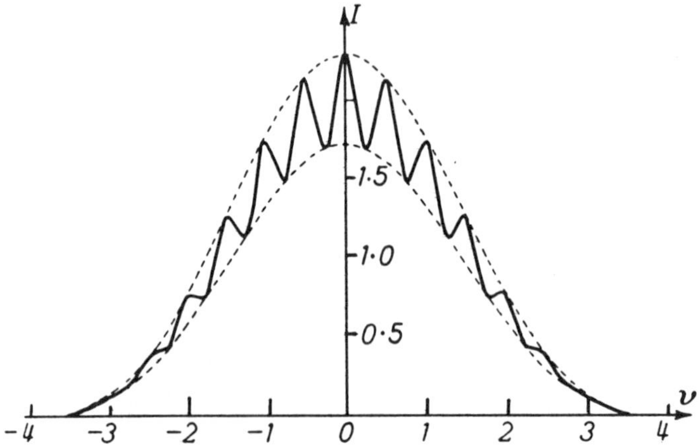

d = 1 cm,
$|\gamma(d)| = 0.146$, $\beta(d) = 0$

Fig. 5-6C

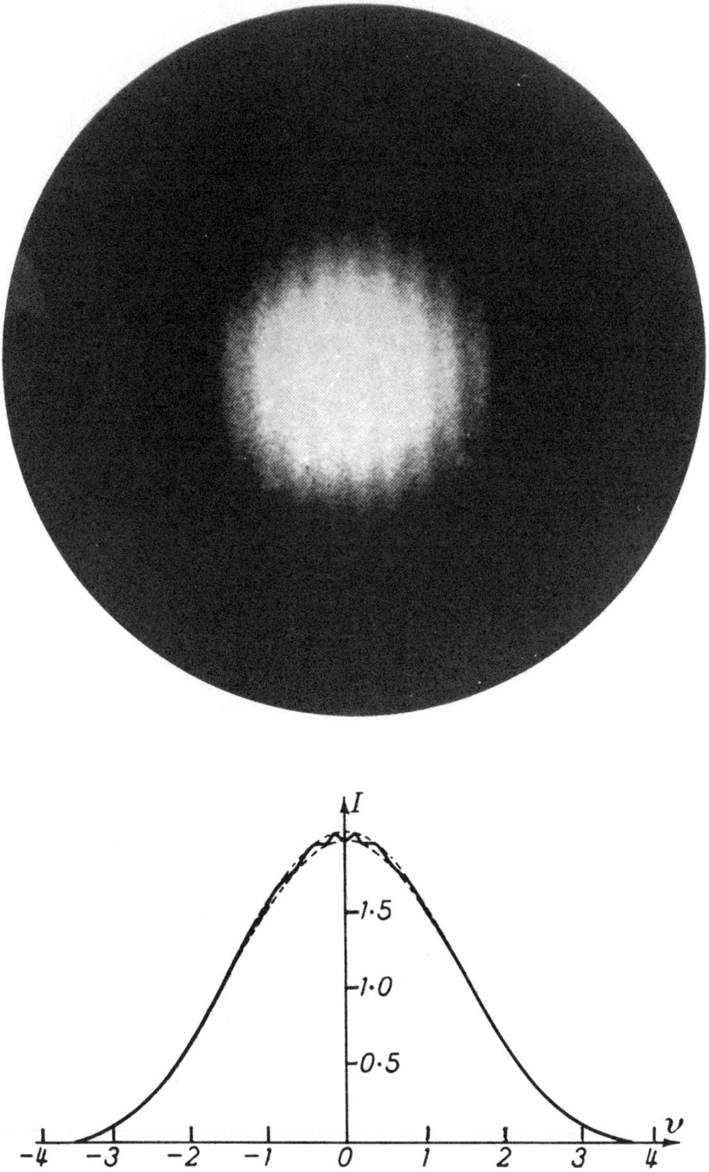

d = 1.2 cm,
$|\gamma(d)| = 0.015$, $\beta(d) = \pi$

Fig. 5-6D

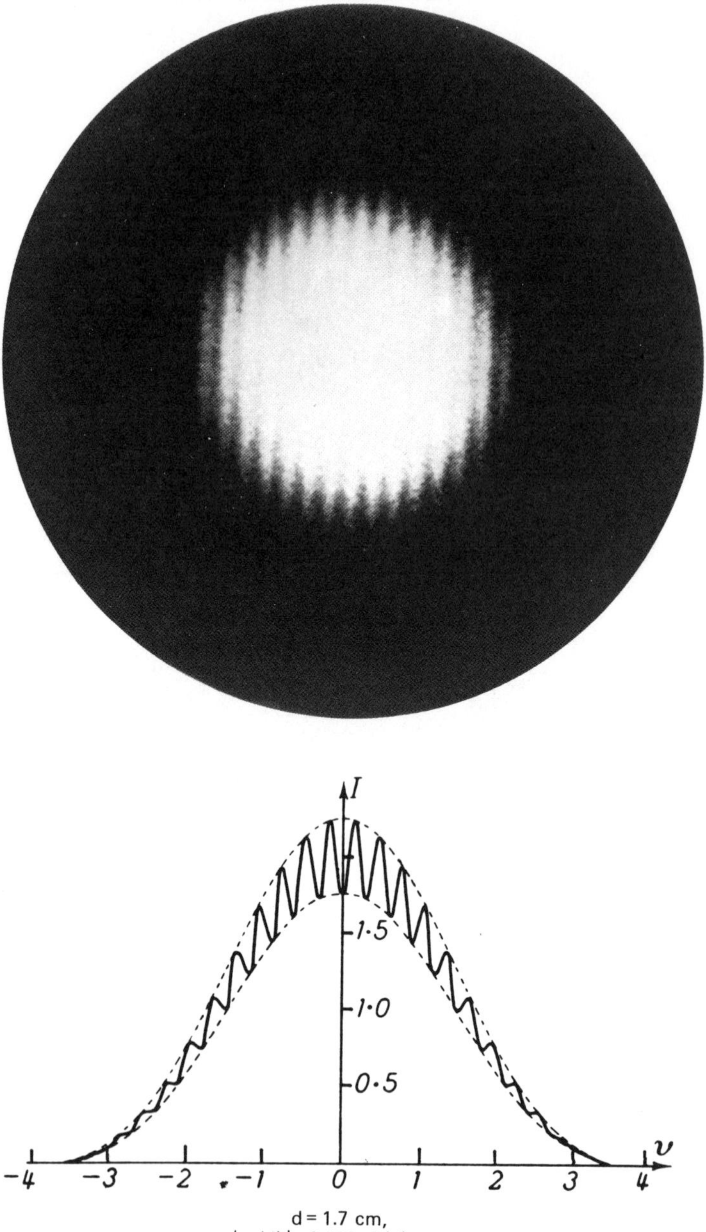

d = 1.7 cm,
$|\gamma(d)| = 0.123$, $\beta(d) = \pi$

Fig. 5-6E

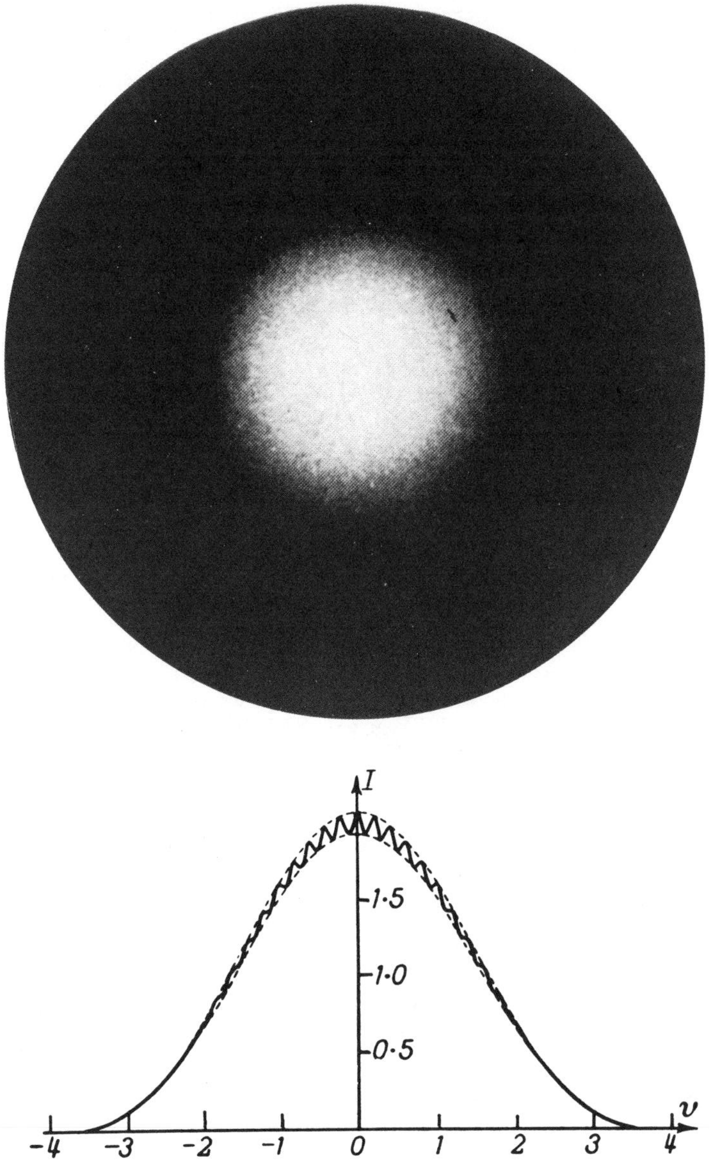

d = 2.3 cm,
$|\gamma(d)| = 0.035$, $\beta(d) = 0$

Fig. 5-6F

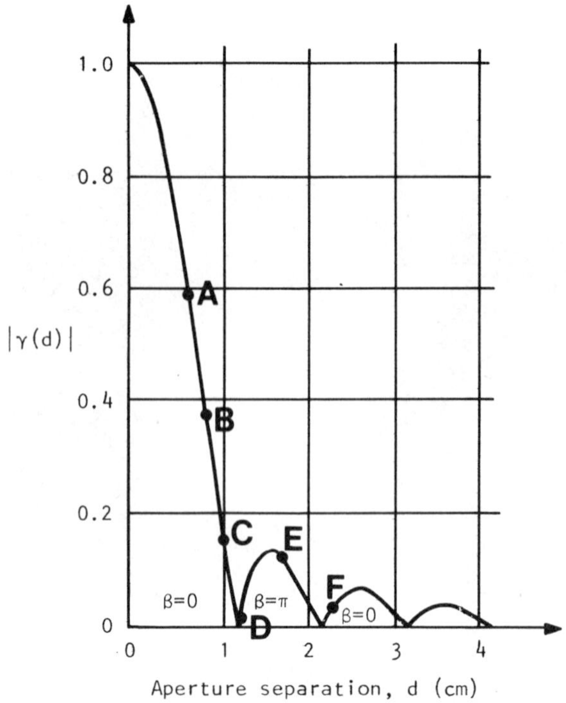

Fig. 5-7. Degree of coherence as a function of the aperature separation d, see Eq. (5-59). Obviously this same equation applies if d is held fixed but the source radius a_0 is changed. [Reproduced by permission from B. J. Thompson and E. Wolf, *J. Opt. Soc. Am.* **47**(10), 895–902 (1957).]

MICHELSON STELLAR INTERFEROMETER

The stellar interferometer experiment performed by Michelson was used to measure the angular diameter of stars and the angular separation of double stars (see Michelson, 1890, 1920; Anderson, 1920). It is described in several texts on optics; we refer to Jenkins and White (1976, Art. 16.8) and Hecht and Zajac (1976, Art. 12.3.1). Since *amplitude correlation* is basic to this experiment, the story can be retold in the language of coherence theory.

Consider a uniform, noncoherent, quasimonochromatic source. Figure 5-8 shows such a circular source of radius a_0 at a distance s from an objective lens. The image is received at a distance s' on the image plane behind the lens. The angular subtense, θ_s, of this source at the lens is

$$\theta_s = \frac{2a_0}{s}. \tag{5-61}$$

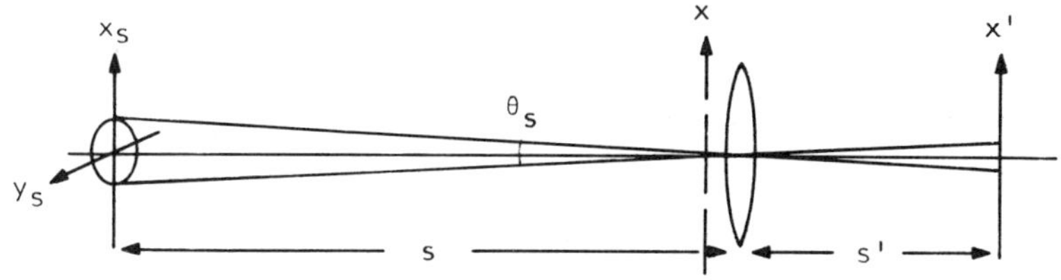

Fig. 5-8. Uniform noncoherent quasimonochromatic source with an imaging lens. Objective lens images a circular source of radius a_0. The source is at a distance s from the lens and the image is at s'. The angular subtense of the source is θ_s. A pair of slits with a variable separation is introduced into the aperture of the objective.

The angular resolving power, θ_{RP}, of the lens is

$$\theta_{RP} = \frac{1.22\bar{\lambda}}{D_{ap}}, \tag{5-62}$$

where D_{ap} is the diameter of the lens aperture.

First let us consider the case of the source disk resolved by the objective, that is, $\theta_s > \theta_{RP}$. By the van Cittert–Zernike theorem the spatial coherence function in the aperture plane is given by Eq. (5-48) of Example 5-2. Let d_{12} be the value of the distance r_{12} in the aperture plane at which the coherence function first goes to zero; then

$$d_{12} = \frac{1.22\bar{\lambda}}{\theta_s}. \tag{5-63}$$

For the present case in which the source disk is resolvable, we find that

$$d_{12} < D_{ap}. \tag{5-64}$$

That is, the coherence function is narrower than the lens aperture.

In Fig. 5-8 a pair of narrow slits with variable separation is introduced into the aperture plane (much like the Thompson and Wolf experiment). This will enable us to study the coherence function, which entails measuring the visibility and phase shift of the interference fringes in the image plane. Knowledge of the coherence function will allow us to determine the optical intensity distribution, $\Gamma(\mathbf{s}, \mathbf{s}, 0)$, across the source disk via an inverse spatial Fourier transformation. In principle, this is the same information we would obtain by using the objective lens in the imaging mode. If, however, only the angular size of the source is desired, then

$$\theta_s = \frac{1.22\bar{\lambda}}{d_{12}}, \tag{5-65}$$

where d_{12} is the distance between the pair of slits in the aperture for which the fringes first disappear. The determination of the angular size in this manner was suggested by Michelson (1890).

Now let us consider the case of the source disk *not* resolved by the objective lens; that is, $\theta_s < \theta_{\mathrm{RP}}$. In this case, the coherence function is broader than the lens aperture,

$$d_{12} > D_{\mathrm{ap}}. \tag{5-66}$$

If the lens is used in the imaging mode, we shall end up with the impulse response of the objective, without any information of the details of the source. Likewise, repeating the above experiment of putting a pair of slits in the aperture, the condition of zero visibility of fringes will be unattainable; hence it will not permit the determination of even the angular size of the source. Such is indeed the case with stellar sources. A typical star whose diameter is on the order of 10^9 m and which is at a distance of 1 light year (about 10^{16} m) has an angular subtense

$$\theta_s \simeq 10^{-7} \text{ rad} \simeq 0.02 \text{ arc second}. \tag{5-67}$$

The first zero of the coherence function of the light from such a stellar source is

$$d_{12} = \frac{1.22\bar{\lambda}}{\theta_s} \simeq 6 \text{ m}, \tag{5-68}$$

where $\bar{\lambda} = 500$ nm, the mean wavelength of the spectral filter used. The aperture diameter of an astronomical telescope will have to be larger than this value in order to determine the angular size. This large a telescope is yet to be made!

At this point, Michelson's extension of the basic theme is pertinent. As shown in Fig. 5-9, a pair of slits at a fixed separation d_{12} is introduced into the telescope aperture. The star light is fed to them by means of a pair of movable mirrors, M_1 and M_2. Interference fringes are studied in the focal plane of the telescope. Since the slits in the aperture are at a fixed separation, the fringe frequency remains unchanged; only the fringe visibility changes as the distance between M_1 and M_2 is changed. The mirror

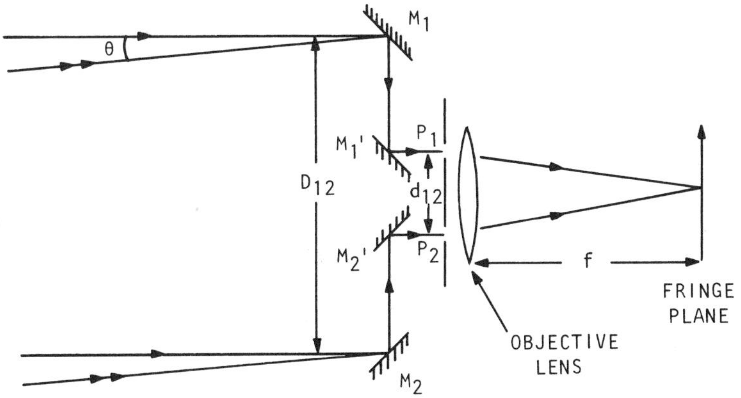

Fig. 5-9. Michelson stellar interferometer. An opaque mask with a pair of slits P_1 and P_2 at a fixed separation d_{12} is in the aperture of the objective lens. Mirrors M_1' and M_2' are fixed, and M_1 and M_2 are movable so that the distance D_{12} can be adjusted for the first disappearance of fringes in the receiving plane. Starlight, depicted as two parallel rays with a single arrowhead, is incident on the mirrors M_1 and M_2. Rays from another point on the star are shown with a double arrowhead at an angle θ to the first pair of parallel rays.

separation D_{12} at which the fringe visibility goes to zero gives the angular size of the star,

$$\theta_s = \frac{1.22\bar{\lambda}}{D_{12}}. \tag{5-69}$$

The maximum mirror separation possible in the instrument is referred to as its baseline length. The fact that stars are not circular disks, but are spherical, necessitates a correction factor. Instead of 1.22 in Eq. (5-69), the factor to be used is 1.4 (see Beran and Parrent, 1964).

Some experimental results are given in Table 5-1. Due to the enormous baseline required, the experiment is not easy to do. To be sure, it calls for Michelson's skill and experience to obtain the results. The general stability of the instrument required is on the order of or smaller than the wavelength of light, Furthermore, the atmospheric phase fluctuations at such large mirror separations degrade the fringes.

Although not generally done, in principle, it is possible to measure the angular separation θ_{ds} of double stars by this method. Since each one of the pair of stars is nonresolvable, it may be treated as a point source (δ function). Application of the van Cittert–Zernike theorem to the pair of stars gives

$$\theta_{ds} = \frac{0.5\bar{\lambda}}{D_{12}}, \tag{5-70}$$

TABLE 5-1. STELLAR DIAMETER MEASUREMENTS BY THE MICHELSON STELLAR INTERFEROMETER[a]

Star	Angular Diameter (arc second)	Approximate Baseline
α Orionis Betelgeuse	0.047	10 ft = 3.05 m
α Tauri	0.02	24 ft = 7.31 m

[a] Michelson and Pease (1921).

where D_{12} is again the mirror separation for the vanishing of the fringes contributed by the double-star source [see Problems 5-10(a) and (b)].

In summary, the Michelson stellar interferometer measures the coherence function in the plane of the telescope aperture. By the van Cittert–Zernike theorem, this function is related to the spatial Fourier transform of the source optical intensity distribution. In this sense the interferometer is a spatial frequency analyzer.

HANBURY-BROWN AND TWISS INTERFEROMETER

Although not quite within the bounds of the second-order correlation, we shall briefly mention the pioneering experiment by Hanbury-Brown and Twiss (1956, 1957a, b), for which we also refer to the book on stellar interferometry by Hanbury-Brown (1974). In their experiment, a new concept was demonstrated, namely, that of the correlation of photocurrents. A proper description of this experiment and the new generation of experiments that followed entails studying the statistics of photoelectrons or Hanbury-Brown's Fourier component model. For this topic, we refer to two review articles, one by Mandel (1963) and the other by Mandel and Wolf (1965). Also, a lucid description of the theory, from both the classical and quantum points of view, is found in the excellent book by Klauder and Sudarshan (1968).

To describe the theoretical basis of the interferometer will take us too far afield. Only a brief discussion of the central quantity measured with the interferometer is given in what is to follow. A photocurrent is proportional to the time-integrated optical intensity falling on the detector during a time interval T. The proportionality constant includes the quantum efficiency of the detector. The question of whether or not the photocurrents are correlated hinges on whether the optical intensities are correlated. This latter correlation is a contracted form of the fourth-order amplitude correlation

$$\langle V_1 V_2 V_3^* V_4^* \rangle \rightarrow \langle I_1 I_2 \rangle,$$

when the point P_3 coincides with P_1 and P_4 with P_2 and the subscripts 1 and 2 on the optical intensity stand for the location of the two respective detectors separated in space like the two outer mirrors of the Michelson stellar interferometer. After this point, the photocurrents are fed to the appropriate electronic circuitry and correlated.

It turns out that, for a completely coherent case like that of the laser, the correlation factors into $\langle I_2 I_2 \rangle = \langle I_1 \rangle \langle I_2 \rangle$, implying thereby that the photocurrents are *not* correlated; the current values simply appear as products without any additional interference term. The factorization is easy to understand (theoretically) if it is calculated via the ensemble average. For this average, the joint probability density $W_2(I_1, I_2)$ is used. If I_1 and I_2 are statistically independent, the density factors into $W_1(I_1)W_1'(I_2)$, which in turn leads to the product of the individual average values.

The story is completely different, however, with thermal light whose amplitude is Gaussian distributed. For this case, by the property of these statistics, the fourth-order correlation of amplitude may be expressed as sums of products of second-order correlations, $\Gamma_{12}(\tau)$. It can be shown that the optical intensity correlation is given by

$$\langle I_1 I_2 \rangle = \Gamma_{11}(0)\, \Gamma_{22}(0)\left[1 + |\gamma_{12}(\tau)|^2\right].$$

The matter is slightly more involved than this since it is the integrated optical intensity that is measured. Whether or not the optical intensities are correlated depends on the ratio τ_c/T, where τ_c is the coherence time and T is the detector integration time. We shall consider only the two limiting cases. In one limit, where $T \ll \tau_c$, it is possible, by correlating photocurrents, to measure $|\gamma_{12}|^2$. In the other limit, $T \gg \tau_c$, the contribution of the term $|\gamma_{12}|^2$ is reduced by a factor τ_c/T. In the Hanbury-Brown and Twiss experiment the ratio τ_c/T was typically $\frac{1}{40}$.

The pioneering contribution of Hanbury-Brown and Twiss was to demonstrate the existence of the correlation of photocurrents and measure $|\gamma_{12}(\tau)|^2$ for both the laboratory and stellar sources. In this way, they were able to extend the baseline length to several hundred feet, a marked improvement over the Michelson stellar interferometer. With a 40-ft (12.2-m) baseline they determined the angular diameter of Sirius to be 0.0063 arc second and with a 390-ft (119-m) baseline they determined the diameter of the star β-Crusis to be 0.0007 arc second. The effect of the atmospheric phase distortion is very serious on amplitude correlation interferometry, whereas for the Hanbury-Brown and Twiss experiment it is negligible.

MICHELSON TWO-BEAM INTERFEROMETER

In the previous two sections we studied spatial coherence, that is, coherence *across* the beam characterized by $\Gamma_{12}(0)$. Now we shall examine coherence *along* the beam, that is, coherence along its direction of propagation. This is the so-called *temporal coherence*, symbolized by $\Gamma_{11}(\tau)$, where the two space points are coincident ($P_1 = P_2$) and where we focus our attention on the time delay τ.

Temporal coherence effects also make their presence felt in a two-pinhole experiment for large enough path differences. An experiment by Thompson

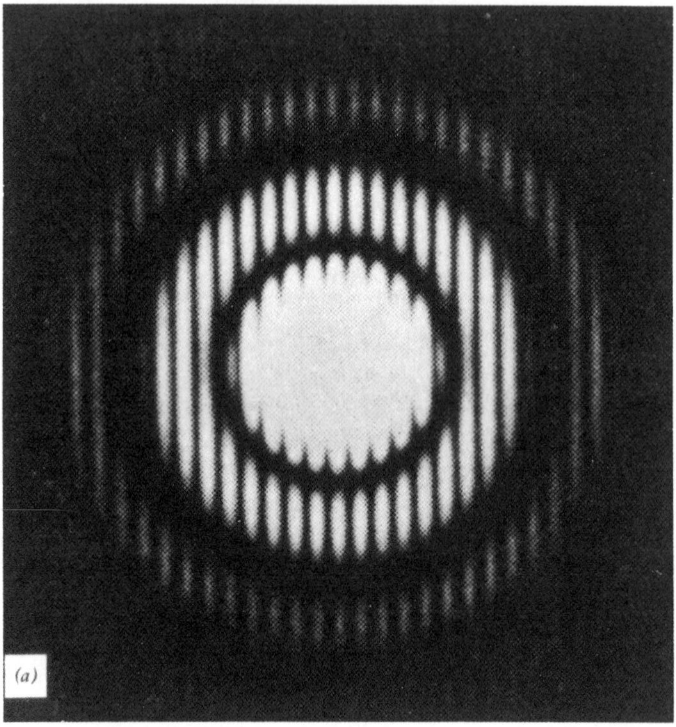

Fig. 5-10A. Thompson's experiment to show temporal coherence effects in two pinhole interference fringes. (*a*) A collimated mercury arc lamp is used to obtain the fringes. (*b*) A plane parallel glass plate is introduced in front of one of the pinholes; the fringes are absent, and only the diffraction pattern of a single hole is visible. (*c*) and (*d*) The same experiment is repeated with He–Ne laser light. The fringes are just as distinct in (*d*) even after introduction of the glass plate (from Thompson, 1965).

(1965) shows this effect on axis by use of a thin parallel glass plate (thickness 0.5 mm) in the light path from one of the pinholes. Its effect is to increase the optical path difference (OPD) in the interfering beams. In the experiment, two circular apertures were illuminated by a mercury arc source filtered to isolate the 546.1-nm line. The experimental results are shown in Fig. 5-10. The familiar double-aperture fringes are seen in Fig. 5-10*a*. With the introduction of the glass plate, no fringes are visible; only the transform of the apertures appears in Fig. 5-10*b*. As we shall see, the extra path difference due to the glass plate was too large compared to the "coherence length" of the filtered mercury arc light. Such was not the case when the experiment was repeated by using the helium–neon laser light of wavelength 632.8 nm. Fringes of good visibility are seen in both pictures Fig. 5-10*c* and *d*, which correspond to the absence of and the presence of the glass plate, respectively.

The Michelson two-beam interferometer (Fig. 5-11) allows us to direct our attention to the temporal coherence effects. Light from a broad noncoherent source S is amplitude divided at the beam divider BD. The divider

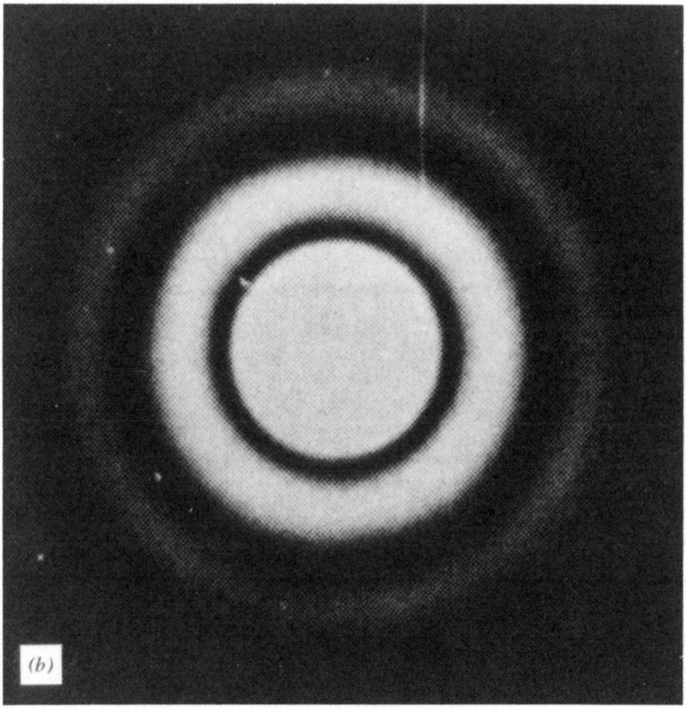

Fig. 5-10B

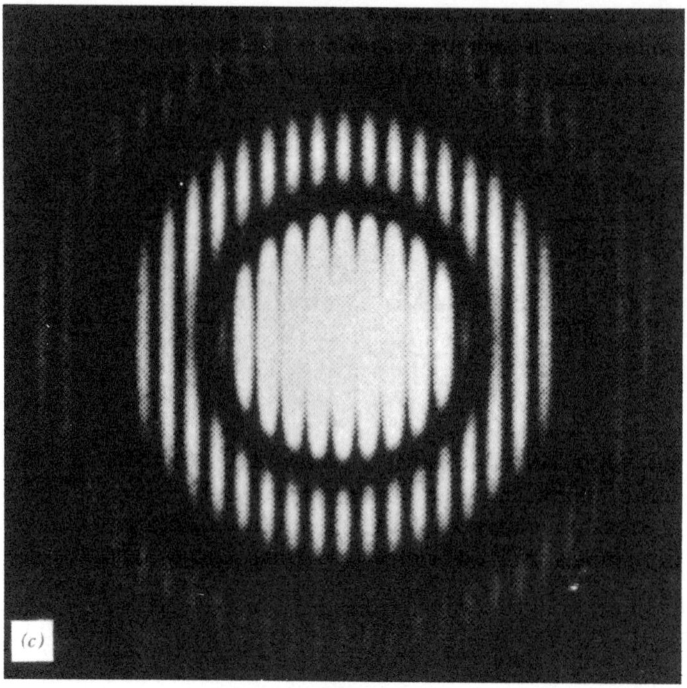

Fig. 5-10C

is partially silvered to achieve equal power in the reflected and transmitted light. The interferometer is adjusted so that the fixed mirror M_1 is exactly perpendicular to the movable mirror M_2. Equivalently, the image M_1' of M_1 in the BD is exactly parallel to the mirror M_2. This arrangement effectively produces two parallel images of the broad source S. The resulting fringes may be thought of as arising from the interference of light from these sources.

In the interferometer, beam 2 travels three times through the glass medium of BD before combining with beam 1, which travels only once through glass. For a monochromatic or sufficiently narrow-band spectrum, the equality of optical path may be achieved by simply moving mirror M_2. For light with a broad spectrum (white light), there can be no completely achromatic fringe, due to the dispersion of the glass. If the OPD for red light is made zero, that of blue light is not. Hence, to observe white light interference with a few bright and dark fringes, one must cause both beams to travel equal distances in glass. This is achieved by introducing a compensator plate C. It is an identical duplicate of the glass plate of BD, except for the silvering. This procedure is called *white light compensation*, various

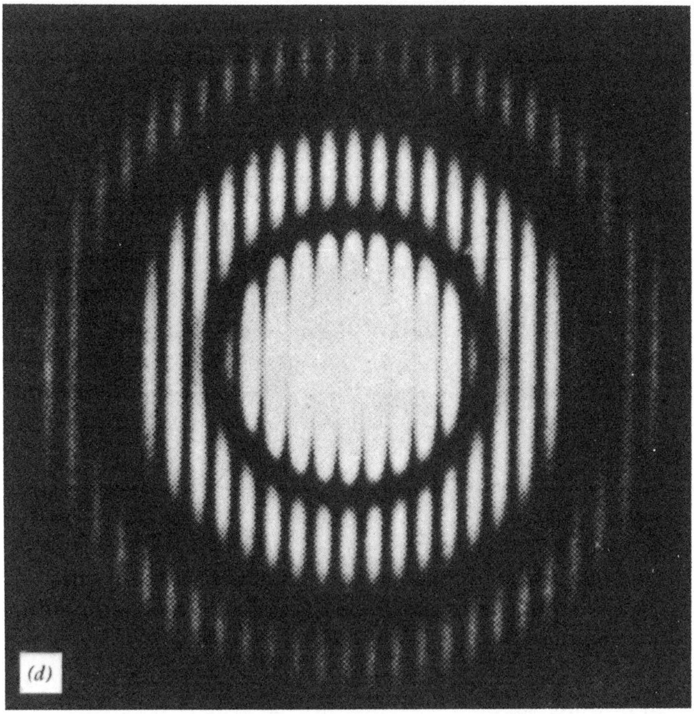

Fig. 5-10D

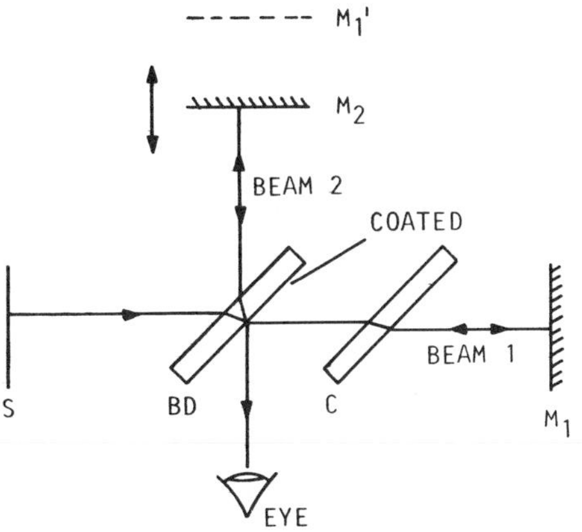

Fig. 5-11. Michelson two-beam interferometer. S, broad noncoherent source; BD, beam divider (lightly silvered, the additional phase introduced due to this coating is omitted from the discussion in the text); C, compensator plate; M_1, fixed mirror with tilting screws; M_2 movable mirror; M'_1, image of M_1 in the BD. Double-headed arrow indicates mirror movement.

aspects of which are described by Longhurst (1967), Françon (1966), Hecht and Zajac (1976), Mertz (1965), and Murty (1978).

If M_2 and M_1' are parallel and separated by d, the OPD condition to be satisfied is

$$2d\cos\theta = m\lambda, \tag{5-71}$$

for a bright fringe of order m observed at angle θ. Since d is fixed, it is θ that varies from one fringe to the next. These are the so-called *fringes of equal inclination*; by symmetry they are circular with center at the foot of the perpendicular from the eye to the mirror M_2. Since the interference is between pairs of parallel rays, the fringes are observed with the unaided eye in the relaxed state or with a suitable telescope.

The difference in the time taken for light to travel from BD to M_1 and back and from BD to M_2 and back is the delay τ. For the point at the center of the fringe pattern, the OPD condition reads

$$c\tau = 2d = m\lambda, \tag{5-72}$$

where the order m is not necessarily an integer. As the mirror M_2 is translated, both τ and d change; hence the light at the center alternates between bright and dark. Across the field of view (FOV) for a fixed d, the variation in θ brings about a variation in τ; that is,

$$c\tau' = 2d\cos\theta = m'\lambda. \tag{5-73}$$

The increase of θ from the center out indicates a decrease of order m'. The time delay τ' thus decreases from the pattern center to the edge of the FOV. For large enough d, there are enough fringes in the FOV for the change in visibility, due to the changing τ, to be noticeable.

Fourier Transform Spectrometry

A better way of studying the visibility as a function of τ is to record (e.g., by use of a photodetector) the variation in light irradiance at the center of the FOV as the mirror M_2 is translated. In Fourier transform spectrometry (FTS), this record is called an *interferogram*; a typical one is reproduced in Fig. 5-12. The horizontal axis may be labeled by the time delay

$$\tau = \frac{2d}{c}, \tag{5-74}$$

where d is the displacement of M_2 from the position of zero OPD.

To describe the interferogram by coherence theory, we let $V_1(t)$ be the field for beam 1 of Fig. 5-11. The field for beam 2 is the same but acquires a time delay τ. The superposition

$$V = V_1(t) + V_1(t+\tau)$$

appears at the FOV center. The observed irradiance (interferogram) is

$$\mathsf{E}(\tau) = \mathsf{C}I(\tau), \tag{5-75}$$

where the optical intensity $I(\tau)$ with the time-average operation is

$$I(\tau) = \langle [V_1(t) + V_1(t+\tau)][V_1^*(t) + V_1^*(t+\tau)] \rangle.$$

Its expression in terms of the coherence function reads

$$I(\tau) = 2I_1 + \Gamma_{11}(\tau) + \Gamma_{11}^*(\tau), \tag{5-76}$$

where $I_1 = \Gamma_{11}(0)$ is the optical intensity in one of the beams and

$$\Gamma_{11}(\tau) = \langle V_1(t+\tau) V_1^*(t) \rangle \tag{5-77}$$

is the self-coherence function. The complex degree of self-coherence is the normalized function

$$\gamma_{11}(\tau) = \frac{\Gamma_{11}(\tau)}{\Gamma_{11}(0)}, \tag{5-78}$$

with the property $0 \leq |\gamma_{11}(\tau)| \leq 1$. The expression for the interferogram in terms of $\gamma_{11}(\tau)$ is

$$\frac{I(\tau)}{2I_1} - 1 = |\gamma_{11}(\tau)| \cos \Phi_{11}(\tau), \tag{5-79}$$

where Φ_{11} is the phase argument of γ_{11}. If the ordinate of the interferogram, Fig. 5-12, is appropriately scaled from zero to unity, the envelope (locus of maxima, not shown in the figure) is the visibility function $\mathfrak{v}(\tau) = |\gamma_{11}(\tau)|$.

The temporal Fourier transform of the analytic signal $\Gamma_{11}(\tau)$ is

$$\Gamma_{11}(\tau) = \int_0^\infty \hat{\Gamma}_{11}(\nu) \exp(-i2\pi\nu\tau)\, d\nu. \tag{5-80}$$

Thus, Eq. (5-76) may be rewritten in the form

$$I(\tau) - 2I_1 = 2\int_0^\infty \hat{\Gamma}_{11}(\nu) \cos 2\pi\nu\tau\, d\nu. \tag{5-81}$$

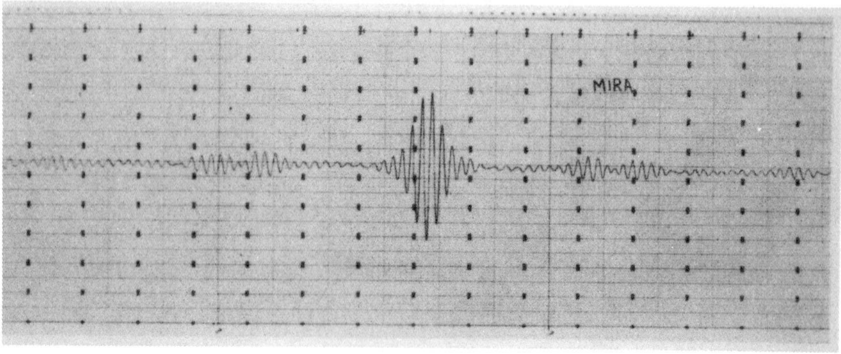

Fig. 5-12. Typical interferogram (from Mertz, 1965).

As is typical of cosine transforms, the inversion leads to the spectrum and its mirror image,

$$\int_{-\infty}^{\infty} [I(\tau) - 2I_1] \cos 2\pi\nu\tau \, d\tau = \hat{\Gamma}_{11}(-\nu) + \hat{\Gamma}_{11}(+\nu). \qquad (5\text{-}82)$$

Of course, in practice there is a limitation on the maximum displacement of mirror M_2 from the zero OPD position; that is, $-\tau_0 \le \tau \le +\tau_0$. Therefore, instead of the true spectrum $\hat{\Gamma}_{11}(\nu)$, one obtains the measured spectrum

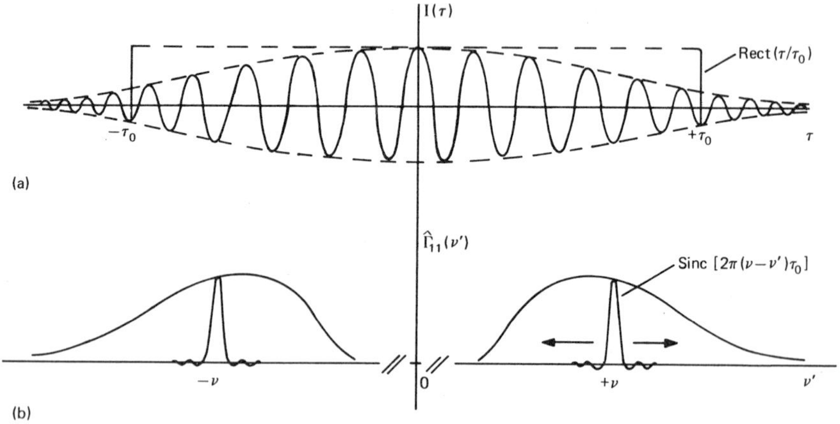

Fig. 5-13. Interferogram and its Fourier transform. (*a*) An interferogram $I(\tau)$ is shown truncated by the Rect function. (*b*) The true spectrum $\hat{\Gamma}_{11}$ and its mirror image. The instrumental profile at $\nu' = \pm\nu$ is shown; the result of the convolution is not shown.

$\hat{\Gamma}'_{11}(\nu)$, given by

$$\hat{\Gamma}'_{11}(-\nu) + \hat{\Gamma}'_{11}(+\nu) = 2\tau_0 \int_{-\infty}^{\infty} \text{Sinc}[2\pi(\nu - \nu')\tau_0] \times [\hat{\Gamma}_{11}(-\nu') + \hat{\Gamma}_{11}(+\nu')]\, d\nu'. \tag{5-83}$$

The interferogram truncated by Rect(τ/τ_0) is sketched in Fig. 5-13*a*. In Fig. 5-13*b* the double (true) spectrum and the instrumental profile, namely, the Sinc function, are sketched. The result of the convolution is not shown.

Resolving Power

The limitation on τ imposes a limit of resolution. As shown in Fig. 5-14, the neighboring spectral ordinates may be said to be just resolved if the maximum of one Sinc centered at ν_1 falls on the minimum of the other centered at ν_2. The spectral interval,

$$\nu_2 - \nu_1 = \delta\nu = \frac{1}{2\tau_0},$$

indicates that the resolving power $\mathcal{R}$ is

$$\mathcal{R} = \left|\frac{\bar{\lambda}}{\delta\lambda}\right| = \left|\frac{\bar{\nu}}{\delta\nu}\right| = 2\bar{\nu}\tau_0, \tag{5-84}$$

where $\bar{\nu}$ is the average value of ν_1 and ν_2. In terms of the OPD value we have

$$\mathcal{R} = \frac{2d_0}{\bar{\lambda}/2}, \tag{5-85}$$

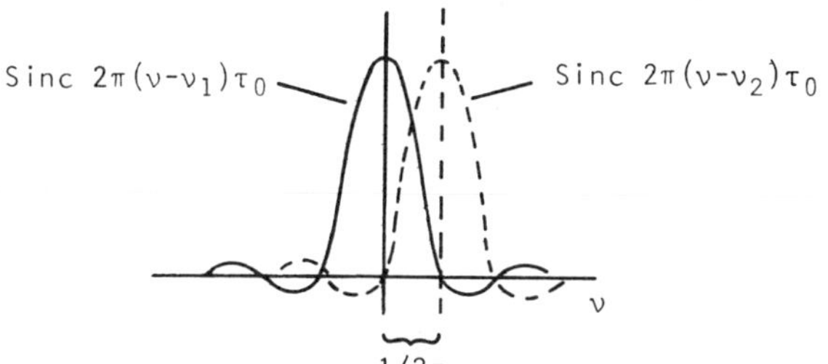

Fig. 5-14. Two just-resolved spectral ordinates. They are separated by $1/2\tau_0$, the distance of the first zero of the instrumental function (Sinc function) from its center.

where $d_0 = c\tau_0/2$ is the maximum displacement of mirror M_2 from the zero OPD position. From Eq. (5-85), $\mathcal{R}$ may be interpreted as the number of half waves that may be fitted between the total travel $(2d_0)$ of the mirror from one side of the zero time-delay position to the other. Thus the resolving power $\mathcal{R}$ at $\bar{\lambda}$ equals the number of maxima that may be counted during the travel distance $(2d_0)$.

The Sinc-function instrumental profile has rather large side lobes; after convolution it produces a false extension beyond the $\nu_{\min}$ and $\nu_{\max}$ values of the true spectrum. This may be avoided in practice at the cost of resolution. The extent of the true spectrum may be found by replacing the $\mathrm{Rect}(\tau/\tau_0)$ by a triangular truncating function from $-\tau_0$ to $+\tau_0$. Although its transform Sinc^2 function has smaller side lobes, it is twice as broad. This procedure for clipping the long tails of the measured spectrum by modifying the raw interferogram data is referred to as *apodization*. For further details, see Mertz (1965).

Coherence Time

When the beams in the two arms are of equal strength, the visibility, $\mathcal{V}$, equals $|\gamma_{11}(\tau)|$. It decreases from its zero OPD value due to the finite width of the spectrum. The detail behavior of $\mathcal{V}$ depends on the structure of the spectrum (see Problem 5-11, and Born and Wolf, 1970, Fig. 7.54). To introduce the concept of *coherence time*, τ_c, we begin with the normalized spectrum defined by

$$\hat{\gamma}_{11}(\nu) = \frac{\hat{\Gamma}_{11}(\nu)}{\int_0^\infty \hat{\Gamma}_{11}(\nu)\,d\nu}. \tag{5-86}$$

It is consistent with the definition of $\gamma_{11}(\tau)$ of Eq. (5-78). The Fourier relationship is

$$\gamma_{11}(\tau) = \int_0^\infty \hat{\gamma}_{11}(\nu)\exp(-i2\pi\nu\tau)\,d\nu, \tag{5-87}$$

and the normalization is such that

$$\gamma_{11}(0) = \int_0^\infty \hat{\gamma}_{11}(\nu)\,d\nu = 1.$$

Example 5-4

Suppose the normalized spectrum is Gaussian in form, with mean frequency $\bar{\nu}$,

$$\hat{\gamma}_{11}(\nu) = \frac{1}{(2\pi\sigma^2)^{1/2}} \exp\left[-\frac{(\nu-\bar{\nu})^2}{2\sigma^2}\right]. \tag{5-88}$$

[Since $\gamma(\tau)$ is an analytic signal, the approximation involved in using a Gaussian was discussed in Chapter 2.] As shown in Fig. 5-15, its spectral width $\Delta\nu$ may be defined in terms of the frequency values $\nu = \bar{\nu} \pm \nu_1$ at which $\hat{\gamma}_{11}$ attains about 4% of its maximum value; thus,

$$\Delta\nu = 2\nu_1 = 2(2\pi\sigma^2)^{1/2}. \tag{5-89}$$

The normalized temporal coherence also has a Gaussian form,

$$\gamma_{11}(\tau) = \exp(-2\pi^2\sigma^2\tau^2)\exp(-i2\pi\bar{\nu}\tau). \tag{5-90}$$

For a narrow spectrum, the visibility of fringes will remain high for larger values of the delay τ or for larger mirror displacements, $d = c\tau/2$. If we take "good" visibility to mean $v \gtrsim 0.46 \simeq \exp(-\pi/4)$, then the coherence time τ_c may be defined by

$$\tau_c = \frac{1}{2(2\pi\sigma^2)^{1/2}}. \tag{5-91}$$

These definitions obey the product relationship

$$\tau_c \Delta\nu = 1.$$

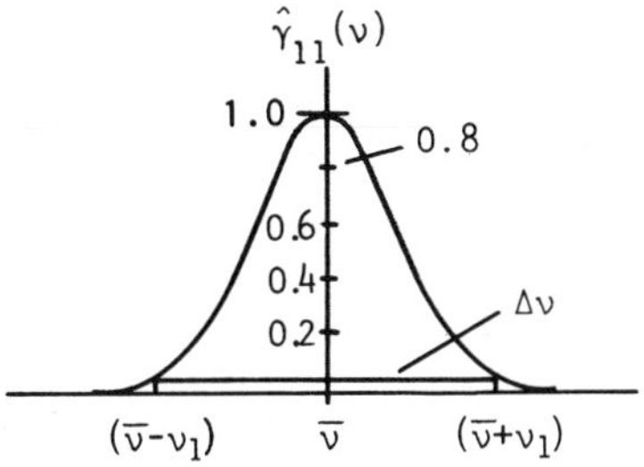

Fig. 5-15. The Gaussian spectral profile. The 4% point is marked by the horizontal line of length equal to the spectral width $\Delta\nu$ used in Eq. (5-89).

The numerical factors may change with the criterion of good visibility. Nevertheless, as a general rule, coherence time is large for narrow $\Delta\nu$ and vice versa. Thus, good visibility fringes are found when the delay τ fulfills the inequality

$$\tau\Delta\nu \lesssim 1. \tag{5-92}$$

More generally, the coherence time τ_c and the bandwidth $\Delta\nu$ may be defined in terms of the squared modulus $|\gamma_{11}(\tau)|^2$. Thus, following Wolf (1958), the coherence time is defined in terms of the "second moment,"

$$\tau_c^2 = \frac{\int_{-\infty}^{\infty} \tau^2 |\gamma_{11}(\tau)|^2 \, d\tau}{\int_{-\infty}^{\infty} |\gamma_{11}(\tau)|^2 \, d\tau}. \tag{5-93}$$

The denominator is related to the normalized spectral density by Parseval's theorem,

$$\int_{-\infty}^{\infty} |\gamma_{11}(\tau)|^2 \, d\tau = \int_0^{\infty} |\hat{\gamma}_{11}(\nu)|^2 \, d\nu. \tag{5-94}$$

The effective bandwidth is defined in terms of the "variance,"

$$(\Delta\nu)^2 = \frac{\int_0^{\infty} (\nu - \bar{\nu})^2 |\hat{\gamma}_{11}(\nu)|^2 \, d\nu}{\int_0^{\infty} |\hat{\gamma}_{11}(\nu)|^2 \, d\nu}, \tag{5-95}$$

where the mean frequency is

$$\bar{\nu} = \frac{\int_0^{\infty} \nu |\hat{\gamma}_{11}(\nu)|^2 \, d\nu}{\int_0^{\infty} |\hat{\gamma}_{11}(\nu)|^2 \, d\nu}. \tag{5-96}$$

With these definitions one obtains the reciprocal inequality (Wolf, 1958; Born and Wolf, 1970),

$$\tau_c \Delta\nu \geq \frac{1}{4\pi}, \tag{5-97}$$

For a narrow Gaussian spectrum, as in Example 5-4, the equality sign holds to a good approximation. It is only approximate because the lower limit of the frequency integral is zero for analytic signals rather than $-\infty$. For more general spectral shapes, the inequality sign holds. For the mathematics of this problem and for alternative definitions, we refer to Mandel and Wolf (1965), where further references are also listed.

The essence of the whole argument is that τ_c increases when $\Delta\nu$ decreases and as such we may regard τ_c to fulfill the order-of-magnitude relationship

$$\tau_c \simeq \frac{1}{\Delta\nu}. \tag{5-98}$$

The corresponding coherence length is defined by

$$(\Delta l)_c = c\tau_c \simeq \frac{c}{\Delta\nu} = \frac{\bar{\lambda}^2}{\Delta\lambda}. \tag{5-99}$$

Recall the small-path-difference condition of the QM approximation, that OPD $< c/\Delta\nu$. The foregoing discussion tells us that the path differences must be held smaller than the coherence length. In Thompson's experiment, Fig. 5-10, the fringes vanished since the path difference, $(n-1)t$, introduced by the plate of thickness t, was larger than $(\Delta l)_c$ (see Problem 5-12).

Observe that τ_c or $(\Delta l)_c$ is defined correctly in terms of the self-coherence function $\gamma_{11}(\tau)$, which is primarily a statistically defined quantity. Both τ_c and $\Delta\nu$ were discussed without mention of the field $V(t)$ itself. Although commonly done in optics textbooks, strictly speaking it is inappropriate to identify $(\Delta l)_c$ with the "length of a single wavetrain" (see Appendix 3.2).

The isolation of the temporal coherence by means of the Michelson two-beam interferometer enables us to study the spectral properties of the source. The interferometer is therefore a temporal frequency analyzer.

COHERENCE VOLUME

In dealing with thermal sources (Gaussian statistics), the second-order function $\Gamma_{12}(\tau)$ is adequate for the description of optical phenomena. The artificial separation into spatial, $\Gamma_{12}(0)$, and temporal, $\Gamma_{11}(\tau)$, coherence is found useful in special applications.

Our calculations of coherence width and area were based on the van Cittert–Zernike theorem, which presupposes a plane noncoherent source. The coherence width is defined in a plane perpendicular to the beam propagation from such a source. As in Eq. (5-51), the width in a particular direction in the plane fulfills the order-of-magnitude relationship

$$w \simeq \frac{\bar{\lambda} z}{2(\text{source width})} = \frac{\bar{\lambda}}{2\theta_s},$$

where the *source width* is the dimension measured parallel to the direction of interest in the plane. The angular subtense of this source dimension is θ_s at the plane of interest. Likewise, the coherence area across the beam is

$$\mathscr{a}_c \simeq \frac{\bar{\lambda}^2 z^2}{4(\text{source area})} \simeq \frac{\bar{\lambda}^2}{4\Omega}, \tag{5-100}$$

where Ω is the approximate solid angle subtended by the source at the plane.

As one moves away from the source, the area $\mathscr{a}_c$ will get slightly larger. This variation may be neglected for a moderate passband $\Delta\nu$ or $\Delta\lambda$ of the spectral filter used. The coherence length along the direction of beam propagation is given by the order-of-magnitude relationship

$$(\Delta l)_c = c\tau_c \simeq \frac{c}{\Delta\nu} = \frac{\bar{\lambda}^2}{\Delta\lambda}. \tag{5-101}$$

Consistent with these considerations, imagine a volume, called *coherence volume* $(\Delta v)_c$, made up of this coherence area and length,

$$(\Delta v)_c = \mathscr{a}_c(\Delta l)_c = \frac{\bar{\lambda}^3}{\Omega}\left(\frac{\bar{\lambda}}{\Delta\lambda}\right) = \frac{\bar{\lambda}^3}{\Omega}\left(\frac{\bar{\nu}}{\Delta\nu}\right), \tag{5-102}$$

marked off in the radiation zone. Radiation coming from points within this volume is capable of interfering to produce fringes of "good" visibility (see Problem 5-13). For further reading, we refer to Mandel and Wolf (1965) for the identification of this coherence volume with a single cell of phase space of photons.

BEAM SPREAD

Light leaving a point source in the plane $z = 0$ spreads into the right half-space, uniformly over spheres of increasing radii with the point source at the center. In general, for a plane source in any state of coherence, the light energy leaving it spreads into the right half-space. We shall use the term *beam spread* to describe how the major portion (fraction $\varepsilon = 0.96$) of the beam power is spread as the distance, z, from the source is increased.

An area integral of the spectral irradiance,

$$\hat{\mathsf{E}}(P, \nu) = \mathsf{C}\hat{\Gamma}(P, P, \nu), \tag{5-103}$$

with units of W m^{-2} Hz^{-1}, gives the spectral power $\hat{\boldsymbol{\Phi}}$ contained in that area. For beam spread, the change in this area for a constant enclosed spectral power (ESP) is studied as the distance z is increased. For a special case of a source with radial symmetry, the area that encloses a constant (96%) ESP will be circular, and its radius may be plotted versus the distance z from the source. Such a plot describes how the light spreads after leaving the source. We shall study an ideal coherent laser and also a completely noncoherent source. The intermediate cases of partial coherence will be studied by using the quasihomogeneous source model of Carter and Wolf (1977). The basic propagation for QM fields, Eq. (5-31), will be used throughout this section. In the context of QM fields, one studies the enclosed power $\mathbf{\Phi}$ in the area integral of the irradiance,

$$\mathsf{E}(P) = \mathsf{C}\Gamma(P, P, 0), \tag{5-104}$$

with units W m^{-2}. The short form EP for enclosed power will be used.

Coherent Source

An ideal laser with a Gaussian optical intensity profile is an example of the coherent source. Its mutual optical intensity function has the factored form as in Eq. (5-33),

$$\Gamma(\mathbf{s}_1, \mathbf{s}_2, 0) = U(\mathbf{s}_1)\, U^*(\mathbf{s}_2),$$

where

$$U(\mathbf{s}_j) = I_L^{1/2} \exp\left(-\frac{r_{sj}^2}{4\sigma_L^2} \right), \qquad j = 1, 2. \tag{5-105}$$

Since the source aperture is in the $z = 0$ plane, r_{sj} is used to denote the radial distance, $r_{sj}^2 = x_{sj}^2 + y_{sj}^2$ in the plane. The zero ordinate of the Gaussian profile is I_L, and σ_L is the standard deviation.

In the coherent limit we may work with the amplitude $U'(\mathbf{r})$, Eq. (5-34), in order to calculate the optical intensity,

$$\Gamma(P, P, 0) = |\, U'(\mathbf{r})\, |^2, \tag{5-106}$$

where P denotes a typical point $\mathbf{r}$ in the plane of observation at z in Fig. 5-16. The use of the quadratic approximation for the normal derivative of the Green's function, Eq. (5-38), leads us to

$$U'(\mathbf{r}) = \frac{-i}{\lambda z} \iint_{\mathcal{A}} U(\mathbf{s}_j) \exp(+ik\rho_j)\, dx_{sj}\, dy_{sj}, \tag{5-107}$$

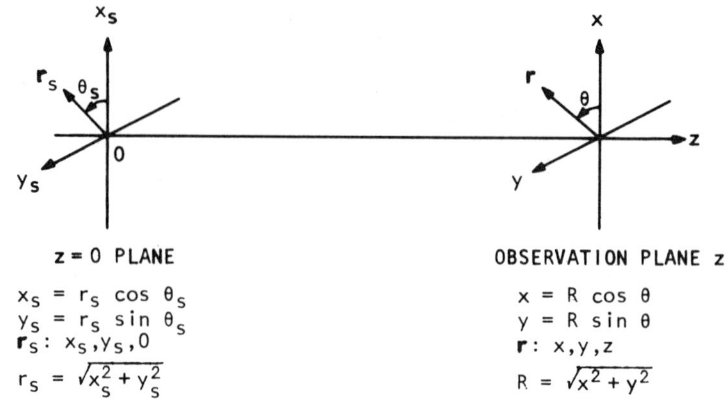

Fig. 5-16. Coordinate system and symbols used.

where λ is the laser wavelength and ρ_j in the approximation is

$$\rho_j = |\mathbf{r} - \mathbf{r}_{sj}|$$
$$\simeq r - \frac{xx_{sj} + yy_{sj}}{z} + \frac{x_{sj}^2 + y_{sj}^2}{2z}.$$

In terms of the polar coordinates described in Fig. 5-16, ρ_j becomes

$$\rho_j \simeq r - \frac{Rr_{sj}\cos(\theta - \theta_{sj})}{z} + \frac{r_{sj}^2}{2z}, \tag{5-108}$$

where R is the radial distance in the plane of observation.

Let the optical aperture of the laser be circular, of radius a_0. We shall suppose that it is very large compared to the standard deviation of the optical intensity profile, $a_0 \gg \sigma_L$. With this condition, the integrals on the source variables may be extended to infinity as though the aperture were absent. The effect of the finite aperture will be studied later. With the circular aperture and the assumed profile, the problem has axial symmetry. The spectral irradiance at P at a distance R from the axis will not be any different from that at any other point at the same distance from the axis for any azimuthal angle θ. For simplicity of algebra, we shall choose $\theta = 0$ without affecting the answer.

Thus the expression for the wave amplitude at $\mathbf{r}$ is

$$U'(\mathbf{r}) = \left(\frac{-iI_L^{1/2}}{\lambda z}\right)\exp(ikr)\int_0^{\infty}\int_0^{2\pi}\exp\left(\frac{-r_{sj}^2}{4\sigma_L^2}\right)$$
$$\times\exp\left(\frac{+ikr_{sj}^2}{2z}\right)\exp\left(\frac{-ikRr_{sj}}{z}\cos\theta_{sj}\right)d\theta_{sj}\,r_{sj}\,dr_{sj}.$$

The integral over θ_{sj} is simply related to the definition of the zero-order Bessel function, which leaves us with the integral on r_{sj},

$$U'(\mathbf{r}) = \left(\frac{-2\pi iI_L^{1/2}}{\lambda z}\right)\exp(ikr)\int_0^{\infty}J_0\left(\frac{-kRr_{sj}}{z}\right)$$
$$\times\exp\left[\left(-\frac{1}{4\sigma_L^2}+\frac{ik}{2z}\right)r_{sj}^2\right]r_{sj}\,dr_{sj}.$$

This integral is in the form of the one reported by Abramowitz and Stegun (1964), which gives us

$$U'(\mathbf{r}) = \left(\frac{-\pi iI_L^{1/2}}{\lambda z}\right)\frac{\exp(ikr)}{1/4\sigma_L^2 - ik/2z}\exp\left[\frac{-k^2R^2}{4z^2}\left(\frac{1}{4\sigma_L^2}-\frac{ik}{2z}\right)^{-1}\right].$$

With a bit more algebraic manipulation as dictated by Eq. (5-106), we may cast the optical intensity in the form

$$\Gamma(P,P,0) = \frac{k^2I_L}{z^2}\left(\frac{4\sigma_L^4z^2}{z^2+4k^2\sigma_L^4}\right)\exp\left(-\frac{2k^2\sigma_L^2R^2}{z^2+4k^2\sigma_L^4}\right) \tag{5-109}$$

(see Problem 5-14). The value of the radius $R = R_L$, for which the argument of the negative exponential equals $-\pi$, defines a circle that encloses almost 96% ($\varepsilon = 0.96$) of the power in the plane of observation at z. This radius is given by

$$R_L = \left[2\pi\sigma_L^2\left(\frac{z^2}{4k^2\sigma_L^4}+1\right)\right]^{1/2}. \tag{5-110}$$

With the choice of the radius a_0 of the aperture to be very large compared to the standard deviation σ_L, the extent of the laser aperture is determined by σ_L. It is convenient to define the diameter d in this situation by

$$d = \left(4\pi\sigma_L^2\right)^{1/2}. \tag{5-111}$$

The far-field distance of the laser aperture may be taken to be

$$z_{\mathrm{FF}} = \frac{d^2}{\lambda} = 2k\sigma_L^2. \tag{5-112}$$

Now the plot of R_L versus z describes the beam spread. Two limiting cases are of interest: (i) Fraunhofer, $z \gg z_{\mathrm{FF}}$,

$$R_L \simeq \left(\frac{\pi}{2k^2\sigma_L^2}\right)^{1/2} z \tag{5-113a}$$

and (ii) Fresnel, $z \ll z_{\mathrm{FF}}$,

$$R_L \simeq \left(2\pi\sigma_L^2\right)^{1/2}. \tag{5-113b}$$

When the plane of observation is well into the far-field region, the beam spreads linearly with z. Well inside the Fresnel region, R_L remains constant; there is almost no spreading of the beam. The distance z_{FF}, which marks off the two regions, depends on both k and σ_L. Its numerical value may change from one laser to another, but the nature of the beam spread of the laser light remains unaltered. This intrinsic property of the light from any laser may be described by using unitless parameters. We use q_L for the beam radius,

$$R_L = q_L\left(2\pi\sigma_L^2\right)^{1/2}, \tag{5-114}$$

and measure the distance z in units of the far-field distance,

$$z = p_L\left(2k\sigma_L^2\right). \tag{5-115}$$

In terms of these unitless parameters, Eq. (5-110) takes the form

$$q_L = \left(p_L^2 + 1\right)^{1/2}. \tag{5-116}$$

It is plotted in Fig. 5-17. In part (a) of this figure, the distance parameter p_L is less than unity. It shows that q_L is very nearly constant throughout this region, indicating almost no beam spread. In this region the laser beam

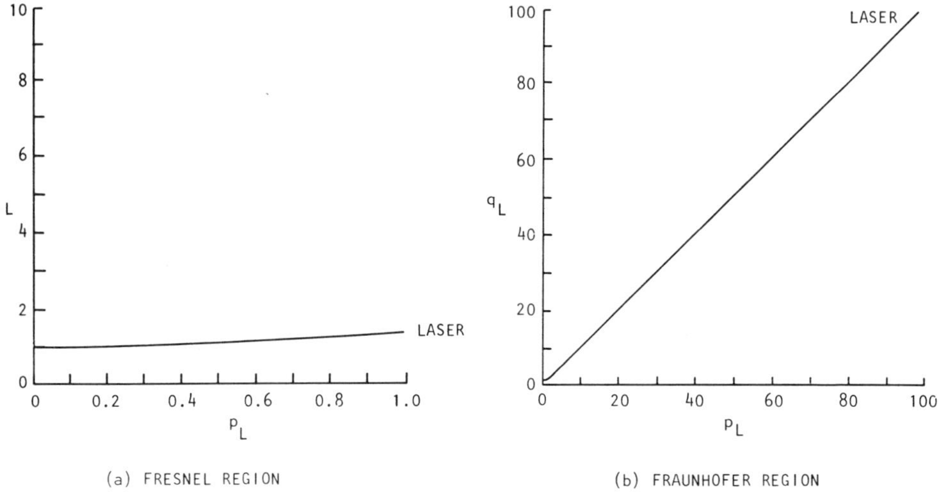

Fig. 5-17. Plots to show how the beam spreads as it leaves the Gaussian intensity profile laser source. The unitless parameter q_L is the ratio of the radial distance from the z axis in the plane of observation to half the width of the source at the 4% point of the Gaussian. The parameter p_L describes the distance z from the source scaled in terms of the far-field distance. (a) Beam spread in the Fresnel region, where the scale of the figure is such that $p_L < 1$ is examined. (b) Beam spread in the Fraunhofer region, where the scale of the abscissa is such that we focus our attention on the large values of p_L, $p_L \gg 1$ (from Marathay and Goring, 1979).

behaves like a "pointer." It loses its pointing capability in the Fraunhofer region, as shown in part (b) of the figure, where q_L increases linearly with p_L. On the scale of this graph the region $p_L \lesssim 1$ is almost imperceptible. In the next subsection we shall study the effect of partial coherence on beam spread.

Partially Coherent Source

We shall use the Carter and Wolf (1977) source as a model for studying the partially coherent source. It is characterized by the factored form of the mutual optical intensity function, see Eq. (5-11),

$$\Gamma(\mathbf{s}_1, \mathbf{s}_2, 0) = I\left(\frac{\mathbf{s}_1 + \mathbf{s}_2}{2}\right) g(\mathbf{s}_1 - \mathbf{s}_2). \tag{5-117}$$

It is also called a quasihomogeneous (QH) source, due to the requirement that the optical intensity I must be slowly varying compared to the spatial coherence described by g. The condition $g(0) = 1$ assures that $\Gamma(\mathbf{s}, \mathbf{s}, 0) = I(\mathbf{s})$. Following Carter and Wolf, we choose them to be Gaussian, see

Eq. (5-17),

$$I\left(\frac{\mathbf{s}_1+\mathbf{s}_2}{2}\right)=I_Q\exp\left[-\frac{(\mathbf{s}_1+\mathbf{s}_2)^2}{8\sigma_Q^2}\right],$$

$$g(\mathbf{s}_1-\mathbf{s}_2)=\exp\left[-\frac{(\mathbf{s}_1-\mathbf{s}_2)^2}{2\sigma_g^2}\right], \qquad (5\text{-}118)$$

with the condition $\sigma_Q \gg \sigma_g$. As in the previous case of a coherent source, we shall assume that the optical aperture of the QH source is circular and its radius $a_0 \gg \sigma_Q$. The area integrals are thus extended to infinite limits without much error. As far as the optical intensity $\Gamma(P, P, 0)$ calculation is concerned there is axial symmetry and we shall put $\theta = 0$ as before. With

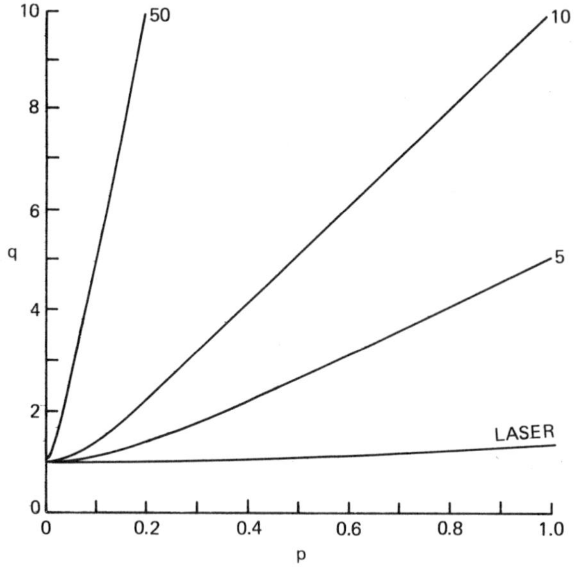

(a) FRESNEL REGION

Fig. 5-18A. Plots to show how the beam spreads as it leaves the quasihomogeneous, partially coherent source, compared with the laser case. The definitions of the unitless parameters [see Eqs. (5-122) and (5-123)] are similar to those used in Fig. 5-17 and are used here without subscripts. The labels 5, 10, 50 refer to the values of the ratio $2\sigma_Q/\sigma_g$, which describes the state of partial coherence. (a) Beam spread in the Fresnel region, $p < 1$. (b) Beam spread in the Fraunhofer region, $p \gg 1$ (from Marathay and Goring, 1979).

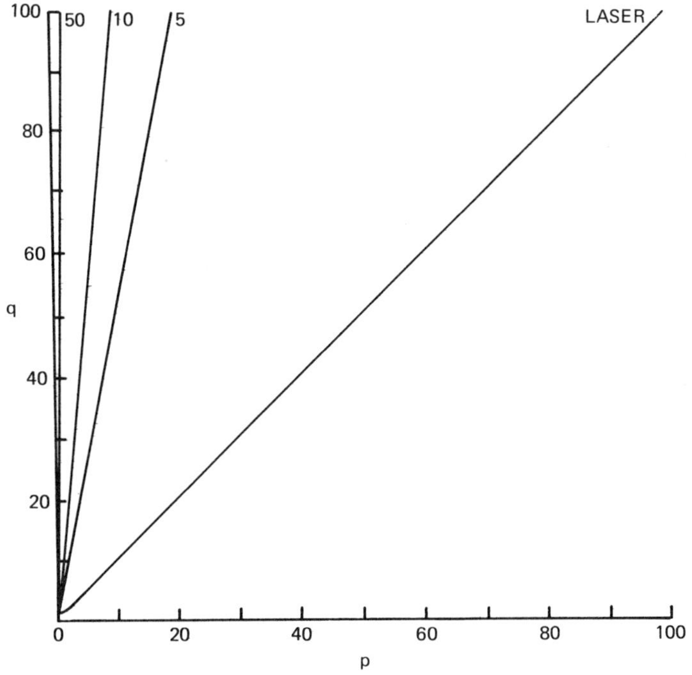

(b) FRAUNHOFER REGION

Fig. 5-18B

the standard quadratic approximation, Eqs. (5-31) and (5-38) give

$$\Gamma(P, P, 0) = \frac{1}{\bar{\lambda}^2 z^2} \iiint\int_{-\infty}^{\infty} \Gamma(\mathbf{s}_1, \mathbf{s}_2, 0) \times \exp[+ik(\rho_1 - \rho_2)]\, dx_{s1}\, dy_{s1}\, dx_{s2}\, dy_{s2}, \quad (5\text{-}119)$$

where $\bar{\lambda}$ is the mean wavelength and $k = 2\pi/\bar{\lambda}$. The algebraic details for the evaluation of the integrals are given in Appendix 5.3. The final expression for the optical intensity at P in the plane at a distance z from the QH source is

$$\Gamma(P, P, 0) = I_Q\left(\frac{k^2\sigma_Q^2\sigma_g^2}{z^2 + k^2\sigma_Q^2\sigma_g^2}\right)\exp\left[-\frac{R^2}{2\sigma_Q^2}\left(\frac{k^2\sigma_Q^2\sigma_g^2}{z^2 + k^2\sigma_Q^2\sigma_g^2}\right)\right]. \quad (5\text{-}120)$$

This same result was obtained in Eq. (5-20) by use of the angular spectrum approach. The circle that encloses almost 96% of the beam power has the radius

$$R_Q = \left[2\pi\sigma_Q^2\left(\frac{z^2}{k^2\sigma_Q^2\sigma_g^2} + 1\right)\right]^{1/2}. \tag{5-121}$$

Under the assumed conditions, the QH source aperture radius is determined by σ_Q. As with the coherent case, we shall scale the radial distance in terms of the effective width of the source

$$R_Q = q_Q\left(2\pi\sigma_Q^2\right)^{1/2} \tag{5-122}$$

and the distance z in terms of what would be a far-field distance for the same size coherent source,

$$z = p_Q\left(2k\sigma_Q^2\right) = p_Q z_{\mathrm{FF}}. \tag{5-123}$$

In terms of the unitless parameters, Eq. (5-121) takes the form

$$q_Q = \left[\left(\frac{2\sigma_Q}{\sigma_g}\right)^2 p_Q^2 + 1\right]^{1/2}. \tag{5-124}$$

Although the mathematical derivation of the result is free of restrictions, we bear in mind that, for the QH source, $\sigma_Q \gg \sigma_g$. The condition delineates the bounds within which the optical interpretation of the result is meaningful. The state of coherence may be described by the ratio $\sigma_g/2\sigma_Q$. The QH source approaches the state of noncoherence when $\sigma_g/2\sigma_Q \ll 1$.

In Eq. (5-121), if $z \gg k\sigma_Q\sigma_g = (\sigma_g/2\sigma_Q)z_{\mathrm{FF}}$, then R_Q increases linearly with z:

$$R_Q \simeq \frac{(2\pi)^{1/2}}{k\sigma_g}z. \tag{5-125}$$

Observe that the slope decreases when the coherence function g becomes broad, which makes the QH source more coherent. In the other limit, $z \ll k\sigma_Q\sigma_g = (\sigma_g/2\sigma_Q)z_{\mathrm{FF}}$, R_Q is reduced to a constant:

$$R_Q \simeq \left(2\pi\sigma_Q^2\right)^{1/2}. \tag{5-126}$$

These general features are seen in Fig. 5-18, where q_Q is plotted as a function of p_Q according to Eq. (5-124). In both Figs. 5-18*a* and 5-18*b*, curves for values 5, 10, and 50 of the ratio $2\sigma_Q/\sigma_g$ are shown. In general, the beam spreads more as the QH source becomes more noncoherent.

For a direct comparison with the coherent case we shall choose both sources of the same dimension, $\sigma_L = \sigma_Q = \sigma$. For both sources the distance z from the source is measured in units of $2k\sigma^2$. The radius parameter is scaled in terms of $(2\pi\sigma^2)^{1/2}$, the same for both sources. The two sources are placed on an equal footing. It is evident that the plots in Fig. 5-18 also apply for this choice of scale. As seen there, the light from a QH source will always spread more than the coherent laser light (see Marathay and Goring, 1979).

Noncoherent Source

The spatial coherence function $\Gamma_{12}(0)$ for a noncoherent source whose aperture is in the $z = 0$ plane may be described by

$$\Gamma(x_1, y_1, x_2, y_2, 0, 0) = \frac{\bar{\lambda}^2}{\pi}\Gamma(x_1, y_1, x_1, y_1, 0, 0)\delta(x_1 - x_2)\,\delta(y_1 - y_2). \tag{5-127}$$

In this limit, the beam spreads through large angles, making the application of scalar diffraction theory suspect. Yet seeking an answer within its confines, we shall avoid the usual approximations. The integrals become more manageable if, instead of a Gaussian profile, we assume the source to be uniform and of finite size. Thus, let the radiant exitance function be

$$\begin{aligned} \mathsf{M}(x_1, y_1, 0) &= \mathsf{C}\Gamma(x_1, y_1, x_1, y_1, 0, 0) \\ &= \mathsf{C} I_0 \,\mathrm{cyl}\left[\frac{(x_1^2 + y_1^2)^{1/2}}{a_0}\right]. \end{aligned}$$

We begin with the basic propagation law, Eq. (5-31), and the expression for the normal derivative of the Green's function, Eq. (5-38), *without* the approximation. The irradiance E on the plane z due to this source is

$$\mathsf{E}(x, y, z) = \frac{\bar{\lambda}^2}{4\pi^3}\int\int \mathsf{M}(x_1, y_1, 0)\left(\frac{z^2}{\rho_1^2}\right)\left(\frac{1 + k^2\rho_1^2}{\rho_1^4}\right)dx_1\,dy_1,$$

where z/ρ_1 is the familiar $\cos\theta_1$ term. Figure 5-19 describes the symbols and the notation of polar coordinates.

For light in the radiation zone, we shall make the approximation that $k^2\rho_1^2 \gg 1$. For a circular source of radius a_0, the calculation is axially symmetric about the z axis. Calculation with $\theta = 0$ requires evaluation of

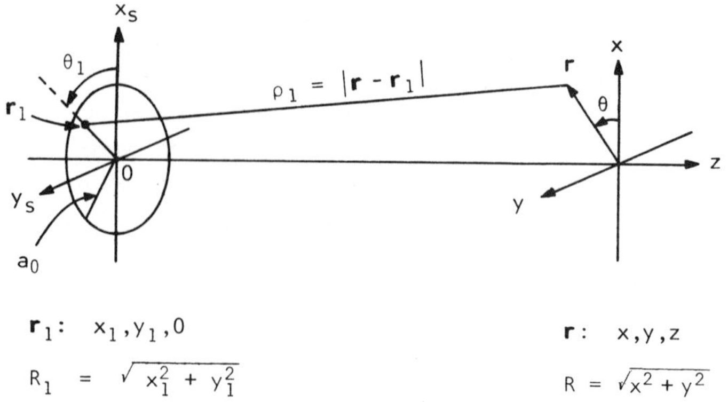

Fig. 5-19. Coordinate system and notation for the beam-spread calculation for a noncoherent circular source of radius a_0.

the expression

$$\mathsf{E}(R, 0, z) = \mathsf{C}\frac{z^2}{\pi} I_0 \int_0^{a_0} \int_0^{2\pi} \frac{R_1\, dR_1\, d\theta_1}{\left(z^2 + R^2 + R_1^2 - 2RR_1 \cos\theta_1\right)^2}.$$

The cancellation of the wavelength $\bar{\lambda}$ suggests a geometrical optical behavior of light in the radiation zone in the limit of a noncoherent source. A look at the table of integrals shows that

$$\mathsf{E}(R, 0, z) = \mathsf{C}\frac{I_0}{2}\left\{1 - \frac{z^2 + R^2 - a_0^2}{\left[R^4 + 2R^2(z^2 - a_0^2) + (z^2 + a_0^2)^2\right]^{1/2}}\right\}. \tag{5-128}$$

The total power enclosed at the source aperture ($z = 0$) is

$$(\mathrm{EP})_0 = \int_0^{a_0} \int_0^{2\pi} \mathsf{M}(R_1 \cos\theta_1, R_1 \sin\theta_1, 0)\, R_1\, dR_1\, d\theta_1$$
$$= \pi a_0^2 \mathsf{C} I_0. \tag{5-129}$$

In the plane $z =$ constant, the power enclosed by a circle of radius R_0 is

$$(\mathrm{EP})_z = \int_0^{R_0} \int_0^{2\pi} \mathsf{E}(R, 0, z)\, R\, dR\, d\theta$$
$$= \frac{\pi}{2}\mathsf{C} I_0\left(R_0^2 - \left\{\left[R_0^4 + 2R_0^2(z^2 - a_0^2) + (z^2 + a_0^2)^2\right]^{1/2} - (z^2 + a_0^2)\right\}\right). \tag{5-130}$$

The fraction $\varepsilon = (\mathrm{EP})_z/(\mathrm{EP})_0$ of the enclosed power is, therefore,

$$\varepsilon = \frac{R_0^2 + z^2 + a_0^2}{2a_0^2} - \left[\left(\frac{R_0^2}{2a_0^2}\right)^2 + \frac{R_0^2(z^2 - a_0^2)}{2a_0^4} + \left(\frac{z^2 + a_0^2}{2a_0^2}\right)^2\right]^{1/2}. \tag{5-131}$$

This expression is one of the roots of the quadratic

$$\varepsilon^2 - \left(\frac{R_0^2 + z^2 + a_0^2}{a_0^2}\right)\varepsilon + \frac{R_0^2}{a_0^2} = 0,$$

from which the radius R_0 of the circle that encloses the fraction ε of the power is found to be

$$R_0 = \left(\frac{\varepsilon}{1-\varepsilon} z^2 + a_0^2 \varepsilon\right)^{1/2}. \tag{5-132}$$

The radius of the circle with similar properties at the source is $a_0\sqrt{\varepsilon}$. If this is used as the effective size of the source, the unitless parameter for the radius is defined through the relationship

$$R_0 = q_0 a_0 \sqrt{\varepsilon}. \tag{5-133}$$

The distance z can be measured in units of the far-field distance of the effective size of the source, namely,

$$z = p_0 \frac{\left(2a_0\sqrt{\varepsilon}\right)^2}{\lambda}. \tag{5-134}$$

With the unitless parameters q_0 and p_0, Eq. (5-132) reads

$$q_0 = \left[\frac{\varepsilon^2}{1-\varepsilon}\left(\frac{4a_0}{\lambda}\right)^2 p_0^2 + 1\right]^{1/2}. \tag{5-135}$$

For an optical source the radius a_0 is generally much larger than λ, $a_0 \gg \lambda$. The coefficient of p_0^2 will thus be very large, indicating a large beam spread.

Comparison of Coherent, Noncoherent, and Quasihomogeneous Sources

For a direct comparison among the coherent, noncoherent, and quasihomogeneous source cases, let all three sources have the same effective size by requiring that

$$a_0\left(\frac{\varepsilon}{2\pi}\right)^{1/2} = \sigma_Q = \sigma_L = \sigma \tag{5-136}$$

(see Marathay et al., 1978). In this situation, the three basic equations of beam spread, Eqs. (5-110), (5-121), and (5-132), may be cast in the form

$$q = \left(Ap^2 + 1\right)^{1/2}, \tag{5-137}$$

where the coefficient A for the three respective cases is:

$$A = \begin{cases} 1, & \text{for coherent laser;} \\ (2\sigma/\sigma_g)^2, & \text{for QH source, } \sigma \gg \sigma_g; \\ [32\pi\varepsilon/(1-\varepsilon)](\sigma/\lambda)^2, & \text{for } \varepsilon \simeq 0.96 \text{, noncoherent.} \end{cases}$$

The subscripts on the unitless parameters have been dropped since their respective definitions are now the same for the three sources of the same effective size. Obviously, the coefficient A for the noncoherent case is much larger than for the other two cases. The calculation applies to free-space propagation; optical devices such as focusing lenses are not present. The QH source, in general, spreads the beam more than the coherent (laser) case. The beam spread increases as the source becomes more noncoherent. The completely noncoherent source is the limiting case. Observe that this conclusion is consistent with the van Cittert–Zernike result that the irradiance in the observation plane parallel to the noncoherent source plane is uniform. This result applies in the forward direction parallel to the z axis. Far off

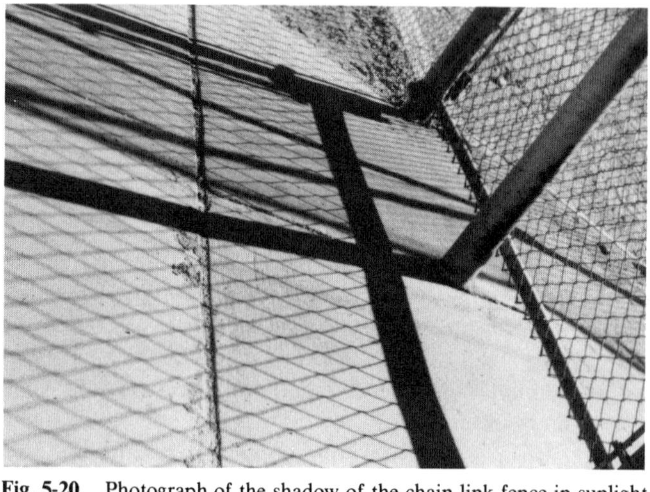

Fig. 5-20. Photograph of the shadow of the chain link fence in sunlight.

axis, the projection of the source area has to be accounted for. This effect is not included in the van Cittert–Zernike theorem, since it was derived with the quadratic approximation. Furthermore, the property of the noncoherent source described is consistent with the *constant spatial frequency spectrum* of the coherence function of such a source.

Before closing this subsection, the reader may find it instructive to refer to Fig. 5-20. It shows the shadow of a chain-link fence in sunlight. The light on the plane of the fence is partially coherent. The width of spatial coherence is approximately the same over all the elements (primitive cell) of the fence. The spreading of the beam with distance is evident in the photograph. The part of the shadow that is near the fence is sharp; as the distance increases, the shadow gets progressively less sharp. An alternative description of this phenomenon is found in an interesting book by Minnaert (1954).

Directionality Theorem of Collett and Wolf

Up to now we have compared the three source types by assigning the *same* size to all [see Eq. (5-136)]. In this section we shall use a different scaling. For a coherent source the angular distribution of the radiation in the far field is related to the two-dimensional Fourier transform of the amplitude in the source aperture. It obeys the reciprocity law that the angular width is inversely proportional to the width of the aperture. If the *same* aperture is illuminated by partially coherent light, the angular distribution is typically broader than in the coherent case and is the broadest for the noncoherent case. In this sense, as is well known, the light from a coherent source is highly directional.

The directionality theorem answers the question: Under what conditions does the radiation from a partially coherent source approach the directionality of radiation from a coherent source? The theorem can be established in relation to a Gaussian amplitude-coherent laser, Eq. (5-105), and a Carter–Wolf source with a Gaussian optical intensity profile, Eq. (5-118). Following Collett and Wolf (1978a, b), we set the standard deviation of the coherence function of the Carter–Wolf source equal to twice that of the coherent laser amplitude,

$$\sigma_g = 2\sigma_L. \tag{5-138}$$

Recall the subsidiary condition, $\sigma_Q \gg \sigma_g$, of the Carter–Wolf source. For the present application, $\sigma_Q \gg 2\sigma_L$, which implies that this source is of very large extent compared to the effective size of the laser source.

We shall continue to assume that the optical aperture of the respective sources is very large compared to the width of the Gaussian profile of the optical spectral intensity in the source aperture. The results of Eqs. (5-110) and (5-121) therefore apply and may be rewritten as

$$R_L = \left(\frac{\pi}{2k^2\sigma_L^2} z^2 + 2\pi\sigma_L^2 \right)^{1/2}$$

and

$$R_Q = \left(\frac{\pi}{2k^2\sigma_L^2} z^2 + 2\pi\sigma_Q^2 \right)^{1/2}.$$

Now we shall scale both radii in terms of the effective radius of the laser source. Thus, put

$$q_L = \frac{R_L}{\left(2\pi\sigma_L^2\right)^{1/2}}$$

and

$$q_{\text{CW}} = \frac{R_Q}{\left(2\pi\sigma_L^2\right)^{1/2}}.$$

The distance z in both cases will be measured in units of the far-field distance of the laser source,

$$p = \frac{z}{2k\sigma_L^2}.$$

In terms of these unitless parameters, the relationships for R_L and R_Q become, respectively,

$$q_L = \left(p^2 + 1 \right)^{1/2};$$

$$q_{\text{CW}} = \left[p^2 + \left(\frac{\sigma_Q}{\sigma_L} \right)^2 \right]^{1/2}, \qquad \sigma_Q \gg 2\sigma_L. \tag{5-139}$$

To examine these expressions, first consider the near-field case,

$$p \ll 1 \ll \frac{\sigma_Q}{\sigma_L}.$$

In this region, we have the approximations

$$q_L \simeq 1 \quad \text{and} \quad q_{\text{CW}} \simeq \frac{\sigma_Q}{\sigma_L}.$$

Both approach constant values, indicating no beam spread and $q_{\text{CW}} \gg q_L$.

As p is increased we may reach a stage where

$$1 \ll p \ll \frac{\sigma_Q}{\sigma_L},$$

in which case

$$q_L \simeq p \quad \text{and} \quad q_{\text{CW}} \simeq \frac{\sigma_Q}{\sigma_L}.$$

In this domain, the beam spread of the Carter–Wolf source is *less* than that of the laser, which increases linearly with p. The directionality theorem is concerned with the radiation in the far field of *both* sources; that is, the limit

$$p \gg \frac{\sigma_Q}{\sigma_L} \gg 1.$$

In this case we find

$$q_L \simeq p \quad \text{and} \quad q_{\text{CW}} \simeq p, \tag{5-140}$$

where both increase linearly with p. We may state the Collett–Wolf theorem as: The radiation from a partially coherent Carter–Wolf source is just as directional as that from a laser if $\sigma_g = 2\sigma_L$ and if the effective size of the Carter–Wolf source is very *large* compared to that of the laser, $\sigma_Q \gg 2\sigma_L$.

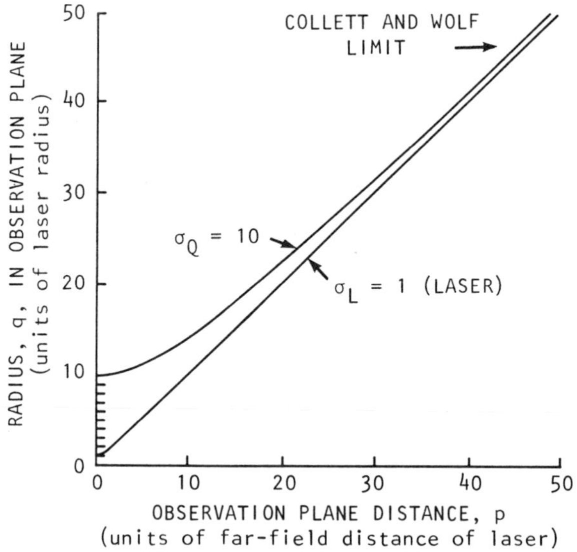

Fig. 5-21. Graph showing the Collett and Wolf limit for large enough distances (from Marathay and Goring, 1979).

These characteristics are shown in Fig. 5-21, drawn for the case $\sigma_Q = 10\sigma_L$. It shows that, in the far field of both, the directionality of the Carter–Wolf source approaches that of the laser.

Originally the theorem was proved by Collett and Wolf (1978a, b) by using the radiant intensity function, which describes the angular distribution of radiation for the coherent as well as for the partially coherent case. For an experimental verification we refer to Farina et al. (1979). For further reading we refer to Foley and Zubairy (1978), Friberg and Wolf (1978), and Rhodes and Webb (1979).

Finite-Aperture Effect

In the analytical calculations up to now, we have assumed the aperture to be very large compared to the width of the Gaussian optical intensity profile.

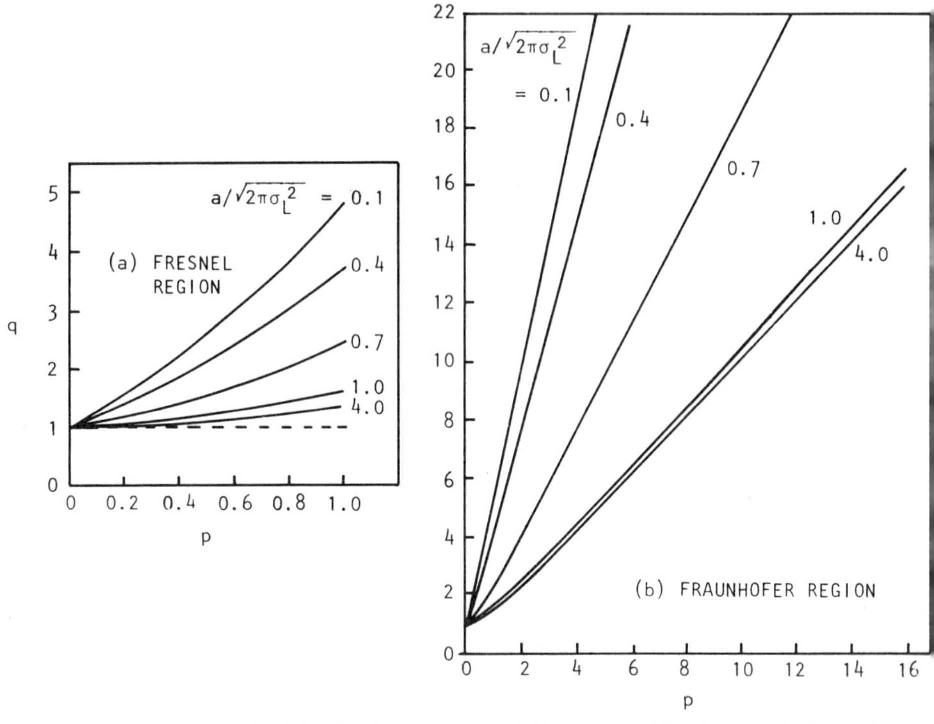

Fig. 5-22. Effect of the finite circular aperture on the beam spread for the case of a laser with a Gaussian intensity profile. The curves are labeled with the values of the ratio $a/(2\pi\sigma_L^2)^{1/2} =$ 4.0, 1.0, 0.7, 0.4, and 0.1. Values of 4 or larger correspond to the infinite-aperture case; values of 0.1 or smaller correspond to the uniformly illuminated aperture: (a) the Fresnel region and (b) the Fraunhofer region. In (a), the broken line is drawn parallel to the abscissa as a reference line.

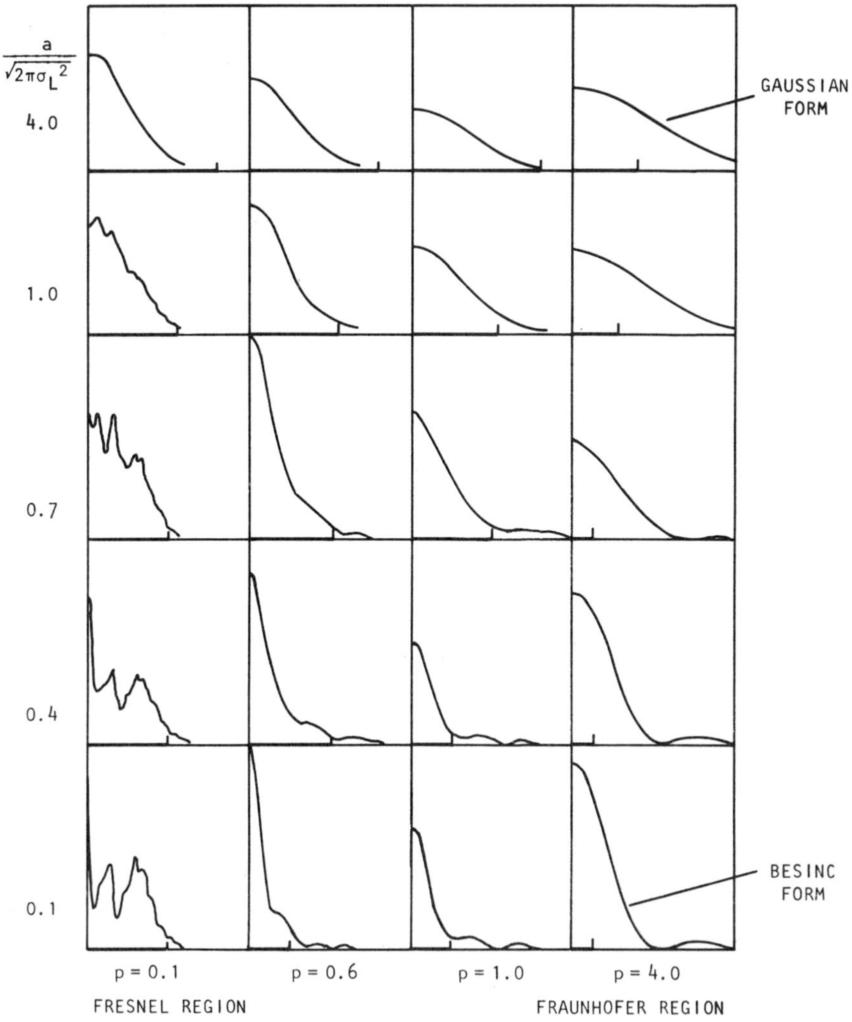

Fig. 5-23. Form of the diffraction patterns in the Fresnel and Fraunhofer regions with a finite aperture on a laser with a Gaussian intensity profile. The scale on the ordinate is not shown. The parameter p is the same as defined in relation to Fig. 5-22.

We may study the influence of the finite aperture by means of a computer calculation.

Consider first the case of coherent light with a Gaussian optical intensity. Suppose the aperture is circular of radius a and centered at the zero ordinate of the Gaussian. The calculation is done as before in the quadratic approximation. Without going into the mathematical details, we mention that the calculation requires the zero-order Hankel transform of the product of the circular aperture function with the Gaussian and the quadratic phase function.

For comparison with the no-aperture case, we need a suitable normalization constant at the source. We choose a_0 as the value of the radius of the circle that encloses about 96% of the power by the formula

$$a_0 = -2\sigma_L^2 \ln\left\{1 - 0.96\left[1 - \exp\left(\frac{-a^2}{2\sigma_L^2}\right)\right]\right\}.$$

Obviously for $a \gg \sqrt{2}\ \sigma_L$ we recover $a_0 = (2\pi\sigma_L^2)^{1/2}$ as used before.

The curves showing the beam spread are plotted in Fig. 5-22. The unitless parameters used are: $q = R/a_0$ for the radius of the circular region of radius R in the field, and $p = z\lambda/2a_0^2$ for the distance from the source. The curves are labeled with the values of the ratio $a/(2\pi\sigma_L^2)^{1/2}$. When this ratio is larger than unity, we regain the no-aperture case. When the ratio is smaller than unity, the beam spreads more than in the laser case shown in Fig. 5-17. When the ratio is 0.1 or smaller, the aperture is effectively uniformly illuminated. Thus, the finite aperture causes an excess beam spread compared to the limiting case of the Gaussian laser.

Figure 5-23 shows several graphs to indicate the nature of the diffraction patterns. The patterns are arranged in a rectangular array. The ordinate scale is different for each pattern. The abscissa shows no scale but only a heavy line with a vertical mark to indicate the extent of half the optical aperture. It was intended to show the manner in which light "spills over" in comparison to the aperture dimension. The plots are labeled with the distance parameter p, which increases as we move to the right. The value $p = 0.1$ takes us well into the Fresnel region, and at $p = 4.0$ we arrive in the Fraunhofer region. As we move along the vertical, the ratio $a/(2\pi\sigma_L^2)^{1/2}$ increases from 0.1 to 4.0. The bottom row thus corresponds to an almost uniformly illuminated aperture. In the top row, the aperture size is ineffective; the patterns are essentially Gaussian.

FREE-SPACE PROPAGATORS

In Chapter 3 we used a factor $K(Q, P)$. It was called a propagator and was used to calculate the field amplitude at the point of observation Q in terms of the amplitude at the initial point P. However, it was left largely undefined. Strictly speaking, the propagation of the light amplitude from P to Q involves an integral relationship. For the light passing through a small aperture (pinhole), we may define a convenient multiplicative "free-space propagator."

Let us assume that the temporal Fourier transform, $\hat{V}$, of the field amplitude, V, exists. [If it does not exist, we could use the time-limited (chopped) version of V over a finite time interval without affecting the discussion in what is to follow.] We begin with the Rayleigh–Sommerfeld formula,

$$\hat{V}(Q,\nu) = \iint_{\mathcal{A}} \hat{V}(P,\nu)\frac{\partial}{\partial n}\left[\hat{G}(Q,P,\nu)\right] d\mathcal{A},$$

where P is a point in the aperture plane. The integration is over the open area $\mathcal{A}$ in the aperture plane, with the area element $d\mathcal{A}$. In the radiation zone, $k\rho \gg 1$, where ρ is the distance between P and Q. In this region, to a good approximation, we have

$$\hat{V}(Q,\nu) = \iint_{\mathcal{A}} \hat{V}(P,\nu)\left[\frac{-ik\cos\theta}{2\pi\rho}\exp(ik\rho)\right] d\mathcal{A}.$$

Since the integrand is almost constant over a small pinhole aperture, we may write

$$\hat{V}(Q,\nu) = \hat{V}(P_1,\nu)\left[\frac{-ik\cos\theta_1}{2\pi\rho_1}\mathcal{A}_1\exp(ik\rho_1)\right], \qquad (5\text{-}141)$$

where $\mathcal{A}_1$ is the area of the pinhole centered at P_1.

In the time domain we have

$$\begin{aligned} V(Q,t) &= \int_0^\infty \hat{V}(Q,\nu)\exp(-i2\pi\nu t)\,d\nu \\ &= \int_0^\infty \hat{V}(P_1,\nu)\left\{\frac{-i\nu\cos\theta_1\mathcal{A}_1}{\rho_1 c}\right\}\exp\left[-i2\pi\nu\left(t-\frac{\rho_1}{c}\right)\right] d\nu. \end{aligned}$$

Now suppose the field has the form $V(t) = U(t)\exp(-2\pi\bar{\nu}t)$, where the complex envelope function $U(t)$ is very slowly varying compared to the carrier at frequency $\bar{\nu}$. This is typically the form of a quasimonochromatic field characterized by $\Delta\nu \ll \bar{\nu}$. Its transform $\hat{V}(P_1,\nu)$ has significant values only over a narrow band of frequencies. If over this band of frequencies the quantity in braces is sensibly constant, then to a good approximation we obtain the product relationship

$$V(Q,t) = \left\{\frac{-i\bar{\nu}\cos\theta_1\mathcal{A}_1}{\rho_1 c}\right\} V\left(P_1, t-\frac{\rho_1}{c}\right).$$

It allows us to define a multiplicative free-space propagator,

$$K(Q,P_1) = \frac{-i\bar{\nu}\cos\theta_1\,\mathcal{A}_1}{\rho_1 c}. \qquad (5\text{-}142)$$

As remarked in Chapter 3, this propagator is purely imaginary, depends on the geometry of the problem, and is time independent. It may be interpreted as the field at the point of observation due to a field of unit amplitude and zero phase at the pinhole.

For more complicated fields such as polychromatic fields, the approximation just made is not valid. The foregoing treatment suggests that for such fields a simple (imaginary) multiplicative propagator does not exist; the relationship involving the integral or Eq. (5-141) of the frequency domain must be used.

MEASUREMENT OF THE MUTUAL COHERENCE FUNCTION

In this section we discuss the experimental determination of the mutual coherence function. Most of the effort in the literature to date seems to have been concentrated on the determination of a spatially stationary MCF for quasimonochromatic fields. Thus, we shall consider the measurement of the mutual optical intensity, $\Gamma(\mathbf{x}_1 - \mathbf{x}_2, z, 0)$, in the plane $z =$ constant and call it simply "coherence function." In the strict sense, the lens apertures used in the measurement of such functions would have to be infinite. In practice, it is sufficient to make the apertures very large compared to the coherence width so that the aperture edge has minimal effect.

This is indeed the form of the coherence function in a plane illuminated by a spatially noncoherent source as given by the van Cittert–Zernike theorem. For the moment, let us ignore the phase factor, $\exp[+i\bar{k}(r_1 - r_2)]$, in Eq. (5-41) and consider only the double spatial Fourier transform. We denote it by

$$\Gamma(x_{12}, y_{12}, z, 0) = \frac{1}{\pi z^2} \int\int I(x_s, y_s) \exp\left(-i2\pi \frac{x_{12}x_s + y_{12}y_s}{\bar{\lambda} z} \right) dx_s\, dy_s. \tag{5-143}$$

In this equation we have put $\Gamma(\mathbf{s}, \mathbf{s}, 0)$ equal to $I(x_s, y_s)$ for the optical intensity distribution of the noncoherent source. This form of the coherence function suggests that it need not be measured for *all pairs* of points in the plane $z =$ constant. It is sufficient to measure it for all *separations* within the region of interest in that plane.

Young's Interference Method

The optical arrangement of the Thompson and Wolf experiment is a convenient realization of Young's interference method. The quality of the fringes is governed by the coherence function. The measurement of the optical intensity distribution in the BFP of the lens L_2 (see Fig. 5-5) will permit us to determine the coherence function in the mask plane for any pinhole separation.

In their experiment, Thompson and Wolf assumed the primary source to be uniform, noncoherent, and circular in shape. The resulting coherence function in the mask plane had the Besinc form, Eq. (5-59). For a noncoherent and nonuniform source of any shape, the complex degree of coherence will be of more general form, $\gamma(d)$. If the other parameters of their experiment are left unchanged, the irradiance distribution in the BFP of L_2 is given by

$$\mathsf{E}(x', y') = \mathsf{E}_1(x', y')\left\{1 + |\gamma(d)|\cos\left[\frac{2\pi x'd}{\bar{\lambda} f} - \phi(d)\right]\right\}, \quad (5\text{-}144)$$

where

$$\mathsf{E}_1(x', y') = \mathsf{E}_0\left[\text{Besinc}\left(\frac{2\pi a r'}{\bar{\lambda} f}\right)\right]^2 \quad (5\text{-}145)$$

is the irradiance due to one circular hole open and the other closed, and, as before, $r' = (x'^2 + y'^2)^{1/2}$. The relation for E contains both the real and the imaginary parts of $\gamma(d)$. If a $\pi/2$ phase difference is introduced in one circular hole relative to the other, as by means of a glass plate of appropriate thickness, the value of E_1 remains unchanged but E changes to say E' to give us a second relationship,

$$\mathsf{E}'(x', y') = \mathsf{E}_1(x', y')\left\{1 + |\gamma(d)|\sin\left[\frac{2\pi x'd}{\bar{\lambda} f} - \phi(d)\right]\right\}.$$

The quantities E, E', and E_1 are accessible to measurement; in terms of these the real and imaginary parts of $\gamma(d)$ are

$$|\gamma(d)|\cos\phi(d) = \left(\frac{\mathsf{E} - \mathsf{E}_1}{\mathsf{E}_1}\right)\cos\left(\frac{2\pi x'd}{\bar{\lambda} f}\right) + \left(\frac{\mathsf{E}' - \mathsf{E}_1}{\mathsf{E}_1}\right)\sin\left(\frac{2\pi x'd}{\bar{\lambda} f}\right)$$

$$|\gamma(d)|\sin\phi(d) = \left(\frac{\mathsf{E} - \mathsf{E}_1}{\mathsf{E}_1}\right)\sin\left(\frac{2\pi x'd}{\bar{\lambda} f}\right) - \left(\frac{\mathsf{E}' - \mathsf{E}_1}{\mathsf{E}_1}\right)\cos\left(\frac{2\pi x'd}{\bar{\lambda} f}\right) \quad (5\text{-}146a)$$

or the absolute value and the phase are given by

$$|\gamma(d)| = \left[\left(\frac{\mathsf{E} - \mathsf{E}_1}{\mathsf{E}_1}\right)^2 + \left(\frac{\mathsf{E}' - \mathsf{E}_1}{\mathsf{E}_1}\right)^2\right]^{1/2}$$

and

$$\phi(d) = \frac{2\pi x' d}{\bar{\lambda} f} - \mathrm{Arctan}\left(\frac{\mathsf{E}' - \mathsf{E}_1}{\mathsf{E} - \mathsf{E}_1}\right). \tag{5-146b}$$

What is measured, of course, is the irradiance averaged over the detector area, namely, the average radiant power $\mathbf{\Phi}$ [W]. The detector size should be small enough that a good representation of the irradiance variation is obtained. This imposes a restriction on how large the separation d can be for a fixed value of the focal length f. Observe that the right-hand side of Eq. (5-146) depends on x' but not the left-hand side. In principle, the real and imaginary parts may be determined by measurements at any point x' (near zero OPD) in the BFP. For practical considerations, the measurements are best made near the interference maxima. Several such measurements at different x' for a fixed d value improve the accuracy of the method. An alternative approach is to determine the E_{max} and E_{min} of Eq. (5-60) to yield the absolute value $|\gamma(d)|$, which is the fringe visibility. The argument $\phi(d)$ is then found by measuring the displacement of the maximum relative to the zero OPD position.

If need be, the coherence may be examined in different azimuths by rotating the line joining the two circular holes to make an azimuthal angle, ψ, with the x axis and maintaining the pair of holes equidistant from the origin. Furthermore, the procedure may be repeated for several mean wavelengths $\bar{\lambda}_1$, $\bar{\lambda}_2$, and so on, to span the polychromatic spectrum of interest. In this way, $\hat{\gamma}(d, \psi, \nu)$ may be found, the temporal Fourier transform of which gives the normalized mutual coherence function.

The phase factor $\exp[+i\bar{k}(r_1 - r_2)]$ that was ignored may be approximated to $\exp[+i\bar{k}(x_1^2 - x_2^2 + y_1^2 - y_2^2)/2z]$. It depends on the distance and on the mean position of the pair of holes. Its presence does not affect the visibility of the fringes but only their position. In principle, by following the procedure outlined in this subsection, one may also determine the nonstationary MCF by allowing the pair of holes to assume positions independently of one another in the mask plane.

Measurement on the Source Image

An indirect but less cumbersome method (see Beran and Parrent, 1964) is to exploit the two-dimensional Fourier transform relationship dictated by the van Cittert–Zernike theorem. Let us return to the Thompson and Wolf experiment and this time imagine that the mask is removed. In its absence, the lenses L_1 and L_2 form the image of the source in the BFP of L_2. To determine the irradiance distribution and the geometrical shape of the image, we may photograph or scan it. The two-dimensional Fourier transform then leads to the spatially stationary coherence function in the plane between the two lenses. The Fourier transformation may be carried out optically (see Goodman, 1968) by an appropriately processed photograph or digitally by using the output of the scan.

The image of the source is scaled with the ratio of the focal lengths of L_2 and L_1. Furthermore, as we shall see in Chapter 6 (see also Goodman, 1968), the source image is a convolution of the impulse response of the imaging lens with the optical intensity distribution of the source. For a large enough lens aperture, the image is a close approximation to the actual source. Inaccuracies in the determination of the irradiance distribution of the image will affect the coherence determination for large point separations.

Wavefront Folding Interferometer

The literature is full of ingenious schemes for the determination of the MCF (see, e.g., Mallick, 1967; Françon and Mallick, 1967; Bradley et al., 1969; Grimes, 1971; Breckinridge, 1972; Wyant, 1975; and Carter, 1977). The Françon–Mallick method uses polarized light and polarization compensators such as a Savart plate. Bradley et al. use multiple-beam interference. The methods described by Wyant and Carter make use of periodic structures such as diffraction gratings or Ronchi rulings. We shall limit our attention to some details of the wavefront folding interferometer due to Mertz (1965), Wessely and Bolstad (1970), and Breckinridge (1972). This sort of interferometer used for coherence measurements is also useful for studying the atmospheric transfer function (see Dainty and Scaddan, 1974, and Burke and Breckinridge, 1978).

We begin by referring to the original wavefront folding interferometer described by Mertz. It is a modified Michelson two-beam interferometer with one of its plane mirrors replaced by a roof mirror. The dihedral angle between the two plane mirrors of the roof is 90°.

Imagine this interferometer, as shown in Fig. 5-24, placed after the lens L_2 of the Thompson–Wolf experimental arrangement without the mask. The interferometer itself need not be very large, since it works in the vicinity of the focus of a converging lens, or say an astronomical telescope. The sketch of Fig. 5-24 is drawn in the x-z plane. The y axis is perpendicular to the plane of the figure and points toward the reader. The point of focus

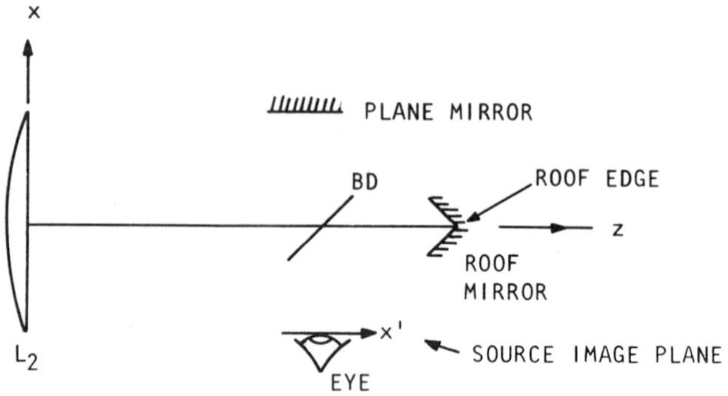

Fig. 5-24. Field-folding interferometer. The interferometer is working at the focus of a converging lens L_2 of the Thompson and Wolf setup (not shown in its entirety) without the mask. BD is the beam divider. The roof edge is along the y axis. The y' axis of the image plane points toward the reader.

(location of the image of the source) is in the plane of the pupil of the observer's eye. The eye is focused on the exit pupil plane of L_2. This viewing arrangement is referred to as the *Maxwellian view* and is used in relation to the wedge fringes and the Twyman–Green interferometer (see Longhurst, 1967).

The plane mirror merely reflects the converging wavefront as it is. The roof mirror, however, reverses or folds it about the y axis (i.e., $+x$ is interchanged with $-x$) while retaining its focusing property. If the plane mirror and the aperture plane of the roof mirror are effectively "parallel" (as dictated by the reflection of one on the other due to the beam divider BD), then the two source images due to the two arms of the interferometer will be coincident. If, however, the plane mirror is tilted by a small angle about an axis common to the mirror and the plane of the figure, the source image due to it will be displaced along the y' axis, slightly above or below the plane of the figure. To the observer, the exit pupil of L_2 now appears crossed with cosine-type straight fringes running parallel to the x axis. For a small point source (unresolved by L_2), the fringes will have good contrast (see Fig. 5-25*a*), and they will run the full extent of the exit pupil. As the source size is increased, so that it is well resolved by L_2, the width of the coherence function becomes smaller than the exit pupil width. For a circular source, the coherence function is of the Besinc type, Eq. (5-48). If its first zero falls well within the exit pupil of L_2, the fringes appear "fragmented," as shown in Fig. 5-25*b*. The part of the fringes in the negative side lobe of the coherence function shows a half-period displacement. The visibility of the fringes does not change because the coherence is constant along the y axis.

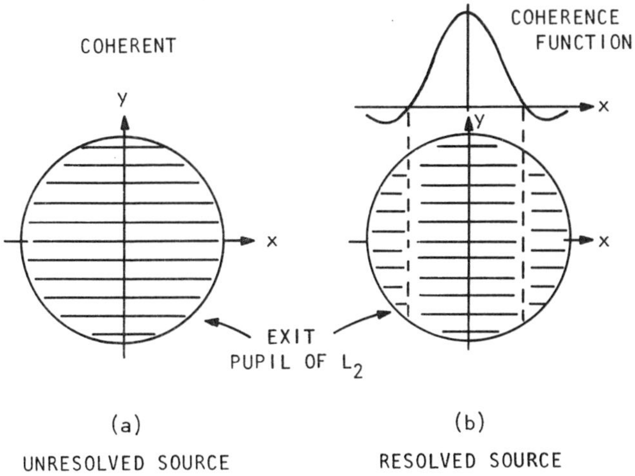

Fig. 5-25. Fringes due to a source (a) not resolved and (b) well resolved by L_2. In (b) the fringes in the region of the side lobes of the coherence function are shifted by half a period.

Tilting the plane mirror to cause a displacement of the source images along the x' axis produces fringes parallel to the y axis in the exit pupil of L_2. Now their visibility will change across the pupil as dictated by the coherence function sketched in Fig. 5-25b. The length of the fringes is always oriented perpendicular to the line joining the two sources. No matter what the orientation of the fringes, the zeros of visibility always remain locked in place.

The folding of the wavefront as discussed above allows us to examine the coherence along the x axis. Use of this interferometer with an astronomical telescope aimed at a star (unresolved) yields fringes whose structure gives the atmospheric transfer function. Typically, the fringes extend over only a limited region of the x axis. The extent of the region is related to the correlation length of the incident wavefront. The paper by Dainty and Scaddan gives an example of a photograph of these fringes.

The reader has undoubtedly guessed the next step, namely, to fold the pupil about the x axis also, so that $+y$ is interchanged with $-y$. For this, the plane mirror is replaced by a second roof mirror as shown in Fig. 5-26. This is the modification of the basic theme by Wessely, Bolstad, and Breckinridge. The roof edge of the second mirror is oriented perpendicular to the edge of the reflection of the first roof mirror by way of the beam divider BD. For the practical realization of this scheme, Breckinridge (1978) constructed a "monolithic" interferometer. In experimental arrangements it is desirable to use auxiliary optics so that the light entering the interferometer is rendered parallel.

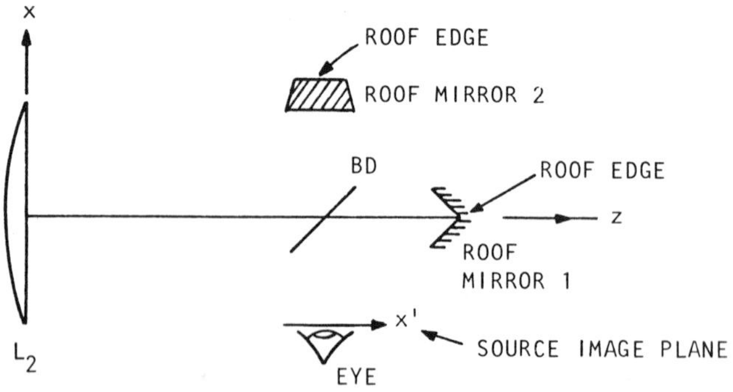

Fig. 5-26. Wavefront double-folding interferometer. The roof edge of mirror 1 is perpendicular to, and that of mirror 2 is parallel to, the plane of the figure. BD is the beam divider. The y' axis of the image plane points toward the reader.

Now let $V(x, y, z_e, t)$ be the field in the exit pupil located at, say, $z = z_e$. Roof 1 of Fig. 5-26 causes a folding, indicated by

$$V(x, y, z_e, t) \to V(-x, y, z_e, t),$$

while roof 2 brings about

$$V(x, y, z_e, t) \to V(x, -y, z_e, t).$$

By an appropriate tilt of one roof relative to the other, the source images may be displaced in any desired direction with components, say a, b, in the x', y' plane. This displacement introduces a relative phase shift, $\exp[i\bar{k}(ax + by)/f]$, between the two folded fields. A superposition of these fields leads to the optical intensity distribution $I(x, y, z_e)$ in the exit pupil of lens L_2. We bear in mind that the statistics are both spatially and temporally stationary and that there is no time delay introduced in one arm of the interferometer relative to the other. Therefore,

$$\begin{aligned} I(x, y, z_e) &= \left\langle \left| V(-x, y, z_e, t) + V(x, -y, z_e, t) \exp\left[\frac{i\bar{k}(ax + by)}{f} \right] \right|^2 \right\rangle \\ &= 2\Gamma(0,0, z_e, 0)\left\{ 1 + |\gamma(-2x, 2y, z_e, 0)| \right. \\ &\quad \left. \times \cos\left[2\pi \left(\frac{ax + by}{\bar{\lambda} f} \right) - \phi(-2x, 2y) \right] \right\}, \end{aligned} \tag{5-147}$$

where f is the focal length of lens L_2. In this equation the mutual optical intensity function is

$$\Gamma(-2x, 2y, z_e, 0) = \langle V(-x, y, z_e, t)\, V^*(x, -y, z_e, t)\rangle, \quad (5\text{-}148)$$

and γ is its normalized form with ϕ as the phase argument. The inclination of the fringes with the x axis is Arccot $(-a/b)$. Hereafter the procedure for the determination of the real and imaginary parts of $\gamma(-2x, 2y, z_e, 0)$ is the same as that used in Young's interference method.

The structure of the fringes shows vividly the effect of the coherence function. The changing visibility in different azimuths and fragmentation (if the function has zeros) of the fringes is seen all at once. These fringes may be photographed by using relay optics to image the exit pupil onto the detector plane. For quantitative measurements it is not essential to have the fringes; we may set $a = b = 0$ so that the source images coincide.

As we shall see in Chapter 6, the points x, y in the exit pupil may be labeled in terms of the spatial frequencies, $(-2x/\bar{\lambda} f)$ and $(2y/\bar{\lambda} f)$. The interferometer is therefore a two-dimensional spatial frequency analyzer. It displays the two-dimensional spatial Fourier transform of the optical intensity distribution of the source.

Comments

The demonstration that the MCF is accessible to measurement is significant for two reasons. The first is that the coherence theoretic formulation of optics is in terms of *observable* quantities. The second reason is related to the diffraction problem. The MCF in the field may be calculated by diffraction theory with known boundary conditions. These conditions are specified in terms of the MCF on the surface, for example, the aperture plane. These boundary values may be experimentally determined and used as inputs to the diffraction calculation. This feature is unique to the MCF, not shared by the wave amplitude. For this latter the usual *ad hoc* boundary conditions (Kirchhoff type) have to be postulated to proceed with the diffraction calculation. For further reading, refer to Skinner (1961).

PROBLEMS

5-1. Verify that the contribution of the sphere $\mathcal{C}$ to the surface integral tends to zero as the radius R tends to infinity. The surface integral in question is

$$\iint_{\mathcal{C}} \left[\hat{G}(\mathbf{r}_1, \mathbf{s}_1, \nu) \frac{\partial \hat{\Gamma}(\mathbf{s}_1, \mathbf{s}_2, \nu)}{\partial n(\mathbf{s}_1)} - \hat{\Gamma}(\mathbf{s}_1, \mathbf{s}_2, \nu) \frac{\partial \hat{G}(\mathbf{r}_1, \mathbf{s}_1, \nu)}{\partial n(\mathbf{s}_1)} \right] d\mathcal{S}_1,$$

where $\mathbf{r}_1$ is a fixed point in the volume and $\mathbf{s}_1$ and $\mathbf{s}_2$ are points on the sphere whose elemental area is denoted by $d\mathcal{S}_1$. A similar surface integral is encountered in dealing with the second variable in $\hat{\Gamma}$. For this problem use the asymptotic form of $\hat{\Gamma}$ as given by the radiation condition. The function $\hat{G}$ is the modified Green's function of Eq. (A5-4).

5-2. For a Carter–Wolf source,

$$\hat{I}_s(x_s, y_s, \nu) = \hat{I}_Q \exp\left(-\frac{x_s^2 + y_s^2}{2\sigma_Q^2} \right)$$

and

$$\hat{g}_s(x_{s12}, y_{s12}, \nu) = \exp\left(-\frac{x_{s12}^2 + y_{s12}^2}{2\sigma_g^2} \right),$$

calculate (a) the MSDF, $\hat{\Gamma}(x_1, y_1, x_2, y_2, z, \nu)$ in the observation plane $z > 0$ by using Eq. (5-15) and (b) the OSI function $\hat{I}(x, y, z, \nu)$ in that plane by using Eq. (5-16).
Hint: Use the fact that a Fourier transform of a Gaussian and the convolution of two Gaussians are themselves Gaussian.

5-3. (a) Consider an optical field illuminating an aperture $\mathcal{A}$. The field is quasimonochromatic with the wavelength condition $\Delta\lambda \ll \bar{\lambda}$. Assume that the MCF in the aperture $\mathcal{A}$ is known. Analyze the free-space propagation problem in view of establishing the appropriate conditions so that the problem may be reduced to the study of the MOI function $\Gamma_{12}(0)$.

We shall discuss the manner of attacking the problem and display the relevant conditions. Their verification is left as a part of the problem. The analysis entails studying the normal derivatives of the Green's functions by use of the quadratic approximation

$$\frac{\partial \hat{G}}{\partial n} \simeq \exp\left[+ik\frac{(x - x')^2 + (y - y')^2}{2z} \right].$$

It is sufficient to study the x part of the exponent. The conclusions also apply to the y part or the radial coordinate.

First, show that the variation of the exponent due to the spectral components contained within $\Delta\lambda$ will be less than $\pi/4$ rad from its value at $\bar{\lambda}$ if the following conditions are satisfied:

(i) The term with the aperture variable x'^2 demands that the plane of observation not be too close to the aperture. The condition is

$$z > \frac{\Delta\lambda}{\bar{\lambda}} \frac{4x'^2_{\max}}{\bar{\lambda}}.$$

It is in terms of the far-field distance of the aperture.

(ii) The term x^2 of the coordinate in the observation plane imposes a limitation

$$x_{\max} < \bar{\lambda}\left(\frac{z}{4\Delta\lambda}\right)^{1/2},$$

where the value of z is consistent with condition (i). For x larger than the maximum x, the spectral components contained in $\Delta\lambda$ begin to show their individuality.

Fulfillment of both conditions assures that the variation of the mixed term xx' is less than $\pi/4$ radian if $\Delta\lambda \ll \bar{\lambda}$. The exponential containing xx' may be likened to cosine-type fringes in the x plane, due to two point sources separated by x' in the aperture. The "fringe" periodicity changes with each wavelength component in the band $\Delta\lambda$. The quasimonochromatic approximation says that this change in periodicity may be ignored over the entire region marked by $x_{\max}$.

(b) It is instructive to use the typical values $\Delta\lambda = 10$ nm and $\bar{\lambda} = 500$ nm for a circular aperture of diameter $d = 2x'_{\max} = 10$ mm, for example. Study the region of validity of the QM approximation and show that at one far-field distance the value of $x_{\max}$ is more than 11 times the diameter d.

5-4. Use the Besinc representation of noncoherence discussed in Chapter 4 and obtain the result of the van Cittert–Zernike theorem. *Hint*: Do it in the spatial frequency domain. Note that the spatial Fourier transform for the Besinc function is a constant over the domain of real waves, and work with the transfer function of free space.

5-5. Discuss the state of the fringes in the interference plane behind the mask when the pinhole P_2 is held fixed on the z axis but P_1 is moved along x. The pinholes are illuminated by a rectangular source as in Fig. 5-3.

5-6. With a criterion for coherence width and area similar to the one used in the text, show that the coherence area for a circular source aperture of radius a_0 is

$$\mathscr{a}_c = \pi\left(\frac{0.61\bar{\lambda} z_0}{2a_0}\right)^2.$$

5-7. The far-field distance for a coherently illuminated aperture of width $2a_0$ is $(2a_0)^2/\bar{\lambda}$. At this distance, show that the coherence width for a noncoherent source of width $2a_0$ is a_0. For a source of width 1 mm and $\bar{\lambda} = 500$ nm, calculate the coherence widths at distances $z_0 = 1$ and 2m.

5-8. Calculate the size (diameter of the central lobe) of the diffraction pattern for a circular aperture of diameter $2a = 1.4$ mm in order to ascertain the extent of the region over which interference fringes may be observed in the BFP of L_2. The focal length of L_2 is $f = 1520$ mm and the wavelength used is $\bar{\lambda} = 579.0$ nm.

5-9. (a) In the Thompson and Wolf experiment, the focal length of L_1 in Fig. 5-5 is $f = 1520$ mm. For a uniform, noncoherent, circular source aperture of radius $a_0 = 0.045$ mm and wavelength $\bar{\lambda} = 579.0$ nm, calculate the coherence width $\mathscr{w}$ and compare it with the circular aperture diameter, $2a = 1.4$ mm, in order to verify the approximation made in deriving Eq. (5-58).

(b) Calculate the location of the first zero of the coherence function for $f = 1520$ mm. Determine the shift in the location of the first zero if the focal length of L_1 is changed to $f = 1710$ mm.

5-10. (a) For a doublet source far away from the telescope aperture, use the van Cittert–Zernike theorem as indicated in the text to verify Eq. (5-70) and also obtain the other possible conditions for the vanishing of the fringes.

(b) Consider the following imaginary experiment. A small telescope objective is fitted with a two-slit mask with variable separation. Consider an automobile moving toward the telescope whose headlights are 122 cm apart. The light filter used is 600 nm wavelength. At a slit separation of 1.8 cm, the interference fringes were found to vanish. Assume the ideal conditions that the road is perfectly smooth and that the automobile is traveling with a uniform velocity, and so on. One hour later, the same slit separation of 1.8 cm again produces a zero fringe visibility. What is the speed of the automobile? (Adapted from Andrews, 1960.)

5-11. The spectrum of sodium has two neighboring lines, $\lambda_1 = 589.0$ nm and $\lambda_2 = 589.6$ nm. Compute the visibility function if such a source is used in a Michelson two-beam interferometer. If a small spectral width is assumed at each of the wavelengths λ_1 and λ_2, what is the effect on the visibility function?

5-12. (a) In the experiment by Thompson (1965), a plane parallel glass plate of thickness 0.5 mm was used. Let the plate index be 1.5. Compute the passband $\Delta\lambda$ of the spectral filter used in the experiment to isolate the line of wavelength 546.1 nm of the mercury source and then compute the coherence length (see Fig. 5-10).

(b) In the same experiment, suppose that the laser of wavelength 115.3 nm is used with a short time frequency stability $\Delta\nu/\nu \simeq 10^{-13}$. Compute the thickness of the glass plate for which no fringes would be seen and also compute the coherence length. (This is the situation close to the frequency stability achieved by Jaseja et al., 1963.)

5-13. The approximate diameter of the sun is 1.40×10^6 km, and its distance from the earth is about 1.50×10^8 km. A spectral filter with mean wavelength 550 nm and with a passband of 1.0 nm is used. Calculate the coherence area at the surface of the earth and the coherence volume.

5-14. For a laser with a Gaussian optical intensity (OI) profile

$$I(x_s, y_s, 0, \nu) = I_L \exp\left(-\frac{x_s^2 + y_s^2}{2\sigma_L^2} \right),$$

calculate the OI function of the light in any plane $z =$ constant, by using the angular spectrum approach to verify Eq. (5-109). *Hint*: Use the Fourier transform of a Gaussian with complex coefficients (see Gaskill, 1978, Table 7-3).

5-15. Consider an xy plane illuminated with light of wavelength λ. The optical intensity distribution on this plane is a constant, $I(x, y) = a$. The spatial coherence function on it has the form

$$\exp\left[-\frac{(x_1 - x_2)^2 + (y_1 - y_2)^2}{2b^2} \right],$$

where b is a parameter. Light from this plane is used to illuminate a parallel plane at a distance z from it. Calculate and describe the optical intensity and the spatial coherence in the plane z. *Hint*: Observe that the coherence function is spatially stationary and make use of the van Cittert–Zernike result of Eq. (5-41) to do this problem.

REFERENCES

Abramowitz, M., and I. A. Stegun, Eds. (1964). *Handbook of Mathematical Functions with Formulas, Graphs, and Mathematical Tables*, National Bureau of Standards, Applied Mathematics Series, 55, U.S. Printing Office, Washington, D.C.

Anderson, J. A. (1920). Application of Michelson's interferometer method to the measurement of close double stars, *Astrophys. J.* **51**:263–275.

Andrews, C. L. (1960). *Optics of the Electromagnetic Spectrum*, Prentice-Hall, Englewood Cliffs, NJ.

Beran, M. J., and G. B. Parrent, Jr. (1964). *Theory of Partial Coherence*, Prentice-Hall, Englewood Cliffs, NJ, 189 pp.

Born, M., and E. Wolf (1970). *Principles of Optics*, 4th ed., Pergamon, New York.

Bradley, D. J., A. W. McCullough, and C. J. Mitchell (1969). A multiple beam interferometer coherence analyser, *Opt. Acta* **16**:735–743.

Breckinridge, J. B. (1972). Coherence interferometer and astronomical applications, *Appl. Opt.* **11** (12):2996–2998.

Breckinridge, J. B. (1978). A white light amplitude interferometer with 180-degree rotational shear, *Opt. Eng.* **17**:156–159.

Burke, J. J., and J. B. Breckinridge (1978). Passive imaging through the turbulent atmosphere: fundamental limits on the spatial frequency resolution of a rotational shearing interferometer, *J. Opt. Soc. Am.* **68** (1):67–77.

Carter, W. H. (1977). Measurement of second-order coherence in a light beam using a microscope and a grating, *Appl. Opt.* **16** (3):558–563.

Carter, W. H., and E. Wolf (1977). Coherence and radiometry with quasihomogeneous planar source, *J. Opt. Soc. Am.* **67** (6):785–796.

Collett, E., and E. Wolf (1978a). Is complete spatial coherence necessary for the generation of highly directional light beams, *Opt. Lett.* **2**:27–29.

Collett, E., and E. Wolf (1978b). New equivalence theorem for radiant intensity from planar sources of different states of spatial coherence (Abstr.), *J. Opt. Soc. Am.* **68** (10):1410.

Dainty, J. C., and R. J. Scaddan (1974). A coherence interferometer for the direct measurement of the atmospheric transfer function *Mon. Not. R. Astron. Soc.* **167**:69–73.

Farina, J. D., L. M. Narducci, and E. Collett (1979). Generation of highly directional beams from quasihomogeneous optical sources (Abstr.), *J. Opt. Soc. Am.* **69** (10):1414.

Foley, J. T., and M. S. Zubairy (1978). Directionality of Gaussian Schell-model beams (Abstr.), *J. Opt. Soc. Am.* **68** (10):1410.

Françon, M. (1966). *Optical Interferometry*, Academic, New York, Chap. IV, pp. 87–100.

Françon, M., and S. Mallick (1967). Measurement of the second order degree of coherence, pp. 71–104, in *Progress in Optics*, Vol. 6, E. Wolf, Ed., North-Holland, Amsterdam.

Friberg, A., and E. Wolf (1978). Reciprocity relations between the far field and the source field generated by a planar source of any state of coherence (Abstr.), *J. Opt. Soc. Am.* **68** (10):1410.

Gaskill, J. D. (1978). *Linear Systems, Fourier Transforms and Optics*, Wiley, New York, 554 pp.

Goodman, J. W. (1968). *Introduction to Fourier Optics*, McGraw-Hill, San Francisco, 287 pp.

Grimes, D. N. (1971). Measurement of the second-order degree of coherence by means of a wavefront shearing interferometer, *Appl. Opt.* **10** (7):1567–1570.

Hanbury-Brown, R. (1974). *The Intensity Interferometer*, Taylor and Francis, London; Halstead Press, New York, 184 pp.

Hanbury-Brown, R., and R. Q. Twiss (1956). A test of a new type of stellar interferometer on Sirius, *Nature* **178**:1046–1048.

Hanbury-Brown, R., and R. Q. Twiss (1957a). Interferometry of the intensity fluctuations in light. I. Basic theory: the correlation between photons in coherent beams of radiation, *Proc. R. Soc. Lond. Ser. A* **242**:300–324.

Hanbury-Brown, R., and R. Q. Twiss (1957b). Interferometry of the intensity fluctuations in light. II. An experimental test of the theory for partially coherent light, *Proc. R. Soc. Lond. Ser. A* **243**:291–319.

Hecht, E., and A. Zajac (1976). *Optics*, Addison-Wesley, Reading, MA, 565 pp.

Jaseja, T. S., A. Javan, and C. R. Townes (1963). Frequency stability of He–Ne masers and measurement of length, *Phys. Rev. Lett.* **10** (5):165–167.

Jenkins, F. A., and H. E. White (1976). *Fundamentals of Optics*, 4th ed., McGraw-Hill, New York, 746 pp.

Klauder, J. R., and E. C. G. Sudarshan (1968). *Fundamentals of Quantum Optics*, W. A. Benjamin, New York, 279 pp.

Longhurst, R. S. (1967). *Geometrical and Physical Optics*, Wiley, New York, 534 pp.

Mallick, S. (1967). Degree of coherence in the image of a quasimonochromatic source, *Appl. Opt.* **6** (8):1403–1405.

Mandel, L. (1963). Fluctuations of light beams, pp. 181–248, in *Progress in Optics*, E. Wolf, Ed., Vol. 2, North-Holland, Amsterdam; Wiley, New York.

Mandel, L., and E. Wolf (1965). Coherence properties of optical fields, *Rev. Mod. Phys.* **37**:231–287.

Marathay, A. S. (1981). Spatial coherence of light from a Carter and Wolf quasihomogeneous source (Abstr.), *J. Opt. Soc. Am.* **71** (12):1625.

Marathay, A. S., and W. P. Goring (1979). Directionality of light beams and spatial coherence, *Phys. Scr.* **19**:40–42.

Marathay, A. S., W. P. Goring, and K. L. Shu (1978). Directionality of light from coherent, partially coherent, and noncoherent sources (Abstr.), *J. Opt. Soc. Am.* **68** (10):1410.

Mertz, L. (1965). *Transformations in Optics*, Wiley, New York.

Michelson, A. A. (1890). On the application of interference methods to astronomical measurements, *Phil. Mag.* **30** (5):1–21.

Michelson, A. A. (1920). On the application of interference methods to astronomical measurements, *Astrophys. J.* **51**:257–262.

Michelson, A. A., and Pease, F. G. (1921). Measurement of the diameter of α Orionis with the interferometer, *Astrophys. J.* **53**:249–259.

Minnaert, M. G. J. (1954). *The Nature of Light and Colour in the Open Air*, Dover, New York, Chap. I, Arts. 2 and 4.

Murty, M. V. R. K. (1978). Lateral shearing interferometer, Chap. 4, pp. 105–148, in *Optical Shop Testing*, D. Malacara, Ed., Wiley, New York.

Rhodes, W. T., and L. L. Webb (1979). Partially coherent sources, highly directional beams, and Fourier optics (Abstr.), *J. Opt. Soc. Am.* **69** (10):1414.

Skinner, T. J. (1961). Ergodic theorem in the solution of the scalar wave equation with statistical boundary conditions, *J. Opt. Soc. Am.* **51** (11):1246–1251.

Sokolnikoff, I. S., and R. M. Redheffer (1958). *Mathematics of Physics and Modern Engineering*, McGraw-Hill, New York, Chap. 3, Sect. 16, pp. 270–277.

Sommerfeld, A. (1972). Optics, Art. 34.C, pp. 199–201, in *Lectures on Theoretical Physics*, Vol. IV, Academic Press, New York.

Thompson, B. J. (1958). Illustration of the phase change in two-beam interference with partially coherent light, *J. Opt. Soc. Am.* **48** (2):95–97.

Thompson, B. J. (1965). Advantages and problems of coherence as applied to photographic situations, *Opt. Eng.* **4**:7-11.

Thompson, B. J., and E. Wolf (1957). Two-beam interference with partially coherent light, *J. Opt. Soc. Am.* **47** (10):895–902.

Wessely, H. W., and J. O. Bolstad (1970). Interferometric technique for measuring the spatial-correlation function of optical radiation fields, *J. Opt. Soc. Am.* **60** (5):678–682.

Wolf, E. (1958). Reciprocity inequalities, coherence time, and bandwidth in signal analysis and optics, *Proc. Phys. Soc. Lond.* **71**:257–269.

Wyant, J. C. (1975). OTF measurements with a white light source: an interferometric technique, *Appl. Opt.* **14** (7):1613–1615.

APPENDIX 5.1. RAYLEIGH–SOMMERFELD DIFFRACTION THEORY

We shall work with the field amplitude itself and assume that its time and space Fourier transform exists. The sources are assumed to be in the left half-space, $z < 0$. They illuminate an aperture $\mathcal{A}$ in the x_s, y_s plane at $z = 0$. The field spectrum $\hat{V}(x_s, y_s, 0, \nu)$ in the aperture plane is assumed to be known. The basic problem is to find the field spectrum $\hat{V}(x, y, z, \nu)$ in the right half-space, $z > 0$.

As discussed in the section "Integral Formulation" of Chapter 4, we begin with a pair of equations,

$$\begin{aligned} \nabla_1^2 \hat{V}(\mathbf{r}_1, \nu) &= -k^2 \hat{V}(\mathbf{r}_1, \nu), \\ \nabla_1^2 \hat{G}_0(\mathbf{r}, \mathbf{r}_1, \nu) &= -k^2 \hat{G}_0(\mathbf{r}, \mathbf{r}_1, \nu) - \delta(\mathbf{r} - \mathbf{r}_1), \end{aligned} \tag{A5-1}$$

where $\hat{G}_0$ is the primitive Green's function,

$$\hat{G}_0(\mathbf{r}, \mathbf{r}_1, \nu) = \frac{\exp(ik\rho)}{4\pi\rho} \tag{A5-2}$$

and

$$\rho = |\mathbf{r} - \mathbf{r}_1| .$$

Following the procedure of that section we are led to the surface integral,

$$\hat{V}(\mathbf{r}, \nu) = \iint_{\mathcal{S}} \left[\hat{G}_0(\mathbf{r}, \mathbf{s}, \nu) \frac{\partial \hat{V}(\mathbf{s}, \nu)}{\partial n(\mathbf{s})} - \hat{V}(\mathbf{s}, \nu) \frac{\partial \hat{G}_0}{\partial n(\mathbf{s})} (\mathbf{r}, \mathbf{s}, \nu) \right] d\mathcal{S} \tag{A5-3}$$

where $\hat{V}(\mathbf{r}, \nu)$ is the field within a volume $\mathcal{V}$ enclosed by the surface $\mathcal{S}$, and **s** with components x_s, y_s, z_x is the point on the surface.

For the problem of the aperture, we regard the surface $\mathcal{S}$ as made of three parts: the aperture $\mathcal{A}$ itself in the x_s, y_s plane at $z = 0$, the portion $\mathcal{B}$ of that plane outside the aperture, and a portion $\mathcal{C}$ of a sphere centered at the point **r** with a very large radius, R, and intersecting the plane $z = 0$.

We seek a modified Green's function $\hat{G}$ such that it vanishes on the plane $z = 0$. This is done by adding to $\hat{G}_0$ any suitable solution of the homogeneous equation

$$[\nabla_1^2 + k^2] \hat{G}_0(\mathbf{r}_m, \mathbf{r}_1, \nu) = 0,$$

where $\mathbf{r}_m$ is a point in the region $z < 0$ to the left of the aperture and $\mathbf{r}_1$ is a point to its right, $z > 0$. Thus, $\mathbf{r}_m$ and $\mathbf{r}_1$ cannot meet and the equation remains homogeneous. We shall pick $\mathbf{r}_m$ as the mirror image of the point **r** with respect to the aperture plane $z = 0$; its components are $x_m = x$, $y_m = y$, and $z_m = -z$. Consider the modified function

$$\hat{G}(\mathbf{r}, \mathbf{r}_1, \nu) = \hat{G}_0(\mathbf{r}, \mathbf{r}_1, \nu) - \hat{G}_0(\mathbf{r}_m, \mathbf{r}_1, \nu). \tag{A5-4}$$

Clearly, $\hat{G} = 0$ whenever the volume point $\mathbf{r}_1$ coincides with any point in the aperture plane $z = 0$, that is, whenever $\mathbf{r}_1 = \mathbf{s}$. The function $\hat{G}$ is a bona fide Green's function since it too is a solution of Eq. (A5-1).

Next, by the so-called Sommerfeld's radiation condition, the contribution of the surface integral on $\mathcal{C}$ may be shown to go to zero as $R \to \infty$. This condition says that far enough away from the aperture the field $\hat{V}$ will have the asymptotic form of an angular distribution $f(\theta, \phi)$ riding on an expanding spherical wave, that is,

$$\hat{V}(\mathbf{r}, \nu) \simeq f(\theta, \phi) \frac{\exp(ikr)}{r}. \tag{A5-5}$$

The above considerations show that the surface integrals may be limited to the parts $\mathscr{A}$ and $\mathscr{B}$ containing only the spectrum $\hat{V}$ itself:

$$\hat{V}(\mathbf{r}, \nu) = -\iint_{\mathscr{A}+\mathscr{B}} \hat{V}(\mathbf{s}, \nu) \frac{\partial \hat{G}(\mathbf{r}, \mathbf{s}, \nu)}{\partial n(\mathbf{s})} \, dx_s \, dy_s .$$

At this point one assumes that the spectrum $\hat{V}(\mathbf{s}, \nu)$ inside the aperture $\mathscr{A}$ is the same as though the aperture were not there (unperturbed field) and that the field in the region $\mathscr{B}$ outside the aperture is zero. (In the Kirchhoff boundary conditions one makes further assumptions regarding the normal derivative of the field, but these are not required in the present context due to the judicious choice of the Green's function.)

The derivative of the Green's function along the normal pointing out of the volume reduces to a partial on $-z_s$ at the point **s**. After this derivative is calculated, it is evaluated at the aperture plane $z_s = 0$ to give

$$\hat{V}(x, y, z, \nu) = +\iint_{\mathscr{A}} \hat{V}(x_s, y_s, 0, \nu) \left[+\frac{1}{2\pi} \frac{z}{\rho} (1 - ik\rho) \frac{\exp(ik\rho)}{\rho^2} \right] dx_s \, dy_s \tag{A5-6}$$

where ρ is the distance between a point in the aperture and the observation point,

$$\rho = |\mathbf{r} - \mathbf{s}| = \left[(x - x_s)^2 + (y - y_s)^2 + z^2 \right]^{1/2}.$$

The ratio z/ρ equals $\cos\theta$ since it is the direction cosine of the difference vector, $\mathbf{r} - \mathbf{s}$. The integral over the aperture $\mathscr{A}$ is the Rayleigh–Sommerfeld answer. It gives the field in the plane $z =$ constant, parallel to the aperture plane $z = 0$. The choice of the boundary conditions (that the spectrum is zero on $\mathscr{B}$ and that it remains unperturbed in $\mathscr{A}$), although an approximation, is useful when the aperture dimensions are very large in comparison to the wavelength λ. Conversely, the boundary conditions are inappropriate if the primary interest is in the field fluctuations over distances on the order of λ in $\mathscr{A}$ or the aperture dimensions are comparable to λ. Such details affect mainly the diffracted field distribution at large angles from the normal in the right half-space. The customary approximations of the quadratic phase (Fresnel) and the linear phase (Fraunhofer) over the aperture follow after Eq. (A5-6).

The Rayleigh–Sommerfeld answer may be rewritten as a convolution integral,

$$\hat{V}(x, y, z, \nu) = \iint_{\mathscr{A}} \hat{V}(x_s, y_s, 0, \nu) \, \mathscr{R}(x - x_s, y - y_s, z, \nu) \, dx_s \, dy_s , \tag{A5-7}$$

where

$$\mathfrak{R}(x - x_s, y - y_s, z, \nu) \equiv \frac{1}{2\pi} \cos\theta\,(1 - ik\rho) \frac{\exp(ik\rho)}{\rho^2}.$$

In the domain of spatial frequencies we have the product relationship

$$\mathring{V}(\kappa p, \kappa q, z, \nu) = \mathring{V}(\kappa p, \kappa q, 0, \nu) \exp(+ikmz), \qquad \text{(A5-8)}$$

where

$$\mathring{V}(\kappa p, \kappa q, z, \nu) = \iint_{-\infty}^{\infty} \hat{V}(x, y, z, \nu) \exp[-i2\pi\kappa(px + qy)]\, dx\, dy,$$

and the third direction cosine is defined in terms of the first two by

$$m = \begin{cases} +\left(1 - p^2 - q^2\right)^{1/2}, & p^2 + q^2 \le 1 \\ +i\left(p^2 + q^2 - 1\right)^{1/2}, & p^2 + q^2 > 1. \end{cases} \qquad \text{(A5-9)}$$

That the function $\mathfrak{R}$ and the $\exp(+ikmz)$ form a Fourier pair may be established by using Weyl's plane-wave decomposition of the spherical wave,

$$\frac{\exp(ikr)}{r} = \frac{i}{\lambda} \iint_{-\infty}^{\infty} \frac{1}{m} \exp(ikmz) \exp[ik(px + qy)]\, dp\, dq. \qquad \text{(A5-10)}$$

For further discussion on this point the reader is referred to Goodman (1968), and for an alternative approach a paper by the author is reproduced in Appendix 5.2. The factor $\exp(+ikmz)$ is called the *transfer function* of free space.

APPENDIX 5.2. FOURIER TRANSFORM OF THE GREEN'S FUNCTION FOR THE HELMHOLTZ EQUATION

This appendix is a reprint of a paper that was published in the *Journal of the Optical Society of America*, Vol. 65, No. 8, August 1975, pp. 964–965.

Fourier transform of the Green's function for the Helmholtz equation

A. S. Marathay*

Optical Sciences Center, University of Arizona, Tucson, Arizona 85721
(Received 7 March 1975; revision received 24 May 1975)

Index Headings: Diffraction; Fourier transform.

In the Rayleigh–Sommerfeld[1,2] formulation of diffraction theory, a Green's function $G(\vec{r}-\vec{r}')$ is constructed so that it vanishes when the point $\vec{r}'$ (or the point $\vec{r}$) lies on the plane (say at $z'=0$) containing the diffracting aperture. The scalar wave amplitude in the diffracted field is found through an integral over the area of the aperture containing the normal derivative of the Green's function. This integral has the form of a convolution. In terms of the plane-wave representation, the convolution integral reduces to a product relation of the corresponding two-dimensional spatial Fourier transforms. In order to establish this relation, the Fourier transform of the normal derivative of the Green's function is required. This is found through the derivative of the transform of the space part of the outgoing spherical wave given by Weyl's formula,[3–5]

$$\frac{\exp(ikr)}{4\pi r}=\frac{i}{4\pi\lambda}\int_{-\infty}^{\infty}\!\!\int \frac{d\vec{p}}{m}\exp(ikmz)\exp(ik\vec{p}\cdot\vec{x}), \tag{1}$$

where $k=2\pi/\lambda$, λ is the wavelength, and r is the magnitude of the vector $\vec{r}$ with components (x,y,z). The vectors $\vec{x}$ and $\vec{p}$ are two dimensional with components (x,y) and (p,q), respectively. The scalar product is $\vec{p}\cdot\vec{x}=px+qy$ and the symbol $d\vec{p}=dp\,dq$. The condition $\vec{p}\cdot\vec{p}\leq 1$ corresponds to real waves for which $m=+(1-\vec{p}\cdot\vec{p})^{1/2}$ and for evanescent waves $\vec{p}\cdot\vec{p}>1$ in which case $m=+i(\vec{p}\cdot\vec{p}-1)^{1/2}$.

The proof of Weyl's formula is much involved, as given in Weyl's original paper[3] and also by Baños[6] and Stratton.[7] For this reason, it generally has to be assumed and taken as a starting point. Pedagogically, it will be advantageous to be able to establish the Fourier-transform relation of Eq. (1) in a simpler way. In this letter, we derive it as a natural consequence of the inhomogeneous Helmholtz equation,

$$(\nabla'^2+k^2)G_0(\vec{r}-\vec{r}')=-\delta(\vec{r}-\vec{r}'), \tag{2}$$

satisfied by the Green's function,

$$G_0(\vec{r}-\vec{r}')=\frac{\exp(ik|\vec{r}-\vec{r}'|)}{4\pi|\vec{r}-\vec{r}'|}. \tag{3}$$

The prime on the del operator indicates operation on the primed variables (x',y',z').

We will assume the existence of the two-dimensional spatial Fourier transform $\tilde{G}_0(\kappa\vec{p},z-z')$ of the Green's function,

$$G_0(\vec{x}-\vec{x}',z-z')=\int_{-\infty}^{\infty}\!\!\int d(\kappa\vec{p})\tilde{G}_0(\kappa\vec{p},z-z')\times\exp[i2\pi\kappa\vec{p}\cdot(\vec{x}-\vec{x}')], \tag{4}$$

where $\kappa=1/\lambda$ and $d(\kappa\vec{p})=\kappa^2dp\,dq$. The delta function in Eq. (2) is three dimensional. Let the x and y parts of this function be represented by the Fourier integral

$$\delta(\vec{x}-\vec{x}')=\int_{-\infty}^{\infty}\!\!\int d(\kappa\vec{p})\exp[i2\pi\kappa\vec{p}\cdot(\vec{x}-\vec{x}')]. \tag{5}$$

When Eqs. (4) and (5) are used in Eq. (2), we get the equation satisfied by $\tilde{G}_0$,

$$\left[\frac{d^2}{dz'^2}+k^2(1-\vec{p}\cdot\vec{p})\right]\tilde{G}_0(\kappa\vec{p},z-z')=-\delta(z-z'). \tag{6}$$

We assume that the unknown function $\tilde{G}_0$ may be constructed by making use of the linearly independent solutions of the corresponding homogeneous equation. Thus, let us write

$$\tilde{G}_0(\kappa\vec{p},z-z')=\begin{cases}A(\kappa\vec{p})\exp[ikm(z-z')], & z\geq z'\\ A(\kappa\vec{p})\exp[-ikm(z-z')], & z\leq z',\end{cases} \tag{7}$$

where m is as defined before. The unknown function $A(\kappa\vec{p})$ is determined such that $\tilde{G}_0(\kappa\vec{p},z-z')$ will satisfy the inhomogeneous equation. This is done by choosing the correct value of the discontinuity in the first derivative of $\tilde{G}_0(\kappa\vec{p},z-z')$. Let us integrate both sides of Eq. (6) with respect to z' within the limits $(z-\epsilon$ to $z+\epsilon)$ and take the limit as $\epsilon\to 0$. The term containing k^2 contributes zero because $\tilde{G}_0(\kappa\vec{p},z-z')$ is continuous at $z=z'$. The integral of the remaining Eq. (6) yields

$$\lim_{\epsilon\to 0}\left\{\frac{d}{dz'}\tilde{G}_0(\kappa\vec{p},z-z')\Big|_{z'=z+\epsilon}-\frac{d}{dz'}\tilde{G}_0(\kappa\vec{p},z-z')\Big|_{z'=z-\epsilon}\right\}=-1.$$

In this equation, we put the appropriate expression for the two conditions $z\leq z'$ and $z\geq z'$ from Eq. (7). The unknown function $A(\kappa\vec{p})$ is found to be

$$A(\kappa\vec{p})=i/2km. \tag{8}$$

Substitution of Eq. (8) back into Eq. (7) gives the desired Fourier transform $\tilde{G}_0(\kappa\vec{p},z-z')$ of the Green's function. The transform relation as given in Eq. (4) can be written

$$G_0(\vec{x}-\vec{x}',z-z')=\begin{cases}\dfrac{i}{4\pi\lambda}\displaystyle\int_{-\infty}^{\infty}\!\!\int\frac{d\vec{p}}{m}\exp[ikm(z-z')]\times\exp[ik\vec{p}\cdot(\vec{x}-\vec{x}')], & z\geq z'\\ \dfrac{i}{4\pi\lambda}\displaystyle\int_{-\infty}^{\infty}\!\!\int\frac{d\vec{p}}{m}\exp[-ikm(z-z')]\times\exp[ik\vec{p}\cdot(\vec{x}-\vec{x}')], & z\leq z'.\end{cases} \tag{9}$$

Equation (9) is Weyl's formula that we wanted to establish. In particular, Eq. (1) is found if we put $x' = y' = z' = 0$ and consider the condition $z > 0$.

In the discussion after Eq. (6), the derivation rests on finding the discontinuity in the first derivative of the function $\bar{G}_0$ at $z = z'$. This fact is concealed in an alternative method that may be followed after Eq. (6). In this second method, the spatial Fourier transforms of $\bar{G}_0$ and the δ function with respect to the variable z' are used in Eq. (6). Then Eq. (9) is established after certain contour integrals in the complex plane are calculated. At first sight, it might be thought that Eq. (9) will also follow by use of the full Fourier transform of G_0 over the variables x', y', and z' in Eq. (2). The contour[8,9] integration in the complex plane in this case, however, leads to the closed-form expression of G_0 given in Eq. (3) but not to Weyl's formula.

The research reported here was supported by Dept. of the Air Force, Space and Missile Systems Organization, Los Angeles, under Contract F04695-67-C-0197.

*Sabbatical leave, Dept. of Electrical Communication Engineering, Indian Institute of Science, Bangalore 560012, India.

[1]A. Sommerfeld, *Optics*, *Vol. IV* (Academic, New York, 1972), p. 199.
[2]J. W. Goodman, *Introduction to Fourier Optics* (McGraw-Hill, New York, 1968), p. 42.
[3]H. Weyl, Ann. Physik **60**, 481 (1919).
[4]G. C. Sherman, J. Opt. Soc. Am. **57**, 546 (1967).
[5]É. Lalor, J. Opt. Soc. Am. 58, 1235 (1968).
[6]Alfredo Baños, Jr., *Dipole Radiation in the Presence of a Conducting Half-Space* (Pergamon, New York, 1966), p. 20.
[7]J. A. Stratton, *Electromagnetic Theory* (McGraw-Hill, New York, 1941), p. 577.
[8]G. Arfken, *Mathematical Methods for Physicists* (Academic, New York, 1970), p. 766.
[9]P. Roman, *Advanced Quantum Theory* (Addison-Wesley, Reading, Mass., 1965), p. 152.

APPENDIX 5.3. QUASIHOMOGENEOUS SOURCE CALCULATION

To derive Eq. (5-120) we start with Eq. (5-119) expressed in polar coordinates:

$$\Gamma(P, P, 0) = \frac{1}{\bar{\lambda}^2 z^2} I_Q \int_0^{\infty}\int_0^{2\pi}\int_0^{\infty}\int_0^{2\pi} \exp\left[-\frac{r_1^2 + r_2^2 + 2r_1 r_2 \cos(\theta_1 - \theta_2)}{8\sigma_Q^2}\right]$$
$$\times \exp\left[-\frac{r_1^2 + r_2^2 - 2r_1 r_2 \cos(\theta_1 - \theta_2)}{2\sigma_g^2}\right]$$
$$\times \exp\left[+ik\left(\frac{r_1^2 - r_2^2}{2z} - \frac{Rr_1\cos\theta_1}{z} + \frac{Rr_2\cos\theta_2}{z}\right)\right]$$
$$\times d\theta_1\, r_1\, dr_1\, d\theta_2\, r_2\, dr_2.$$

The integrals are to be performed in the same order as the display of differentials. For economy of subscripts we have dropped the subscript s from the polar coordinates r_{s1}, θ_{s1}, r_{s2}, and θ_{s2}, which refer to the pair of points of the source (see Fig. 5-16). It is convenient to define the symbols

$$u = \left(\frac{2}{\sigma_g}\right)^2 + \left(\frac{1}{\sigma_Q}\right)^2,$$

$$v = \left(\frac{2}{\sigma_g}\right)^2 - \left(\frac{1}{\sigma_Q}\right)^2,$$

and

$$w = u + i\left(\frac{4k}{z}\right).$$

The integration on θ_1, to be denoted by I_1, is

$$I_1 = \int_0^{2\pi} \exp\left[\left(-\frac{r_1 r_2 v \cos\theta_2}{4}\right)\cos\theta_1 + \left(-\frac{r_1 r_2 v \sin\theta_2}{4}\right)\sin\theta_1\right] \times \exp\left(-\frac{ikRr_1\cos\theta_1}{z}\right) d\theta_1.$$

We may express this integral as the sum of two integrals by splitting the complex exponential into its real and imaginary parts. Each one of these may be identified with those listed by Gröbner and Hoffreiter (1966) to get the result

$$I_1 = 2\pi J_0\left[ir_1(C' - iD')^{1/2}\right]$$

where J_0 is the zero-order Bessel function. The symbols C' and D' are

$$C' = \left(\frac{r_2 v}{4}\right)^2 - \left(\frac{kr}{z}\right)^2 \quad \text{and} \quad D' = \frac{kRr_2 v\cos\theta_2}{2z}.$$

The primes are used so as not to confuse them with the symbols used by Gröbner and Hoffreiter.

With this evaluation, $\Gamma(P, P, 0)$ becomes

$$\Gamma(P, P, \dot{0}) = \frac{2\pi}{\bar{\lambda}^2 z^2} I_Q \int_0^\infty \int_0^{2\pi} \exp\left(-\frac{r_2^2 u}{8}\right) \exp\left[ik\left(-\frac{r_2^2}{2z} + \frac{Rr_2\cos\theta_2}{z}\right)\right] \times \left[\int_0^\infty \exp\left(-\frac{r_1^2 w^*}{8}\right) J_0\left[ir_1(C' - iD')^{1/2}\right] r_1\, dr_1\right] d\theta_2\, r_2\, dr_2.$$

The integral on r_1, to be denoted by I_2, is found to be a special case of the general form listed in the *Handbook of Mathematical Functions* (Abramowitz and Stegun, 1964). We find that

$$I_2 = \frac{4}{w^*} \exp\left[\frac{2(C' - iD')}{w^*}\right].$$

This evaluation modifies the expression of Γ to read

$$\Gamma(P, P, 0) = \frac{2\pi}{\bar{\lambda}^2 z^2} I_Q \frac{4w}{|w|^2} \int_0^\infty \exp\left(-\frac{r_2^2 w}{8}\right) \exp\left[\left(r_2^2 v^2 - \frac{16k^2R^2}{z^2}\right)\frac{w}{8|w|^2}\right] \times \left\{\int_0^{2\pi} \exp\left[\frac{ikRr_2}{z}\left(1 - \frac{vw}{|w|^2}\right)\cos\theta_2\right] d\theta_2\right\} r_2\, dr_2.$$

The θ_2 integral is, in fact, the integral representation of the zero-order Bessel function. Therefore,

$$\Gamma(P,P,0) = \frac{4\pi^2}{\lambda^2 z^2} I_Q \frac{4w}{|w|^2} \exp\left(-\frac{2k^2R^2w}{z^2|w|^2}\right)$$
$$\times \int_0^\infty \exp\left[-\left(1-\frac{v^2}{|w|^2}\right)\frac{wr_2^2}{8}\right] J_0\left[\left(1-\frac{vw}{|w|^2}\right)\frac{kR}{z}r_2\right] r_2\,dr_2.$$

The last integral on r_2 may be evaluated in closed form by appealing once more to the integral from the *Handbook of Mathematical Functions*. The evaluation coupled with considerable algebra gives the expression

$$\Gamma(P,P,0) = \frac{k^2}{z^2} I_Q \frac{16}{|w|^2 - v^2} \exp\left[\frac{-4k^2R^2(u-v)}{z^2(|w|^2 - v^2)}\right].$$

Substitution for the symbols u, v, and w leads to the expression of Eq. (5-120) given in the text. (For evaluation of these integrals, the help of W. P. Goring is acknowledged.)

REFERENCES

Abramowitz, M., and I. A. Stegun, Eds. (1964). *Handbook of Mathematical Functions with Formulas, Graphs, and Mathematical Tables*, National Bureau of Standards, Applied Mathematics Series, 55, U.S. Government Printing Office, Washington, D.C.

Gröbner, W., and N. Hoffreiter (1966). *Integraltafel: Bestimmte Integrale*, Vol. 2, 4th revised ed., Springer-Verlag, Vienna and New York.

6

Image Formation

The theory of partial coherence provides a unique framework for describing image formation by an optical system with lenses or mirrors. It is convenient to regard the optical system as a "black box" with a known input, namely, the object, and to study the output, namely, the image. The only way to "know" what is in the object plane is to measure it or determine it through experiments. In principle, the mutual coherence function is such an observable. The light distribution in the object plane may be described by the MCF or its associated mutual spectral density function (MSDF). Only in the extreme case of coherence need the wave amplitude be mentioned, and in that case, the MCF can be factored and hence uniquely provide us with a function of a single space variable. The purpose of this chapter is to show how the basic equations of image formation and the associated spatial frequency characteristics of the optical system follow in a unified manner starting from the propagation law of the MCF.

NOTATION, SYMBOLS, AND SIGN CONVENTION

Let the optical system be rotationally symmetric with a common optical axis for all its elements. It has an entrance pupil and an exit pupil. The location of these conjugate planes on the optical axis is known through paraxial optics. The object and image planes are also conjugate planes. These four planes along with the pairs of coordinates are displayed in Fig. 6-1.

For simplicity and where there is no fear of confusion we shall denote the two-dimensional vectors, for example, in the object plane, by

$$\boldsymbol{\xi} = \hat{\mathbf{i}}\xi + \hat{\mathbf{j}}\eta. \tag{6-1}$$

The distance of any point from the origin will be stated explicitly, $(\xi^2 + \eta^2)^{1/2}$, while ξ will be used only to show the component of $\boldsymbol{\xi}$. A function, say $f(\xi, \eta)$, will be abbreviated by

$$f(\boldsymbol{\xi}) \equiv f(\xi, \eta), \tag{6-2}$$

and the area element by

$$d\boldsymbol{\xi} = d\xi\, d\eta, \tag{6-3}$$

where two-dimensional integrals are involved. A similar scheme will be used

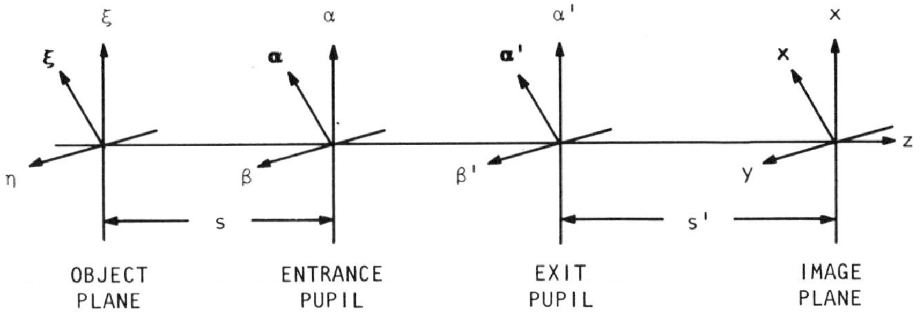

Fig. 6-1. Notation used for image formation. The optical system with its components is not explicitly shown; it is simply represented by its entrance and exit pupils. The horizontal line (z axis) is the optical axis. The distances s and s' between the entrance pupil and object plane and between the exit pupil and image plane are measured parallel to the optical axis. Refer to the text for the sign convention used.

for the entrance pupil, $\boldsymbol{\alpha}$, the exit pupil, $\boldsymbol{\alpha}'$, and the image plane, $\mathbf{x}$, as indicated in Fig. 6-1. Frequently we shall refer to circular pupils with radii a and a' for the entrance and exit pupils, respectively. Light leaves the object plane and arrives at the image plane via the optical system. It enters and exits the system through the pupils. The light amplitude transmittance of the system may be represented by

$$\text{Optical System:}\quad \mathfrak{L}(\boldsymbol{\alpha}')\exp\left[-ik\frac{(\boldsymbol{\alpha}'\cdot\boldsymbol{\alpha}')}{2z_F}\right].$$

The part $\mathfrak{L}(\boldsymbol{\alpha}')$ describes the amplitude and phase transmittance of the system in terms of the exit pupil coordinates α' and β'. It is a function that includes the physical size of the exit pupil, and its phase part incorporates the wavefront aberration terms, $W(\boldsymbol{\alpha}')$, with respect to the Gaussian reference sphere. A power series expansion for $W(\boldsymbol{\alpha}')$ in terms of the powers of normalized exit pupil coordinates $(\boldsymbol{\alpha}'/a')$ is used. The expansion coefficients correspond to different aberration terms and have units of length measured in terms of the wavelength λ. The function W is a measure of the departure of the wavefront from the Gaussian reference sphere. It is assumed that W is the same for all points of the extended object. An optical system with this property is referred to as *isoplanatic*. For further reading on this point we refer to the excellent book by Welford (1974). As an example, a thin lens with a clear circular aperture of radius a', $\mathfrak{L}(\boldsymbol{\alpha}')$, has the form

$$\mathfrak{L}(\boldsymbol{\alpha}') = \text{cyl}\left[\frac{(\alpha'^2+\beta'^2)^{1/2}}{a'}\right]\exp[ikW(\boldsymbol{\alpha}')]. \tag{6-4}$$

The quadratic phase factor in the optical system transmittance contains $\boldsymbol{\alpha}'\cdot\boldsymbol{\alpha}' = \alpha'^2+\beta'^2$ and it describes the light-focusing ability of the lens (see Goodman, 1968). It converts the incoming expanding spherical wave into an outgoing converging spherical wave in the quadratic (paraxial optics) approximation. The exponent contains the distance z_F instead of the familiar focal length F. This is because the object and image distances are measured with reference to the entrance and exit pupils, respectively. If s and s' were to be measured with reference to the principal planes, then of course $z_F = F$. Furthermore, in terms of s and s' the paraxial image-forming equation takes the form

$$\frac{1}{s(m_p)^2}+\frac{1}{s'}=\frac{1}{z_F} \tag{6-5}$$

where m_p is the pupil magnification. This is the form in which the focusing condition will appear in our calculations. The parameter z_F is found to be

$$z_F = +Fm_p, \tag{6-6}$$

and the ratio of the distances is

$$\frac{s'}{s} = -mm_p, \tag{6-7}$$

where m is the magnification in the image. By using Eq. (6-6) we may put the focusing condition into another form,

$$\frac{1}{sm_p} + \frac{m_p}{s'} = \frac{1}{F}.$$

We observe that when the pupil magnification is set equal to $+1$, as it is for the principal planes, the above equations reduce to the familiar form (see Problem 6-1).

It is important to point out that we have *not* used the coordinate sign convention used in lens design because that would entail making changes in the light propagation equations that we have been studying up to now. Instead we have used the following sign convention for the paraxial optical equations given above (see Jenkins and White, 1976, Art. 3.5):

1. All figures are drawn with the light traveling from left to right.
2. All object distances are considered positive when they are measured to the left of the lens and negative when they are measured to the right.
3. All image distances are positive when they are measured to the right of the lens and negative when measured to the left.
4. Both focal lengths are positive for a converging lens and negative for a diverging lens.
5. Transverse directions are positive when measured upward from the axis and negative when measured downward.

In this sign convention the image magnification is negative for a real inverted image.

OPTICAL INTENSITY AND COHERENCE IN THE IMAGE

We begin by assuming that the MCF $\Gamma_{ob}(\xi_1, \xi_2, \tau)$, or equivalently the MSDF $\hat{\Gamma}_{ob}(\xi_1, \xi_2, \nu)$, is known for all pairs of points over the object plane. The MSDF $\hat{\Gamma}_{ob}$ contains the illumination and the object characteristics. As an example, consider a transilluminated object with amplitude transmittance $\hat{t}_{ob}(\xi, \nu)$. Let $\hat{\Gamma}(\xi_1, \xi_2, \nu)$ be the MSDF of the illuminating beam. Then

$$\hat{\Gamma}_{ob}(\xi_1, \xi_2, \nu) = \hat{\Gamma}(\xi_1, \xi_2, \nu)\, \hat{t}_{ob}(\xi_1, \nu)\, \hat{t}^*_{ob}(\xi_2, \nu). \tag{6-8}$$

If the beam is due to a uniform, noncoherent source, $|\hat{\Gamma}(\xi_1, \xi_2, \nu)|$ will be a function of the coordinate difference, $\xi_1 - \xi_2$, as seen in Examples 5-1 and 5-2. For theoretical considerations the Carter–Wolf quasihomogeneous source may also be used for this purpose. The phase part of $\hat{\Gamma}_{ob}$ thus contains the phase structure of the object plus a contribution due to the light statistics. The optical intensity over the object plane is given by $\Gamma_{ob}(\xi, \xi, 0)$.

Light leaves the object plane and arrives at the entrance pupil. By the generalized van Cittert–Zernike theorem discussed in Chapter 5, the MSDF $\hat{\Gamma}_{ent}(\boldsymbol{\alpha}_1, \boldsymbol{\alpha}_2, \nu)$ of the entrance pupil plane is found in terms of $\hat{\Gamma}_{ob}$. Light then passes through the optical system and reaches the exit pupil plane. The MSDF $\hat{\Gamma}_{exit}(\boldsymbol{\alpha}'_1, \boldsymbol{\alpha}'_2, \nu)$ of the exit pupil is found through the product of $\hat{\Gamma}_{ent}$ and the amplitude transmittance functions of the optical system with $\boldsymbol{\alpha}'$ as the argument. The pupil variables $\boldsymbol{\alpha}$ and $\boldsymbol{\alpha}'$ are related by the pupil magnification, $\boldsymbol{\alpha} = \boldsymbol{\alpha}'/m_p$. Finally, we propagate the MSDF of the exit pupil and relate it to the MSDF $\hat{\Gamma}_{im}(\mathbf{x}_1, \mathbf{x}_2, \nu)$ of the image plane. A Fourier transformation yields the MCF $\Gamma_{im}(\mathbf{x}_1, \mathbf{x}_2, \tau)$, which gives us the optical intensity $\Gamma_{im}(\mathbf{x}, \mathbf{x}, 0)$ across the image plane.

These words when translated into mathematics yield the following:

Object Plane:

$$\hat{\Gamma}_{ob}(\xi_1, \xi_2, \nu) = \int_{-\infty}^{\infty} \Gamma_{ob}(\xi_1, \xi_2, \tau) \exp(i2\pi\nu\tau)\, d\tau. \tag{6-9}$$

Entrance Pupil:

$$\hat{\Gamma}_{ent}(\boldsymbol{\alpha}_1, \boldsymbol{\alpha}_2, \nu) = \int\int d\xi_1\, d\xi_2\, \hat{\Gamma}_{ob}(\xi_1, \xi_2, \nu) \times \frac{\partial \hat{G}(\boldsymbol{\alpha}_1, \xi_1, \nu)}{\partial n(\xi_1)} \frac{\partial \hat{G}^*(\boldsymbol{\alpha}_2, \xi_2, \nu)}{\partial n(\xi_2)}. \tag{6-10}$$

Exit Pupil:

$$\hat{\Gamma}_{exit}(\boldsymbol{\alpha}'_1, \boldsymbol{\alpha}'_2, \nu) = \hat{\Gamma}_{ent}\left(\frac{\boldsymbol{\alpha}'_1}{m_p}, \frac{\boldsymbol{\alpha}'_2}{m_p}, \nu\right) \mathfrak{L}(\boldsymbol{\alpha}'_1)\, \mathfrak{L}^*(\boldsymbol{\alpha}'_2) \times \exp\left[-\frac{ik}{2z_F}(\boldsymbol{\alpha}'_1 \cdot \boldsymbol{\alpha}'_1 - \boldsymbol{\alpha}'_2 \cdot \boldsymbol{\alpha}'_2)\right]. \tag{6-11}$$

Image Plane:

$$\hat{\Gamma}_{\text{im}}(\mathbf{x}_1, \mathbf{x}_2, \nu) = \int\!\!\int d\boldsymbol{\alpha}'_1\, d\boldsymbol{\alpha}'_2\, \hat{\Gamma}_{\text{exit}}(\boldsymbol{\alpha}'_1, \boldsymbol{\alpha}'_2, \nu) \times \frac{\partial \hat{G}(\mathbf{x}_1, \boldsymbol{\alpha}'_1, \nu)}{\partial n(\boldsymbol{\alpha}'_1)} \frac{\partial \hat{G}^*(\mathbf{x}_2, \boldsymbol{\alpha}'_2, \nu)}{\partial n(\boldsymbol{\alpha}'_2)}, \tag{6-12}$$

and

$$\Gamma_{\text{im}}(\mathbf{x}_1, \mathbf{x}_2, \tau) = \int_0^{\infty} \hat{\Gamma}_{\text{im}}(\mathbf{x}_1, \mathbf{x}_2, \nu) \exp(-i2\pi\nu\tau)\, d\nu. \tag{6-13}$$

In Eqs. (6-10) and (6-12) the differentials stand for the elemental areas as in Eq. (6-3). Only one integral sign is shown per area integral. The limits are not explicitly displayed; formally they are to be taken from $-\infty$ to $+\infty$. This is to be understood for all such area integrals throughout this chapter unless otherwise indicated.

We shall make the customary approximation, as, for example, in Eq. (5-38), for the normal derivatives of $\hat{G}$. Thus,

$$\frac{\partial \hat{G}(\mathbf{x}_1, \boldsymbol{\alpha}'_1, \nu)}{\partial n(\boldsymbol{\alpha}'_1)} \simeq \frac{-i}{\lambda s'} \exp(ik\rho'_1), \tag{6-14}$$

wherein the distance ρ'_1 between $\mathbf{x}_1$ and $\boldsymbol{\alpha}'_1$ is approximated up to the quadratic terms. The expression for ρ'_1 is

$$\begin{aligned}\rho'_1 &= \left[(x_1 - \alpha'_1)^2 + (y_1 - \beta'_1)^2 + s'^2\right]^{1/2} \\ &\simeq s' + \frac{(x_1 - \alpha'_1)^2 + (y_1 - \beta'_1)^2}{2s'}.\end{aligned}$$

Expressed in terms of the two-dimensional vectors, it reads

$$\rho'_1 = s' + \frac{\mathbf{x}_1 \cdot \mathbf{x}_1}{2s'} + \frac{\boldsymbol{\alpha}'_1 \cdot \boldsymbol{\alpha}'_1}{2s'} - \frac{\mathbf{x}_1 \cdot \boldsymbol{\alpha}'_1}{s'}. \tag{6-15}$$

Similar considerations apply to the other normal derivatives in Eqs. (6-10) and (6-12). Now we have all the tools necessary to obtain the MSDF $\hat{\Gamma}_{\text{im}}$ in the image plane to a good approximation.

Thus for the entrance pupil we find

$$\hat{\Gamma}_{\text{ent}}(\boldsymbol{\alpha}_1, \boldsymbol{\alpha}_2, \nu) = \left(\frac{1}{\lambda s}\right)^2 \int\int d\xi_1\, d\xi_2\, \hat{\Gamma}_{\text{ob}}(\xi_1, \xi_2, \nu) \times \exp\left[ik\left(\frac{\boldsymbol{\alpha}_1 \cdot \boldsymbol{\alpha}_1}{2s} + \frac{\xi_1 \cdot \xi_1}{2s} - \frac{\boldsymbol{\alpha}_1 \cdot \xi_1}{s}\right)\right] \times \exp\left[-ik\left(\frac{\boldsymbol{\alpha}_2 \cdot \boldsymbol{\alpha}_2}{2s} + \frac{\xi_2 \cdot \xi_2}{2s} - \frac{\boldsymbol{\alpha}_2 \cdot \xi_2}{s}\right)\right].$$

This expression is to be used in Eq. (6-11) to obtain $\hat{\Gamma}_{\text{exit}}$, which in turn is to be put into Eq. (6-12) to obtain the MSDF $\hat{\Gamma}_{\text{im}}$ in the image plane. Following this procedure we can get

$$\hat{\Gamma}_{\text{im}}(\mathbf{x}_1, \mathbf{x}_2, \nu) = \left(\frac{1}{\lambda s}\right)^2 \left(\frac{1}{\lambda s'}\right)^2 \int\int d\boldsymbol{\alpha}_1'\, d\boldsymbol{\alpha}_2' \int\int d\xi_1\, d\xi_2\, \hat{\Gamma}_{\text{ob}}(\xi_1, \xi_2, \nu)$$
$$\times \exp\left[ik\left(\frac{\boldsymbol{\alpha}_1' \cdot \boldsymbol{\alpha}_1'}{2sm_p^2} + \frac{\xi_1 \cdot \xi_1}{2s} - \frac{\boldsymbol{\alpha}_1' \cdot \xi_1}{sm_p} - \frac{\boldsymbol{\alpha}_2' \cdot \boldsymbol{\alpha}_2'}{2sm_p^2} - \frac{\xi_2 \cdot \xi_2}{2s} + \frac{\boldsymbol{\alpha}_2' \cdot \xi_2}{sm_p}\right)\right]$$
$$\times \mathfrak{L}(\boldsymbol{\alpha}_1')\, \mathfrak{L}^*(\boldsymbol{\alpha}_2') \exp\left[-\frac{ik}{2z_F}(\boldsymbol{\alpha}_1' \cdot \boldsymbol{\alpha}_1' - \boldsymbol{\alpha}_2' \cdot \boldsymbol{\alpha}_2')\right]$$
$$\times \exp\left[ik\left(\frac{\mathbf{x}_1 \cdot \mathbf{x}_1}{2s'} + \frac{\boldsymbol{\alpha}_1' \cdot \boldsymbol{\alpha}_1'}{2s'} - \frac{\mathbf{x}_1 \cdot \boldsymbol{\alpha}_1'}{s'} - \frac{\mathbf{x}_2 \cdot \mathbf{x}_2}{2s'} - \frac{\boldsymbol{\alpha}_2' \cdot \boldsymbol{\alpha}_2'}{2s'} + \frac{\mathbf{x}_2 \cdot \boldsymbol{\alpha}_2'}{s'}\right)\right] \tag{6-16}$$

We immediately observe that all the quadratic terms of the type $(\boldsymbol{\alpha}' \cdot \boldsymbol{\alpha}')$ vanish, due to the focusing condition of Eq. (6-5).

Now consider, for example, the integral on $\boldsymbol{\alpha}_1'$, namely,

$$\int d\boldsymbol{\alpha}_1'\, \mathfrak{L}(\boldsymbol{\alpha}_1') \exp\left[-ik\left(\frac{\mathbf{x}_1}{s'} + \frac{\xi_1}{sm_p}\right) \cdot \boldsymbol{\alpha}_1'\right].$$

The Fourier kernel in it may be rearranged by using Eq. (6-7) for the ratio s'/s. Then the integral becomes a function of $\mathbf{x}_1 - m\xi_1$. This is true for an isoplanatic system for which the pupil function $\mathfrak{L}$ does not change as $\mathbf{x}$ and ξ vary over the image and the object, respectively. Therefore, it is convenient to define a function A of this argument by

$$A(\mathbf{x}_1 - m\xi_1) = \frac{1}{\lambda^2 ss'} \int d\boldsymbol{\alpha}_1'\, \mathfrak{L}(\boldsymbol{\alpha}_1') \exp\left[-i2\pi(\mathbf{x}_1 - m\xi_1) \cdot \left(\frac{\boldsymbol{\alpha}_1'}{\lambda s'}\right)\right]. \tag{6-17}$$

Let us define a pair of spatial frequency variables in relation to the image space by

$$f' \equiv \frac{\alpha'}{\lambda s'} \quad \text{and} \quad g' \equiv \frac{\beta'}{\lambda s'} \tag{6-18}$$

and a spatial frequency vector by

$$\mathbf{f}' = \hat{\mathbf{i}}\, f' + \hat{\mathbf{j}}\, g' = \frac{\boldsymbol{\alpha}'}{\lambda s'}. \tag{6-19}$$

The writing of the differentials in the integrals over f' and g' may be shortened to

$$d\mathbf{f}' \equiv df'\, dg' = \frac{d\boldsymbol{\alpha}'}{(\lambda s')^2}. \tag{6-20}$$

Although not of immediate interest, we mention for completeness that a spatial frequency vector in relation to the object space may also be defined, $f = +\alpha/\lambda s$, and $g = \beta/\lambda s$, such that $\mathbf{f} = -m\mathbf{f}'$. In terms of the spatial frequency vector the definition of A becomes

$$A(\mathbf{x}_1 - m\boldsymbol{\xi}_1) = \frac{s'}{s} \int d\mathbf{f}'_1\, \mathcal{L}(\lambda s' \mathbf{f}'_1) \exp[-i2\pi(\mathbf{x}_1 - m\boldsymbol{\xi}_1) \cdot \mathbf{f}'_1]. \tag{6-21}$$

The optical meaning and the usefulness of the function A will be established in what is to follow. For now we observe that A is defined as a two-dimensional spatial Fourier transform of the lens aperture function $\mathcal{L}(\boldsymbol{\alpha}')$ with reference to the exit pupil. The units of A are $(\text{area})^{-1}$, and $\mathcal{L}$ is unitless.

In terms of the function A, the description of the image given in Eq. (6-16) may be cast into a more compact form,

$$\begin{aligned}\hat{\Gamma}_{\text{im}}(\mathbf{x}_1, \mathbf{x}_2, \nu) = {} & \exp\left[\frac{ik}{2s'}(\mathbf{x}_1 \cdot \mathbf{x}_1 - \mathbf{x}_2 \cdot \mathbf{x}_2)\right] \int\int d\boldsymbol{\xi}_1\, d\boldsymbol{\xi}_2\, \hat{\Gamma}_{\text{ob}}(\boldsymbol{\xi}_1, \boldsymbol{\xi}_2, \nu) \\ & \times \exp\left[\frac{ik}{2s}(\boldsymbol{\xi}_1 \cdot \boldsymbol{\xi}_1 - \boldsymbol{\xi}_2 \cdot \boldsymbol{\xi}_2)\right] A(\mathbf{x}_1 - m\boldsymbol{\xi}_1)\, A^*(\mathbf{x}_2 - m\boldsymbol{\xi}_2).\end{aligned} \tag{6-22}$$

This is our final general answer for *polychromatic partially coherent* conditions of illumination. It consists of a linear relationship between the object and image characteristics. Without displaying it explicitly, we observe that a linear relationship also connects the MCFs of the object and the image.

Either one may be used to obtain the optical intensity in the image,

$$\Gamma_{\text{im}}(\mathbf{x},\mathbf{x},0)=\int_0^{\infty}d\nu\int_{-\infty}^{\infty}d\tau'[\exp(+i2\pi\nu\tau')]\iint d\xi_1\,d\xi_2\,\Gamma_{\text{ob}}(\xi_1,\xi_2,\tau')$$
$$\times\exp\left[+\frac{ik}{2s}(\xi_1\cdot\xi_1-\xi_2\cdot\xi_2)\right]A(\mathbf{x}-m\xi_1)\,A^*(\mathbf{x}-m\xi_2)$$
$$=\int_0^{\infty}d\nu\iint d\xi_1\,d\xi_2\,\hat{\Gamma}_{\text{ob}}(\xi_1,\xi_2,\nu)\exp\left[+\frac{ik}{2s}(\xi_1\cdot\xi_1-\xi_2\cdot\xi_2)\right]$$
$$\times A(\mathbf{x}-m\xi_1)\,A^*(\mathbf{x}-m\xi_2). \tag{6-23}$$

The image optical intensity may be derived through either the full MCF $\Gamma_{\text{ob}}(\xi_1,\xi_2,\tau)$ or the MSDF $\hat{\Gamma}_{\text{ob}}(\xi_1,\xi_2,\nu)$ of the object. That $\Gamma_{\text{im}}(\mathbf{x},\mathbf{x},0)$ is real and nonnegative may be established, as asked for in Problem 6-2. We shall now undertake to study the details of the various limiting cases of interest.

IMAGE FORMATION IN THE COHERENT LIMIT

In the coherent limit, the mutual spectral density in the object plane factors in the form

$$\hat{\Gamma}_{\text{ob}}(\xi_1,\xi_2,\nu)=U_{\text{ob}}(\xi_1)\,U^*_{\text{ob}}(\xi_2)\,\delta(\nu-\nu_0), \tag{6-24}$$

where ν_0 is the monochromatic frequency. When this factored form is put into Eq. (6-22), it is obvious that it also factors. Thus in the image plane, $\hat{\Gamma}_{\text{im}}$ takes the form

$$\hat{\Gamma}_{\text{im}}(\mathbf{x}_1,\mathbf{x}_2,\nu)=U_{\text{im}}(\mathbf{x}_1)\,U^*_{\text{im}}(\mathbf{x}_2)\,\delta(\nu-\nu_0), \tag{6-25}$$

where the function U_{im} of a single variable $\mathbf{x}$ is related to U_{ob} by

$$U_{\text{im}}(\mathbf{x})=\exp\left[\frac{ik_0}{2s'}(\mathbf{x}\cdot\mathbf{x})\right]\iint d\xi\,U_{\text{ob}}(\xi)\exp\left[\frac{ik_0}{2s}(\xi\cdot\xi)\right]A(\mathbf{x}-m\xi), \tag{6-26}$$

where $k_0=2\pi\nu_0/c$. Thus in the coherent limit, the object–image relationship may be described by a linear operation on an amplitude function $U(\xi)$. For this linear system we would like to find the amplitude impulse response S_{coh} and the amplitude transfer function T_{coh} for the optical system.

The impulse in the object plane implies a point of light at some location ξ_0. Consider an

$$\text{Impulse-Object:} \quad U_{\text{ob}}(\xi) = \mathcal{A} U_0 \delta(\xi - \xi_0). \tag{6-27}$$

The delta function is used for mathematical convenience. In practice, a small opening of area $\mathcal{A}$ is used in an otherwise opaque object plane. The dimensions of the opening are taken so small that the function A and the quadratic phase factor in the integrand of Eq. (6-26) are sensibly constant over it. The image, $U_{\text{im}}(\mathbf{x})$, of a unit impulse ($\mathcal{A} U_0 = 1$) object of Eq. (6-27) is called the *amplitude impulse response* S_{coh} of the optical system operating with coherent illumination. The expression for S_{coh} reads

$$S_{\text{coh}}(\mathbf{x}, \xi_0) = \exp\left[\frac{ik_0}{2}\left(\frac{\mathbf{x}\cdot\mathbf{x}}{s'} + \frac{\xi_0 \cdot \xi_0}{s}\right)\right] A(\mathbf{x} - m\xi_0). \tag{6-28}$$

Here the quadratic phase factor in $\mathbf{x}$ implies that the response S_{coh} is residing on a spherical surface with radius s' with the center at the midpoint of the exit pupil. But S_{coh} depends on the $\mathbf{x}$ coordinate and the location ξ_0 of the impulse in the object plane. That is, the response of the lens is different for different locations of the impulse in the object plane. Such a system is said to be *spatially nonstationary*.

We shall now discuss the conditions under which the lens with coherent illumination may be regarded as *approximately* spatially stationary. For stationarity, S_{coh} must not depend on $\mathbf{x}$ and ξ_0 separately but rather on the combination $\mathbf{x} - m\xi_0$. Now the quadratic phase factors of Eq. (6-28) will be very nearly unity; that is,

$$\exp\left[\frac{ik_0}{2}\left(\frac{\mathbf{x}\cdot\mathbf{x}}{s'} + \frac{\xi_0 \cdot \xi_0}{s}\right)\right] \simeq 1 \tag{6-29}$$

if we demand that the object distance s and the image distance s' satisfy

$$s \gg \frac{\xi_0^2 + \eta_0^2}{\lambda_0} \quad \text{and} \quad s' \gg \frac{x^2 + y^2}{\lambda_0}.$$

This implies that the lens must be in the far field of both the object and image size—requirements that are almost never satisfied in practice for reasonably sized objects and their images.

We must therefore look for a weaker set of conditions that will allow us to regard the coherent imaging system as approximately stationary. For this purpose, let us make a temporary change to a pair of variables u and v, defined by the two-dimensional vector

$$\mathbf{u} \equiv \mathbf{x} - m\boldsymbol{\xi}_0 \quad \text{or} \quad u = x - m\xi_0 \quad \text{and } v = y - m\eta_0.$$

Thus $\boldsymbol{\xi}_0$ may be expressed in terms of $\mathbf{u}$ and $\mathbf{x}$ in Eq. (6-28) to give

$$S_{\text{coh}}(\mathbf{x}, \mathbf{u}) = \exp\left[\frac{ik_0}{2s'}\left(1 - \frac{m_p}{m}\right)\mathbf{x}\cdot\mathbf{x}\right]\exp\left[\frac{-ik_0 m_p}{2s'm}(\mathbf{u}\cdot\mathbf{u} - 2\mathbf{x}\cdot\mathbf{u})\right]A(\mathbf{u}). \tag{6-30}$$

Now the approximation consists of asking for the conditions under which the quadratic and linear phase factors in $\mathbf{u}$ may be regarded as unity. To understand how the necessary conditions come about, let $\mathcal{R}_2$ and $\mathcal{R}_1$ be the regions in u, v space over which the quadratic and linear phase factors may be approximated to unity, respectively. Let $\mathcal{R}$ be the region over which the function A has significant values. $\mathcal{R}$ is sometimes called a *resolved element*. The sought for weaker conditions of stationarity may be simply derived by asking that both the regions $\mathcal{R}_2$ and $\mathcal{R}_1$ satisfy

$$\mathcal{R}_2 \gg \mathcal{R} \tag{6-31}$$

and

$$\mathcal{R}_1 \gg \mathcal{R}. \tag{6-32}$$

To go any further we need to know the behavior of $A(\mathbf{u})$. From the definition of this function in Eq. (6-17), we observe that it is a spatial Fourier transform of the lens aperture function $\mathcal{L}(\boldsymbol{\alpha}')$. For an ideal, unaberrated lens with a clear aperture of radius a', the function A is a Besinc function given by

$$A(\mathbf{u}) = \frac{\pi a'^2}{\lambda_0^2 ss'}\frac{2J_1(\sigma)}{\sigma} \equiv \frac{\pi a'^2}{\lambda_0^2 ss'}\text{Besinc}(\sigma) \tag{6-33}$$

where J_1 is the Bessel function of order 1, and the argument σ is

$$\sigma \equiv \frac{2\pi a'(u^2 + v^2)^{1/2}}{\lambda_0 s'}. \tag{6-34}$$

We now define the region $\mathcal{R}$ as the area of the circle whose radius is equal to the distance of the first zero of $A(\mathbf{u})$ from the origin,

$$\mathcal{R} = \pi\left(\frac{1.22\lambda_0 s'}{2a'}\right)^2. \tag{6-35}$$

For this representative case, it can be shown that the following set of weaker conditions leads one to stationarity. The conditions will be numbered as (C1), (C2), and (C3). Thus the inequality in Eq. (6-31) is fulfilled if

(C1) The image distance s' is much *smaller* than the far-field distance $4a'^2/\lambda_0$ of the exit pupil, and the object distance s is much *smaller* than the far-field distance of the entrance pupil, $4a^2/\lambda_0$.

The inequality in Eq. (6-32) imposes a restriction on the size of the object and image as given in condition (C2):

(C2) The maximum dimensions of the object and image are, respectively, much smaller than the diameter of the entrance and exit pupils of the lens system.

To be sure, there are certain factors of magnification (m, m_p) and numerical factors of order less than unity that enter into the derivation of conditions (C1) and (C2). Since these factors are of lesser importance to our present discussion, we have chosen to display the condition in words and save the algebra. The appropriate calculation is delegated to Problem 6-3.

We first observe that condition (C1) is almost always satisfied in practice, and it is indeed possible to arrange for condition (C2) to be fulfilled. Hence the expression for the amplitude impulse response S_{coh} of Eq. (6-30) may be approximated to

$$S_{\text{coh}}(\mathbf{x}, \boldsymbol{\xi}_0) = \exp\left[\frac{ik_0}{2s'}\left(1 - \frac{m_p}{m}\right)\mathbf{x} \cdot \mathbf{x}\right] A(\mathbf{x} - m\boldsymbol{\xi}_0). \tag{6-36}$$

The remaining quadratic phase factor in $\mathbf{x}$ cannot be removed, even for a special case. At first sight it might appear that it may be removed by choosing $m = m_p$. This implies, however, that the pupils themselves be the object and image, to which case we are not addressing ourselves. Furthermore, let us denote by $\mathcal{R}_2'$ the region over which this phase factor is nearly unity, and let $\mathcal{R}_{\text{im}}$ be the region over which the image is defined by condition (C2). It can be shown that, for most cases of practical interest,

$$\mathcal{R} \ll \mathcal{R}_2' \ll \mathcal{R}_{\text{im}}, \tag{6-37}$$

which further emphasizes that the phase factor cannot be approximated to unity over the entire area of the image. The first part of the inequality indicates, however, that it may be equated to unity over an area the size of several resolved elements $\mathcal{R}$.

The form of S_{coh} of Eq. (6-36) suggests that the image equation may be rewritten in the form

$$U_{\text{im}}(\mathbf{x}) = \exp\left[\frac{ik_0}{2s'}\left(1 - \frac{m_p}{m}\right)\mathbf{x}\cdot\mathbf{x}\right]\int d\boldsymbol{\xi}\, U_{\text{ob}}(\boldsymbol{\xi})\, A(\mathbf{x} - m\boldsymbol{\xi}). \quad (6\text{-}38)$$

Finally, we shall adopt the physical condition:

(C3) As far as irradiance measurements are concerned, the quadratic phase factor in $\mathbf{x}$ of Eq. (6-38) is not important and may be dropped. Alternatively, for questions of resolution one frequently considers only a small portion (on the order of several resolved elements, $\mathcal{R}$) of the image. Over this region the phase factor reduces to a constant.

Adoption of the alternative given in condition (C3) amounts to the fulfillment of the conditions demanded by Eq. (6-29).

Thus for an isoplanatic coherent imaging system we may write

$$U_{\text{im}}(\mathbf{x}) = \int d\boldsymbol{\xi}\, U_{\text{ob}}(\boldsymbol{\xi})\, A(\mathbf{x} - m\boldsymbol{\xi}), \quad (6\text{-}39)$$

provided conditions (C1) to (C3) are fulfilled. Such a system is referred to as *linear* and *spatially stationary*; it is also called *shift invariant* (Gaskill, 1978). Evidently, we may interpret the function A of Eqs. (6-17) and (6-21) as the amplitude impulse response and put

$$S_{\text{coh}}(\mathbf{x} - m\boldsymbol{\xi}) = A(\mathbf{x} - m\boldsymbol{\xi}). \quad (6\text{-}40)$$

For the rest of this subsection we assume that the required conditions are fulfilled.

Now the transfer function $T_{\text{coh}}(\mathbf{f}')$ of this system is the spatial Fourier transform of the amplitude impulse response

$$T_{\text{coh}}(\mathbf{f}') = \mathcal{F}[S_{\text{coh}}(\mathbf{f}')] = \int_{-\infty}^{\infty} d\mathbf{x}\, S_{\text{coh}}(\mathbf{x}) \exp(-i2\pi\mathbf{x}\cdot\mathbf{f}'). \quad (6\text{-}41)$$

For $\boldsymbol{\xi}_0 \neq 0$, an inconsequential linear phase appears in T_{coh}, implying a shift of the image as a whole by $m\boldsymbol{\xi}_0$. From the definition of A in Eq. (6-21), it is obvious that

$$T_{\text{coh}}(\mathbf{f}') = \frac{s'}{s}\mathcal{L}(-\lambda_0 s'\mathbf{f}'). \quad (6\text{-}42)$$

Thus the light amplitude transmittance $\mathcal{L}$ of the exit pupil, including its aberrations, properly scaled plays the role of the coherent transfer function.

In the spatial frequency domain the coherent imaging equation, Eq. (6-39), takes the form

$$\tilde{U}_{im}(\mathbf{f}') = \tilde{U}_{ob}(m\mathbf{f}')\, T_{coh}(\mathbf{f}'). \tag{6-43}$$

The frequency variable in the object spectrum is scaled with the magnification, as expected. The spatial Fourier transform of the object is defined by

$$U_{ob}(\boldsymbol{\xi}) = \int d\mathbf{f}\, \tilde{U}_{ob}(\mathbf{f}) \exp(+i2\pi\boldsymbol{\xi} \cdot \mathbf{f}),$$

where $\mathbf{f}$ is the spatial frequency in object space. A similar definition of $\tilde{U}_{im}$ is used in terms of the image-space frequency $\mathbf{f}'$.

To recapitulate, the coherent amplitude impulse response, S_{coh}, is the response of the system to a unit impulse. The function A is defined as the spatial Fourier transform of the system's exit pupil function $\mathfrak{L}$. Under the conditions of spatial stationarity, S_{coh} is A itself. The system transfer function T_{coh} is defined as the spatial Fourier transform of S_{coh}. Thus T_{coh} is proportional to $\mathfrak{L}$ itself with the scaled coordinates $\boldsymbol{\alpha}' = \lambda_0 s'\mathbf{f}'$.

For a circular lens of radius a' the maximum spatial frequency passed is the same for all azimuths. The maximum frequency is

$$f'_{max} = g'_{max} = \frac{a'}{\lambda_0 s'}. \tag{6-44}$$

For an unaberrated lens of radius a' the relevant functions are sketched in Fig. 6-2. In terms of the image-forming angle θ' (see Fig. 6-3), we find that

$$f'_{max} = \frac{\theta'}{2\lambda_0}, \tag{6-45}$$

where $\theta' = 2a'/s'$ is the angular subtense of the exit pupil to the image plane.

In general, we observe that for a fixed image distance s' a larger aperture lens will pass higher spatial frequencies, leading to better resolution. Furthermore, the maximum spatial frequency passed by the lens increases as the image-forming angle θ' increases and/or the wavelength λ decreases.

Since T_{coh} goes to zero beyond $a'/\lambda_0 s'$, it allows passage of only low spatial frequencies; the system is therefore a *low-pass filter*. The frequencies that it passes do so without any change for the case of an unaberrated lens with a clear aperture. In general, the amplitude attenuation and/or phase distortion of the spatial frequency components of the object is dictated by the complex amplitude transmittance of the exit pupil of the system. We shall return to this discussion in the next section.

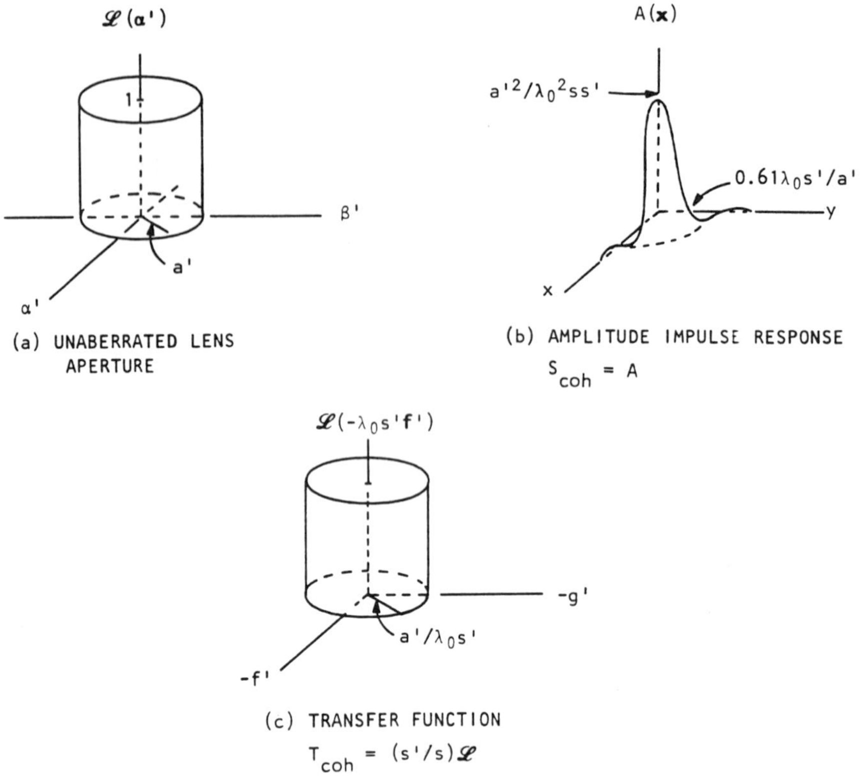

Fig. 6-2. Sketch of the functions $\mathfrak{L}$, S_{coh}, and T_{coh} for the case of an unaberrated lens of radius a'.

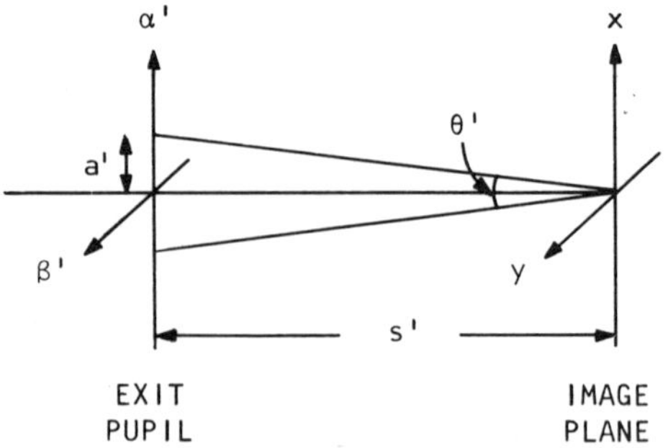

Fig. 6-3. Image-forming angle θ'.

IMAGE FORMATION IN THE NONCOHERENT LIMIT

In the noncoherent limit the mutual spectral density function in the object plane has the form

$$\hat{\Gamma}_{\text{ob}}(\xi_1, \xi_2, \nu) = \frac{4\pi}{k^2}\hat{I}_{\text{ob}}(\xi_1, \nu)\,\delta(\xi_1 - \xi_2), \tag{6-46}$$

where $\hat{I}_{\text{ob}}$ is the optical spectral intensity (OSI) in the object. When this is put into Eq. (6-22), the MSDF in the image plane is found to be

$$\hat{\Gamma}_{\text{im}}(\mathbf{x}_1, \mathbf{x}_2, \nu) = \frac{4\pi}{k^2}\exp\left[\frac{ik}{2s'}(\mathbf{x}_1\cdot\mathbf{x}_1 - \mathbf{x}_2\cdot\mathbf{x}_2)\right] \times \int d\boldsymbol{\xi}\,\hat{I}_{\text{ob}}(\boldsymbol{\xi}, \nu)\,A(\mathbf{x}_1 - m\boldsymbol{\xi})\,A^*(\mathbf{x}_2 - m\boldsymbol{\xi}). \tag{6-47}$$

This equation reveals the acquisition of coherence through propagation, since a noncoherent object produces a partially coherent image. Even in an imaging situation the coherence characteristics of the object are not necessarily carried over to the image plane. If we ask for the OSI $\hat{I}_{\text{im}}(\mathbf{x}, \nu) = \hat{\Gamma}_{\text{im}}(\mathbf{x}, \mathbf{x}, \nu)$ in the image we find that

$$\hat{I}_{\text{im}}(\mathbf{x}, \nu) = \frac{4\pi}{k^2}\int d\boldsymbol{\xi}\,\hat{I}_{\text{ob}}(\boldsymbol{\xi}, \nu)\,|A(\mathbf{x} - m\boldsymbol{\xi})|^2. \tag{6-48}$$

Thus an isoplanatic noncoherent imaging system is not only linear but also spatially stationary in terms of $\hat{I}$. Unlike the coherent case, this one involves no added restrictions, but the ones ensuing from the approximations on the Green's functions [see Eqs. (6-14) and (6-15)] of course apply.

Now the significance of the function $|A|^2$ in Eq. (6-48) is realized by asking for the impulse response S_{ncoh} of the noncoherent imaging system. Thus, let the object plane contain a pinhole at $\boldsymbol{\xi} = \boldsymbol{\xi}_0$ of area $\mathcal{A}$, fed with OSI $\hat{I}_0(\nu)$. That is, let

$$\hat{I}_{\text{ob}}(\boldsymbol{\xi}, \nu) = \mathcal{A}\hat{I}_0(\nu)\,\delta(\boldsymbol{\xi} - \boldsymbol{\xi}_0). \tag{6-49}$$

When this impulse is put into Eq. (6-48), it leads to

$$\hat{I}_{\text{im}}(\mathbf{x}, \nu) = \frac{4\pi}{k^2}\mathcal{A}\hat{I}_0(\nu)\,|A(\mathbf{x} - m\boldsymbol{\xi}_0)|^2.$$

Now the impulse response S_{ncoh} is defined as the image obtained for a unit impulse, $\mathcal{Q}\hat{I}_0(\nu) = 1$, at the origin, $\boldsymbol{\xi}_0 = 0$. Therefore,

$$S_{\text{ncoh}}(\mathbf{x}) = \frac{4\pi}{k^2} | A(\mathbf{x}) |^2. \tag{6-50}$$

Thus $| A |^2$ is proportional to the impulse response of the noncoherent imaging system. For $\boldsymbol{\xi}_0 \neq 0$, the response is simply bodily shifted without change of form. When the optical system was used in coherent light, the impulse response S_{coh} was proportional to A itself. In the previous section, it was shown in Eq. (6-33) that for a perfect unaberrated lens, A has the form of a Besinc function. Hence, for the same system with noncoherent illumination, this impulse response is

$$S_{\text{ncoh}}(\mathbf{x}) = \frac{4\pi}{k^2} | A(\mathbf{x}) |^2 = \frac{1}{\pi\lambda^2}\left[\frac{\pi a'^2}{ss'} \operatorname{Besinc}(\sigma)\right]^2 \tag{6-51}$$

with the argument σ given by

$$\sigma = \frac{2\pi a'(x^2 + y^2)^{1/2}}{\lambda s'}.$$

For a clear circular pupil of radius a' as shown in Fig. 6-4a, the form of $| A |^2$ is sketched in Fig. 6-4b. We return to this example for further discussion later, in relation to Fig. 6-4.

In terms of the impulse response, the relationship for imaging, Eq. (6-48), takes the form

$$\hat{I}_{\text{im}}(\mathbf{x}, \nu) = \int d\boldsymbol{\xi}\, \hat{I}_{\text{ob}}(\boldsymbol{\xi}, \nu)\, S_{\text{ncoh}}(\mathbf{x} - m\boldsymbol{\xi}). \tag{6-52}$$

The impulse response S_{ncoh} for the noncoherent imaging system is called the *point spread function* (PSF).

The transform domain description begins with the spatial Fourier transform of $\hat{I}_{\text{ob}}(\boldsymbol{\xi}, \nu)$, denoted by $\mathring{I}_{\text{ob}}(\mathbf{f}, \nu)$, namely,

$$\hat{I}_{\text{ob}}(\boldsymbol{\xi}, \nu) = \int_{-\infty}^{\infty} d\mathbf{f}\, \mathring{I}_{\text{ob}}(\mathbf{f}, \nu) \exp[+i2\pi(\boldsymbol{\xi} \cdot \mathbf{f})], \tag{6-53}$$

where, as before, $\mathbf{f}$ is the object-space frequency vector. The noncoherent transfer function T_{ncoh} is the spatial Fourier transform of the impulse response S_{ncoh}:

$$T_{\text{ncoh}}(\mathbf{f}') = \int_{-\infty}^{\infty} d\mathbf{x}\, S_{\text{ncoh}}(\mathbf{x}) \exp[-i2\pi(\mathbf{x} \cdot \mathbf{f}')]. \tag{6-54}$$

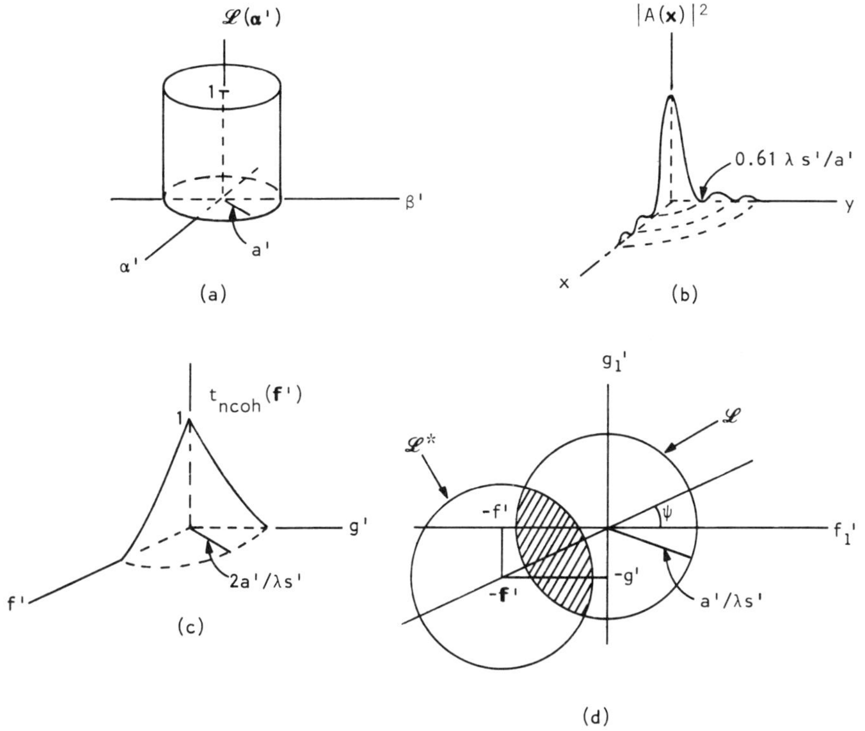

Fig. 6-4. Case of the unaberrated lens of clear aperture of radius a'. (a) The lens function $\mathcal{L}(\boldsymbol{\alpha}')$. ($b$) Impulse response $S_{\text{ncoh}} = (4\pi/k^2)|A|^2 \propto \text{Besinc}^2$. ($c$) Optical transfer function t_{ncoh}. (d) Geometry for calculating the complex autocorrelation function of Eq. (6-55). The lens is shown in the frequency domain $\mathbf{f}_1'$; the region of integration is shown shaded.

Observe that, by definition [Eq. (6-50)], S_{ncoh} has units of $(\text{area})^{-1}$ and T_{ncoh} is unitless. Since S_{ncoh} is related to $|A|^2$, the transfer function may be expressed in terms of the exit pupil function $\mathcal{L}$,

$$T_{\text{ncoh}}(\mathbf{f}') = \frac{4\pi}{k^2}\left(\frac{s'}{s}\right)^2 \int d\mathbf{f}_1'\, \mathcal{L}(\lambda s'\mathbf{f}_1')\, \mathcal{L}^*[\lambda s'(\mathbf{f}_1' + \mathbf{f}')]. \qquad (6\text{-}55)$$

Or in terms of the exit pupil variable $\boldsymbol{\alpha}' = \lambda s'\mathbf{f}'$, we may write

$$T_{\text{ncoh}}\left(\frac{\boldsymbol{\alpha}'}{\lambda s'}\right) = \frac{1}{\pi s^2}\int d\boldsymbol{\alpha}_1'\, \mathcal{L}(\boldsymbol{\alpha}_1')\, \mathcal{L}^*(\boldsymbol{\alpha}_1' + \boldsymbol{\alpha}'). \qquad (6\text{-}56)$$

This integration is over the common area shown shaded in Fig. 6-4d. In this form, T_{ncoh} is a complex autocorrelation of the exit pupil function $\mathcal{L}$. As such, it assumes its maximum value at the origin, $\mathbf{f}' = 0$. Also, since the function $\mathfrak{L}$ is zero outside the physical extent of the pupil, T_{ncoh} goes to zero when the shift variable $\boldsymbol{\alpha}'$ becomes larger than, for example, the diameter of the circular pupil. With these definitions of the spatial Fourier transforms of the relevant functions, the imaging equation in the spatial frequency domain reads

$$\mathring{I}_{im}(\mathbf{f}', \nu) = \mathring{I}_{ob}(m\mathbf{f}', \nu)\, T_{ncoh}(\mathbf{f}'). \tag{6-57}$$

In frequency space the equation of image formation reduces to a simple product relationship. Each spatial frequency component in the object is modified in amplitude and phase by the transfer function T_{ncoh}. Its behavior is such that it favors the low frequencies in the immediate vicinity of $\mathbf{f}' = 0$ and cuts off the object spectrum beyond a certain $\mathbf{f}'_{max}$. Thus the high-frequency structure is absent in the image and the low-frequency components are weighted down by T_{ncoh} even for a system with a clear aperture and free of aberrations. The image is thus a smoothed-out version of the object. On the other hand, in the coherent case the frequencies that arrive in the image pass the unaberrated system with a clear aperture without change.

Since light detectors (including the human eye) respond to optical intensity, a more interesting interpretation of T_{ncoh} is possible. Since, like optical intensity, the OSI [e.g., $\hat{I}_{im}(\mathbf{x}, \nu)$] is not only real but also nonnegative, each spatial frequency structure in the object contains the omnipresent constant background, or what is sometimes called the "dc component." The spatial frequency structure constitutes a variation of $\hat{I}_{ob}$ relative to this background. Consider a high-contrast spatial frequency structure as an object. That is, its dark areas are perfectly dark [zeros of $\hat{I}_{ob}(\boldsymbol{\xi}, \nu)$]. The system transfer function changes the strength of the frequency component relative to the value $T_{ncoh}(0)$ at the dc component. Visually, such a spatial frequency that appears in the image will show loss of contrast. The dark areas will no longer be as dark and the frequency component will undergo a shift proportional to the phase distortion. Within the system passband, the zeros of T_{ncoh} correspond to frequencies that are absent in the image; visually they will show no contrast. Therefore, the *normalized* transfer function, t_{ncoh}, is generally more relevant and is defined by

$$t_{ncoh}(\mathbf{f}') = \frac{T_{ncoh}(\mathbf{f}')}{T_{ncoh}(0)}, \tag{6-58}$$

such that $t_{ncoh}(0) = 1$. It is called the *optical transfer function* (OTF). Its absolute value, $|t_{ncoh}|$, is called the *modulation transfer function* (MTF). Figure 6-4 shows a sketch of the impulse response and the OTF for an unaberrated lens with a clear aperture. For a discussion of different shaped exit pupils, we refer to Gaskill (1978) and Goodman (1968). For the calculation and plots of the transfer function for the special case of defocusing and in the presence of astigmatism, we refer to the pioneering work of Hopkins (1955a) and De (1955). For these and several other cases of Seidel and fifth-order aberrations in both the physical and geometrical optical treatment, we refer to Goodbody (1958), Bromilow (1958), Marathay (1959), and O'Neill (1956, 1963).

Since the complex autocorrelation of Eq. (6-56) involves a finite lens aperture (e.g., circular of radius a'), the transfer function T_{ncoh} will go to zero and stay zero after a maximum spatial frequency given by

$$f'_{max} = g'_{max} = \frac{2a'}{\lambda s'} = \frac{\theta'}{\lambda}. \tag{6-59}$$

The angle θ' is the image-forming angle defined in Fig. 6-3. The maximum spatial frequency passed by the lens working with noncoherent illumination is twice that of the similar quantity defined for the same lens used with coherent illumination. We shall return to this point later.

As in the coherent case, it is also true in the noncoherent case that for a fixed s' a larger lens aperture will pass higher spatial frequencies, leading to better resolution. The maximum spatial frequency passed by the lens increases with increasing image-forming angle θ' and/or decreasing wavelength λ. For an especially interesting case of photographic objectives or telescopes, if the distance s' is very nearly equal to the focal length f (if $s' \simeq f$), then the image-forming angle θ approximates $1/F\#$, where $F\#$ is (focal length)/diameter. Then we arrive at the familiar relationship: maximum spatial frequency $= 1/\lambda F\#$. It does not have general validity. It is the image-forming angle θ' that is of primary importance.

The statement that the maximum spatial frequency passed by the lens in the noncoherent case is twice that of the coherent case calls for a comment. Careful consideration must be given to attempting a comparison of the two cases. For one thing, the coherent case is linear in wave amplitude, which is not an observable. Thus, a direct interpretation of T_{coh} in terms of the observed contrast of spatial frequencies is not possible as done with T_{ncoh}. On the other hand, in terms of the optical intensity or OSI, the coherent system is highly nonlinear, in which case T_{coh} loses significance.

To understand the effect of this nonlinearity, consider two objects transilluminated by coherent light with amplitude transmittance: (1) $t_{ob} = \cos 2\pi f_{ob}\xi$ and (2) $t_{ob} = |\cos 2\pi f_{ob}\xi|$. The first object may be thought of as derived from the second by use of $\lambda/2$ phase strips at alternate maxima. The spatial frequency contents of the two objects are markedly different. In the first there are only two frequencies, $\pm f_{ob}$, and no dc component. The second object has a background level of $2/\pi$ and a multitude of frequencies with $2f_{ob}$ as the fundamental frequency. If the illumination is changed to the noncoherent type, the relevant quantity for image formation is the optical intensity transmittance, $|t_{ob}|^2 = \cos^2 2\pi f_{ob}\xi = (1 + \cos 2\pi 2 f_{ob}\xi)/2$, for either object. It contains a dc component and two frequencies, $\pm 2f_{ob}$.

Now consider an unaberrated lens system with a clear aperture which passes a maximum frequency f'_{max} with coherent light such that

$$f_{ob} < f'_{max} < 2f_{ob}. \tag{6-60}$$

The first object is clearly admitted with either coherent or noncoherent illumination although the contrast in the noncoherent case will be lower than that in the coherent case. Under these circumstances, the coherent mode may be regarded as better than the noncoherent mode of operation. The second object, however, is not resolved ($2f_{ob} > f'_{max}$) with coherent light but is visible ($2f_{ob} < 2f'_{max}$) with some contrast in noncoherent illumination. In this instance the noncoherent illumination is clearly better. Thus although mathematically the cutoff frequency differs by a numerical factor of 2 between the two illumination types, what we mean by a "better" system is object dependent, due to the nonlinearity (see Problems 6-4 and 6-5).

The use of step-function objects or sine-wave objects reveals further differences between the two modes of operation of the system. We shall postpone further comparison, however, until we deal with the partially coherent mode of operation in the next section.

IMAGE FORMATION IN PARTIALLY COHERENT POLYCHROMATIC LIGHT

We always seek to describe optical systems in terms of a linear theory. In the case of coherent illumination we found that the system was linear in terms of the wave amplitude $U_{ob}(\boldsymbol{\xi})$, whereas with noncoherent light we found that linearity was maintained by using the OSI $\hat{I}_{ob}(\boldsymbol{\xi}, \nu)$ to describe the object and a similar quantity for the image. With partially coherent polychromatic light, the system is linear in neither of these two quantities. But as already seen in Eq. (6-22), namely,

$$\begin{aligned}\hat{\Gamma}_{im}(\mathbf{x}_1, \mathbf{x}_2, \nu) = \exp\left[\frac{ik}{2s'}(\mathbf{x}_1 \cdot \mathbf{x}_1 - \mathbf{x}_2 \cdot \mathbf{x}_2)\right] \\ \times \int\int d\boldsymbol{\xi}_1\, d\boldsymbol{\xi}_2\, \hat{\Gamma}_{ob}(\boldsymbol{\xi}_1, \boldsymbol{\xi}_2, \nu) \exp\left[\frac{ik}{2s}(\boldsymbol{\xi}_1 \cdot \boldsymbol{\xi}_1 - \boldsymbol{\xi}_2 \cdot \boldsymbol{\xi}_2)\right] \\ \times A(\mathbf{x}_1 - m\boldsymbol{\xi}_1)\, A^*(\mathbf{x}_2 - m\boldsymbol{\xi}_2),\end{aligned}$$

linearity is achieved in terms of the MSDF $\hat{\Gamma}_{\text{ob}}(\xi_1, \xi_2, \nu)$ to describe the object and a similar function for the image.

Conditions (C1), (C2), and (C3) discussed in relation to the coherent case may be used to advantage to limit our attention to spatially stationary (shift-invariant) image formation,

$$\hat{\Gamma}_{\text{im}}(\mathbf{x}_1, \mathbf{x}_2, \nu) = \int\int d\xi_1\, d\xi_2\, \hat{\Gamma}_{\text{ob}}(\xi_1, \xi_2, \nu)\, A(\mathbf{x}_1 - m\xi_1)\, A^*(\mathbf{x}_2 - m\xi_2). \tag{6-61}$$

In the coherent and noncoherent cases the object and image were described by functions of a single variable. In the partially coherent case the description is with two variables through the MSDF. Two steps are needed to find the impulse response. First, we set the MSDF proportional to $\delta(\xi_1 - \xi_2)$, which corresponds to a noncoherent or a purely noisy object. Second, we isolate the radiation from a point, say at ξ_0, of this noncoherent object by use of $\delta(\xi_1 - \xi_0)$. The procedure adopted in the noncoherent case, Eq. (6-49), suggests that the double impulse necessary in the present case be written in the form

$$\hat{\Gamma}_{\text{ob}}(\xi_1, \xi_2, \nu) = \frac{4\pi}{k^2}\mathcal{Q}\hat{I}_0(\nu)\, \delta(\xi_1 - \xi_0)\, \delta(\xi_1 - \xi_2), \tag{6-62}$$

where $\mathcal{Q}$ is the area of the pinhole and $\hat{I}_0(\nu)$ is the spectral dependence of the MSDF. This same form may be arrived at by asking for the double impulse in terms of the average, $(\xi_1 + \xi_2)/2$, and difference, $\xi_1 - \xi_2$, variables. The impulse response S_{pcoh} is defined as the image obtained for a unit impulse, $(4\pi/k^2)\mathcal{Q}\hat{I}_0 = 1$, at the origin ($\xi_0 = 0$), for a shift-invariant system:

$$S_{\text{pcoh}}(\mathbf{x}_1, \mathbf{x}_2, \nu) = A(\mathbf{x}_1)\, A^*(\mathbf{x}_2). \tag{6-63}$$

The frequency variable is displayed in the response but is implicit in A through its definition. The factored form of the response implies coherence in the image plane. Thus, as is well known, it shows that the radiation from an isolated point of a noncoherent source is coherent with itself.

In terms of the impulse response S_{pcoh}, the imaging equation assumes the form

$$\hat{\Gamma}_{\text{im}}(\mathbf{x}_1, \mathbf{x}_2, \nu) = \int\int d\xi_1\, d\xi_2\, \hat{\Gamma}_{\text{ob}}(\xi_1, \xi_2, \nu)\, S_{\text{pcoh}}(\mathbf{x}_1 - m\xi_1, \mathbf{x}_2 - m\xi_2, \nu). \tag{6-64}$$

The transfer function T_{pcoh} of this system is defined through the *double* Fourier transform

$$T_{\text{pcoh}}(\mathbf{f}'_1, \mathbf{f}'_2, \nu) = \int\int d\mathbf{x}_1\, d\mathbf{x}_2\, S_{\text{pcoh}}(\mathbf{x}_1, \mathbf{x}_2, \nu) \exp[-i2\pi(\mathbf{f}'_1 \cdot \mathbf{x}_1 - \mathbf{f}'_2 \cdot \mathbf{x}_2)], \tag{6-65}$$

which, due to the definition of A in Eq. (6-21), reduces to

$$T_{\text{pcoh}}(\mathbf{f}'_1, \mathbf{f}'_2, \nu) = \left(\frac{s'}{s}\right)^2 \mathfrak{L}(-\lambda s'\mathbf{f}'_1)\, \mathfrak{L}^*(-\lambda s'\mathbf{f}'_2). \tag{6-66}$$

Now the spatial frequency description of the object $\hat{\Gamma}_{\text{ob}}$ is known through its total Fourier transform $\mathring{\Gamma}_{\text{ob}}$, defined by

$$\hat{\Gamma}_{\text{ob}}(\boldsymbol{\xi}_1, \boldsymbol{\xi}_2, \nu) = \int\int d\mathbf{f}_1\, d\mathbf{f}_2\, \mathring{\Gamma}_{\text{ob}}(\mathbf{f}_1, \mathbf{f}_2, \nu) \exp[+i2\pi(\boldsymbol{\xi}_1 \cdot \mathbf{f}_1 - \boldsymbol{\xi}_2 \cdot \mathbf{f}_2)], \tag{6-67}$$

by using the object-space frequency vectors $\mathbf{f}_1$ and $\mathbf{f}_2$. Then, as would be expected, image formation in the domain of spatial frequencies reduces to a product relationship

$$\mathring{\Gamma}_{\text{im}}(\mathbf{f}'_1, \mathbf{f}'_2, \nu) = \mathring{\Gamma}_{\text{ob}}(m\mathbf{f}'_1, m\mathbf{f}'_2, \nu)\, T_{\text{pcoh}}(\mathbf{f}'_1, \mathbf{f}'_2, \nu), \tag{6-68}$$

where $\mathring{\Gamma}_{\text{im}}$ describes the spatial frequency content of the image analogous to Eq. (6-67).

Thus we have succeeded in casting the theory of image formation with partially coherent polychromatic light into much the same form as we did in the coherent and noncoherent modes of operation. However, the description contains a pair of spatial frequency vectors, $\mathbf{f}'_1$ and $\mathbf{f}'_2$. They are coupled through the total spectrum of the object, $\mathring{\Gamma}_{\text{ob}}(m\mathbf{f}'_1, m\mathbf{f}'_2, \nu)$. The system forming the image modifies this coupling to $\mathring{\Gamma}_{\text{im}}(\mathbf{f}'_1, \mathbf{f}'_2, \nu)$ through the use of the transfer function T_{pcoh}. On the other hand, in coordinate space, Eq. (6-64), light from a pair of points at $\boldsymbol{\xi}_1$ and $\boldsymbol{\xi}_2$ in the object plane is coupled through $\hat{\Gamma}_{\text{ob}}(\boldsymbol{\xi}_1, \boldsymbol{\xi}_2, \nu)$. The statistics of the object and its illumination are contained in $\hat{\Gamma}_{\text{ob}}$; the process of image formation enters through the functions A and A^* in a completely deterministic manner. In the coherent limit the system decouples due to the factorization of $\hat{\Gamma}_{\text{ob}}$ [see Eq. (6-24)], whereas in the noncoherent limit the two space variables coalesce into one variable $\boldsymbol{\xi}$ with a corresponding single spatial frequency.

Hopkins' Effective Source

Applications of coherence theory in practice call for approximations. The idea of the effective source is one useful example. Hopkins (1955a, b, 1957) used it extensively to study the influence of partial coherence in optical image formation and interferometry (see also Baker, 1955).

The study of image formation is most conveniently carried out with reference to the optical system of a microscope, mainly because its illumination system is readily adjustable. The object, namely, the microscope slide, is transilluminated; or for opaque objects a suitable reflection system is used. The illumination is kept uniform and the coherence width of the spatially stationary coherence function is varied over a certain range. The image is studied as the condition of partial coherence is adjusted over this range.

There are basically two types of illumination systems (see Born and Wolf, 1970). In one, called the *critical* illumination, a noncoherent primary source is imaged by a condenser system on the object plane, namely, the microscope slide. The nonuniformities of the source, however, get transferred to the object. This difficulty is removed in the second system, called *Köhler* illumination. In this the primary source is imaged by an auxiliary lens onto the front focal plane of a condenser. The microscope slide is situated in the back focal plane of the condenser. The state of partial coherence of light illuminating the microscope slide is equally controllable in either system.

In order to introduce the concept of the effective source, let us consider the critical illumination system. Let $\mathsf{C}\hat{I}_s(\boldsymbol{\xi}, \nu)$ be the spectral radiant exitance [W m^{-2} Hz^{-1}] of the primary source. Now Eq. (6-46) may be used to describe the state of noncoherence with $\hat{I}_{\text{ob}}$ replaced by $\hat{I}_s$. The state of coherence $\hat{\Gamma}$ on the microscope slide is then found through Eq. (6-47), namely,

$$\hat{\Gamma}(\mathbf{x}_1, \mathbf{x}_2, \nu) = \frac{4\pi}{k^2} \exp\left[\frac{ik}{2s'}(\mathbf{x}_1 \cdot \mathbf{x}_1 - \mathbf{x}_2 \cdot \mathbf{x}_2)\right]$$
$$\times \int d\boldsymbol{\xi}\, \hat{I}_s(\boldsymbol{\xi}, \nu)\, A(\mathbf{x}_1 - m\boldsymbol{\xi})\, A^*(\mathbf{x}_2 - m\boldsymbol{\xi}), \qquad (6\text{-}69)$$

which is, in general, a spatially nonstationary MSDF. The function A now refers to the condenser of the illumination system. The expression of A in terms of the system function $\mathfrak{L}$, Eq. (6-17), allows us to isolate the integral on $\boldsymbol{\xi}$ as follows:

$$\hat{\Gamma}(\mathbf{x}_1, \mathbf{x}_2, \nu) = \frac{\lambda^2}{\pi} \exp\left[\frac{ik}{2s'}(\mathbf{x}_1 \cdot \mathbf{x}_1 - \mathbf{x}_2 \cdot \mathbf{x}_2)\right] \frac{1}{(\lambda^2 ss')^2}$$
$$\times \iint d\boldsymbol{\alpha}_1'\, d\boldsymbol{\alpha}_2'\, \mathfrak{L}(\boldsymbol{\alpha}_1')\, \mathfrak{L}^*(\boldsymbol{\alpha}_2')$$
$$\times \exp\left[-i\frac{2\pi}{\lambda s'}(\mathbf{x}_1 \cdot \boldsymbol{\alpha}_1' - \mathbf{x}_2 \cdot \boldsymbol{\alpha}_2')\right]$$
$$\times \int d\boldsymbol{\xi}\, \hat{I}_s(\boldsymbol{\xi}, \nu) \exp\left[+i\frac{2\pi m}{\lambda s'}(\boldsymbol{\alpha}_1' - \boldsymbol{\alpha}_2') \cdot \boldsymbol{\xi}\right]. \qquad (6\text{-}70)$$

It contains the condenser exit pupil variables, $\boldsymbol{\alpha}'$, which may be converted to those in the entrance pupil by $\boldsymbol{\alpha}' = m_p\boldsymbol{\alpha}$; then the $\boldsymbol{\xi}$ integral becomes

$$\int d\boldsymbol{\xi}\, \hat{I}_s(\boldsymbol{\xi}, \nu) \exp\left[-i\frac{2\pi}{\lambda s}(\boldsymbol{\alpha}_1 - \boldsymbol{\alpha}_2) \cdot \boldsymbol{\xi}\right],$$

where $s' = -mm_p s$ was also used. This is just the form of the coherence function, MSDF, in the entrance pupil, apart from some quadratic phase factors, that would be obtained by applying the van Cittert–Zernike theorem, Eq. (5-41), to a single spectral component ν of the primary noncoherent source. Depending on the physical extent of the source, the integral will be a rather narrow function of the coordinate difference $\boldsymbol{\alpha}_1 - \boldsymbol{\alpha}_2$. For a large enough primary source, we may make the coherence area in the entrance pupil as small as we please. The essential nature of the approximation is to regard the primary source as very large so that the entrance pupil and hence also the exit pupil may be regarded as effectively noncoherently illuminated. A source of 1-cm diameter at a distance of 10 cm has an angular subtense θ of 0.1 at the entrance pupil, which offers a coherence width [Eq. (5-51)] of only 5λ. Consider for the moment that the source is also uniform, $\hat{I}_s(\boldsymbol{\xi}, \nu) = \hat{I}_s(\nu)$ independent of $\boldsymbol{\xi}$, with whatever shape as long as it is large. We shall account for its possible nonuniformity later on. Thus with this "large uniform source" we may approximate the $\boldsymbol{\xi}$ integral in Eq. (6-70) by a delta function, namely,

$$\int d\boldsymbol{\xi}\, \hat{I}_s(\boldsymbol{\xi}, \nu) \exp\left[+i\frac{2\pi m}{\lambda s'}(\boldsymbol{\alpha}_1' - \boldsymbol{\alpha}_2') \cdot \boldsymbol{\xi}\right] \simeq \left(\frac{\lambda s'}{m}\right)^2 \hat{I}_s(\nu)\, \delta(\boldsymbol{\alpha}_1' - \boldsymbol{\alpha}_2'). \tag{6-71}$$

With this approximation the MSDF in the plane of the microscope slide takes the form

$$\hat{\Gamma}(\mathbf{x}_1, \mathbf{x}_2, \nu) = \frac{1}{\pi s'^2} \exp\left[\frac{ik}{2s'}(\mathbf{x}_1 \cdot \mathbf{x}_1 - \mathbf{x}_2 \cdot \mathbf{x}_2)\right] \times \int d\boldsymbol{\alpha}'\, m_p^2 \hat{I}_s(\nu)\, |\mathfrak{L}(\boldsymbol{\alpha}')|^2 \exp\left[-i\frac{2\pi}{\lambda s'}(\mathbf{x}_1 - \mathbf{x}_2) \cdot \boldsymbol{\alpha}'\right]. \tag{6-72}$$

By applying the van Cittert–Zernike theorem this is precisely what we would obtain if the illumination system were to be replaced by an *effective source*,

$$\hat{I}_{\text{eff}}(\boldsymbol{\alpha}', \nu) = m_p^2 \hat{I}_s(\nu) \,|\, \mathfrak{L}(\boldsymbol{\alpha}') \,|^2, \tag{6-73}$$

at a distance of s' in front of the microscope slide and with size and shape the same as the exit pupil of the condenser. Observe that the state of coherence is independent of the condenser aberrations, a result first established by Zernike (1938). Furthermore, observe that the effective source need not be defined exclusively in terms of the exit pupil, since we may locate it anywhere on the optical axis, even at the entrance pupil of the microscope objective (Hopkins, 1957), to obtain the same state of coherence, provided we scale it appropriately to achieve the same angular subtense. The van Cittert–Zernike theorem, however, implies a uniform optical spectral intensity (OSI) distribution on the microscope slide. To account for the possible nonuniformity of the illumination (as found with critical illumination), we simply imagine, following Hopkins, that a nonuniform absorber is inserted in front of the slide. However, for the discussion in what is to follow we shall assume the object is uniformly illuminated.

The layout of the optical system of the microscope is shown in Fig. 6-5. It shows the coordinate symbols and the refractive indexes of the intervening spaces to be used in the following analysis.

The presence of the refractive index other than unity brings about minor changes in the previous equations. For one thing, the propagation constant k gets multiplied by the refractive index, whereas λ is retained for the vacuum wavelength. The definition of A in Eq. (6-17) acquires the factor nn' outside the integral. The spatial frequency variables with the appropriate indexes are listed in the caption of Fig. 6-5.

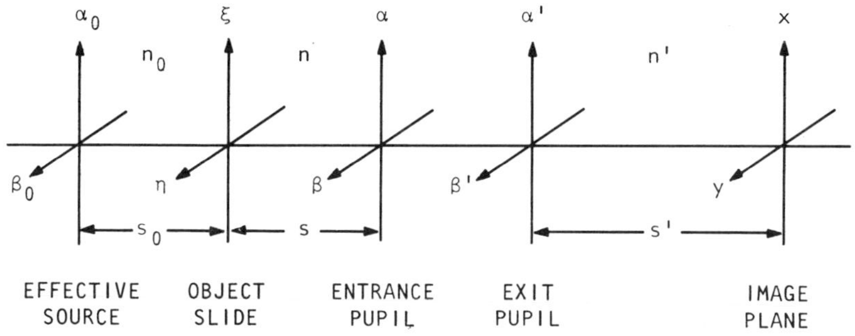

Fig. 6-5. Optical system layout of the microscope. n_0, n, and n' are the refractive indexes of the intervening spaces. Radii of the effective source, entrance pupil, and exit pupil, respectively, are a_0, a, and a'. Pupil magnification: $m_p = \alpha'/\alpha = \beta'/\beta = a'/a$. Image magnification: $m = x/\xi = y/\eta$. Spatial frequency variables: $\mathbf{f}_0 = \boldsymbol{\alpha}_0 n_0/\lambda s_0$, $\mathbf{f} = \boldsymbol{\alpha} n/\lambda s$, $\mathbf{f}' = \boldsymbol{\alpha}' n'/\lambda s'$. Since $ns'/n's = -mm_p$, we retain $\mathbf{f} = -m\mathbf{f}'$.

With reference to Fig. 6-5, if the medium has refractive index n_0, the MSDF on the slide may be expressed by

$$\hat{\Gamma}(\boldsymbol{\xi}_1 - \boldsymbol{\xi}_2, \nu) = \frac{1}{\pi s_0^2} \int d\boldsymbol{\alpha}_0 \, \hat{I}_{\text{eff}}(\boldsymbol{\alpha}_0, \nu) \exp\left[-i2\pi(\boldsymbol{\xi}_1 - \boldsymbol{\xi}_2) \cdot \frac{\boldsymbol{\alpha}_0 n_0}{\lambda s_0} \right]. \tag{6-74}$$

The quadratic phase factor has been approximated to unity. In this relationship the $\boldsymbol{\alpha}_0$ are coordinates in the plane of the effective source situated at a distance s_0.

The MSDF $\hat{\Gamma}$ and $\hat{I}_{\text{eff}}$ form a Fourier pair—a fact that enables us to indulge in more abstraction. To any state of coherence, as described by a function of the coordinate difference, which has crossing symmetry $\hat{\Gamma}(\boldsymbol{\xi}_1 - \boldsymbol{\xi}_2, \nu) = \hat{\Gamma}^*(\boldsymbol{\xi}_2 - \boldsymbol{\xi}_1, \nu)$, there corresponds an effective source distribution $\hat{I}_{\text{eff}}$ by virtue of the Fourier inversion. Because of the crossing symmetry, $\hat{I}_{\text{eff}}$ is real but there is no assurance that it is nonnegative. For this reason we also ask that the mathematical form of $\hat{\Gamma}$ be that of a complex autocorrelation function, so that its spatial Fourier transform is guaranteed to be nonnegative. The effective source is then optically realizable and has an interesting interpretation. Each point of the noncoherent effective source radiates spherical waves, which we have approximated by plane waves. They are mutually noncoherent. All problems in optics for which the MSDF is spatially stationary may be reduced to the noncoherent superposition of these plane waves, each weighted by the appropriate ordinate of the $\hat{I}_{\text{eff}}$ function as dictated by the Fourier integral of Eq. (6-74). This interpretation suggests that the contribution of each plane wave in the problem at hand may be obtained and the individual contributions may be noncoherently superimposed (addition of intensities) with the appropriate weighting in order to understand the effect of an extended noncoherent source without reference to the intermediate spatial coherence function. In fact, this is precisely how such problems were dealt with prior to the formulation of the theory of partial coherence (see, e.g., Jenkins and White, 1976, Art. 16.7). Thus the method of the summation of OSI values, each contributed by a point on the effective source, shows that *a wide class of optical problems may be solved without the explicit use of statistical averaging.*

Alternatively, the effective source concept discussed with reference to Eq. (6-74) may be interpreted in the domain of spatial frequencies, $\mathbf{f}_0 = \boldsymbol{\alpha}_0 n_0 / \lambda s_0$. The coherent limit may be achieved by use of a point source, from which we may derive, by collimation, a single plane wave to illuminate the object. It has only a *single* direction and, in particular, if it is propagating along the optical axis, we have $\hat{I}_{\text{eff}}(\lambda s_0 \mathbf{f}_0 / n_0, \nu) \simeq \delta(\mathbf{f}_0)$, leading to

$|\hat{\Gamma}(\xi_1 - \xi_2, \nu)| \simeq$ constant. The limit of noncoherence, $\hat{\Gamma}(\xi_1 - \xi_2, \nu) \simeq \delta(\xi_1 - \xi_2)$, is reached when $\hat{I}_{\text{eff}}(\lambda s_0 \mathbf{f}_0 / n_0, \nu) \simeq$ constant. It corresponds to a *flat* spatial frequency spectrum for noncoherence as discussed in Chapter 4. Due to the one-to-one correspondence between $\boldsymbol{\alpha}_0$ and $\mathbf{f}_0$, the calculations may be performed just as effectively in coordinate space $\boldsymbol{\alpha}_0$ as in frequency space $\mathbf{f}_0$.

Effective Source and Imaging

To appreciate the usefulness of the effective source concept, we shall consider image formation in partially coherent light with the optical spectral intensity as the measured quantity in both the object and image. With this approach, the property of linearity [Eq. (6-64)] is lost. Nevertheless, we shall see how elegantly the approach reveals the spatial frequency content of the image.

We refer back to Fig. 6-5. Let us suppose that the object has an amplitude transmittance $\hat{t}_{\text{ob}}(\xi, \nu)$ and is transilluminated uniformly with the MSDF, $\hat{\Gamma}(\xi_1 - \xi_2, \nu)$, given in Eq. (6-74). In the object plane, we have [see, e.g., Eq. (6-8)]

$$\hat{\Gamma}_{\text{ob}}(\xi_1, \xi_2, \nu) = \hat{\Gamma}(\xi_1 - \xi_2, \nu)\, \hat{t}_{\text{ob}}(\xi_1, \nu)\, \hat{t}^*_{\text{ob}}(\xi_2, \nu), \tag{6-75}$$

and the OSI, $\hat{I}_{\text{ob}}(\xi, \nu)$, transmitted by it is the zero ordinate,

$$\hat{I}_{\text{ob}}(\xi, \nu) = \hat{\Gamma}(0, \nu)\, |\hat{t}_{\text{ob}}(\xi, \nu)|^2. \tag{6-76}$$

The associated spatial frequencies $\mathbf{f} = \boldsymbol{\alpha} n / \lambda s$ in object space may be used to show its frequency content,

$$\hat{I}_{\text{ob}}(\xi, \nu) = \hat{\Gamma}(0, \nu) \iint d\mathbf{f}_1\, d\mathbf{f}_2\, \mathring{t}_{\text{ob}}(\mathbf{f}_1, \nu)\, \mathring{t}^*_{\text{ob}}(\mathbf{f}_2, \nu) \exp[+i2\pi \xi \cdot (\mathbf{f}_1 - \mathbf{f}_2)]. \tag{6-77}$$

The OSI $\hat{I}_{\text{im}}(\mathbf{x}, \nu)$ in the image is found through Eq. (6-61) by using the form of the object in Eq. (6-75), namely,

$$\hat{I}_{\text{im}}(\mathbf{x}, \nu) = \iint d\xi_1\, d\xi_2\, \hat{\Gamma}(\xi_1 - \xi_2, \nu)\, \hat{t}_{\text{ob}}(\xi_1, \nu)\, \hat{t}^*_{\text{ob}}(\xi_2, \nu) \times A(\mathbf{x} - m\xi_1)\, A^*(\mathbf{x} - m\xi_2). \tag{6-78}$$

We express the coherence function of the illumination in terms of the effective source distribution of Eq. (6-74) and the A functions in terms of the image-space frequencies $\mathbf{f}'$ from Eq. (6-21) to get

$$\hat{I}_{\text{im}}(\mathbf{x}, \nu) = \int\int d\boldsymbol{\xi}_1\, d\boldsymbol{\xi}_2\, \hat{t}_{\text{ob}}(\boldsymbol{\xi}_1, \nu)\, \hat{t}^*_{\text{ob}}(\boldsymbol{\xi}_2, \nu)$$
$$\times \left\{ \frac{1}{\pi s_0^2} \int d\boldsymbol{\alpha}_0\, \hat{I}_{\text{eff}}(\boldsymbol{\alpha}_0, \nu) \exp[-i2\pi(\boldsymbol{\xi}_1 - \boldsymbol{\xi}_2) \cdot \mathbf{f}_0] \right\}$$
$$\times \left\{ \frac{ns'}{n's} \int d\mathbf{f}'_3\, \mathfrak{L}\left(\frac{\lambda s' \mathbf{f}'_3}{n'} \right) \exp[-i2\pi(\mathbf{x} - m\boldsymbol{\xi}_1) \cdot \mathbf{f}'_3] \right\}$$
$$\times \left\{ \frac{ns'}{n's} \int d\mathbf{f}'_4\, \mathfrak{L}^*\left(\frac{\lambda s' \mathbf{f}'_4}{n'} \right) \exp[+i2\pi(\mathbf{x} - m\boldsymbol{\xi}_2) \cdot \mathbf{f}'_4] \right\}, \tag{6-79}$$

wherein the frequencies associated with the effective-source variables are $\mathbf{f}_0 = \boldsymbol{\alpha}_0 n_0/\lambda s_0$. Now the $\boldsymbol{\xi}$ integrals are, in fact, the spatial Fourier transforms over the object structure; for example,

$$\int d\boldsymbol{\xi}_1\, \hat{t}_{\text{ob}}(\boldsymbol{\xi}_1, \nu) \exp[-i2\pi\boldsymbol{\xi}_1 \cdot (\mathbf{f}_0 - m\mathbf{f}'_3)] = \mathring{t}_{\text{ob}}(\mathbf{f}_0 - m\mathbf{f}'_3, \nu), \tag{6-80}$$

with which we may convert Eq. (6-79) to

$$\hat{I}_{\text{im}}(\mathbf{x}, \nu) = \frac{1}{\pi s_0^2} \left(\frac{ns'}{n's} \right)^2 \int d\boldsymbol{\alpha}_0 \int\int d\mathbf{f}'_3\, d\mathbf{f}'_4\, \hat{I}_{\text{eff}}(\boldsymbol{\alpha}_0, \nu)\, \mathfrak{L}\left(\frac{\lambda s' \mathbf{f}'_3}{n'} \right) \mathfrak{L}^*\left(\frac{\lambda s' \mathbf{f}'_4}{n'} \right)$$
$$\times \mathring{t}_{\text{ob}}(\mathbf{f}_0 - m\mathbf{f}'_3, \nu)\, \mathring{t}^*_{\text{ob}}(\mathbf{f}_0 - m\mathbf{f}'_4, \nu) \exp[-i2\pi\mathbf{x} \cdot (\mathbf{f}'_3 - \mathbf{f}'_4)].$$

A change of variables, $m\mathbf{f}'_1 = \mathbf{f}_0 - m\mathbf{f}'_3$ and $m\mathbf{f}'_2 = \mathbf{f}_0 - m\mathbf{f}'_4$, enables us to separate the object structure from the system functions as in

$$\hat{I}_{\text{im}}(\mathbf{x}, \nu) = \int\int d\mathbf{f}'_1\, d\mathbf{f}'_2\, \mathring{t}_{\text{ob}}(m\mathbf{f}'_1, \nu)\, \mathring{t}^*_{\text{ob}}(m\mathbf{f}'_2, \nu) \exp[+i2\pi\mathbf{x} \cdot (\mathbf{f}'_1 - \mathbf{f}'_2)]$$
$$\times \frac{1}{\pi s_0^2} \left(\frac{ns'}{n's} \right)^2 \int d\boldsymbol{\alpha}_0\, \hat{I}_{\text{eff}}(\boldsymbol{\alpha}_0, \nu)\, \mathfrak{L}\left[\frac{\lambda s'}{n'm} (\mathbf{f}_0 - m\mathbf{f}'_1) \right]$$
$$\times \mathfrak{L}^*\left[\frac{\lambda s'}{n'm} (\mathbf{f}_0 - m\mathbf{f}'_2) \right]. \tag{6-81}$$

The integral over $\boldsymbol{\alpha}_0$ is a function of a pair of image-space frequency vectors $\mathbf{f}'_1$ and $\mathbf{f}'_2$. We plan to define a normalized function, $C(\mathbf{f}'_1, \mathbf{f}'_2, \nu)$, in terms of it. But first let's rewrite the argument of $\mathfrak{L}$ in terms of the exit pupil variables $\boldsymbol{\alpha}'$,

$$\mathfrak{L}\left[\frac{\lambda s'}{n'm}(\mathbf{f}_0 - m\mathbf{f}_1')\right] = \mathfrak{L}\left(\frac{s'n_0}{s_0 n'm}\boldsymbol{\alpha}_0 - \boldsymbol{\alpha}_1'\right).$$

The argument now contains a scaled version of the alien parameter $\boldsymbol{\alpha}_0$, which belongs to the effective source. The scaling constant may be expressed in terms of the ratio σ of the numerical aperture of the condenser to that of the microscope objective,

$$\sigma = \frac{(\mathrm{N.A.})_c}{(\mathrm{N.A.})_o} = \frac{n_0 \sin(\theta_0/2)}{n \sin(\theta/2)}, \tag{6-82}$$

where θ_0 and θ are, respectively, the angles subtended by the condenser and the objective to the origin of the ξ-η plane. Up to the quadratic approximation, we have

$$\sigma \simeq \frac{n_0 a_0 s}{n a s_0} = -\frac{a_0}{a'}\frac{s'n_0}{s_0 n'm}.$$

Thus the argument of $\mathfrak{L}$ may be shown in terms of the normalized coordinates,

$$\mathfrak{L}\left[\frac{\lambda s'}{n'm}(\mathbf{f}_0 - m\mathbf{f}_1')\right] = \mathfrak{L}\left[-a'\left(\frac{\boldsymbol{\alpha}_1'}{a'} + \sigma\frac{\boldsymbol{\alpha}_0}{a_0}\right)\right]. \tag{6.83}$$

Now we are ready to define the normalized function $C(\mathbf{f}_1', \mathbf{f}_2', \nu)$. A suitable normalization constant is

$$\hat{\Gamma}'(0,\nu) = \frac{1}{\pi s_0^2}\left(\frac{ns'}{n's}\right)^2 \int d\boldsymbol{\alpha}_0\, \hat{I}_{\mathrm{eff}}(\boldsymbol{\alpha}_0,\nu)\left|\mathfrak{L}\left(\frac{-a'\sigma\boldsymbol{\alpha}_0}{a_0}\right)\right|^2, \tag{6-84}$$

so we may define $C(\mathbf{f}_1', \mathbf{f}_2', \nu)$ by the relationship

$$\begin{aligned}\hat{\Gamma}'(0,\nu)\, C(\mathbf{f}_1', \mathbf{f}_2', \nu) &= \hat{\Gamma}'(0,\nu)\, C\left(\frac{\boldsymbol{\alpha}_1' n'}{\lambda s'}, \frac{\boldsymbol{\alpha}_2' n'}{\lambda s'}, \nu\right) \\ &= \frac{1}{\pi s_0^2}\left(\frac{ns'}{n's}\right)^2 \int d\boldsymbol{\alpha}_0\, \hat{I}_{\mathrm{eff}}(\boldsymbol{\alpha}_0,\nu)\, \mathfrak{L}\left(-\boldsymbol{\alpha}_1' - \frac{a'\sigma\boldsymbol{\alpha}_0}{a_0}\right) \\ &\quad \times \mathfrak{L}^*\left(-\boldsymbol{\alpha}_2' - \frac{a'\sigma\boldsymbol{\alpha}_0}{a_0}\right).\end{aligned} \tag{6-85}$$

It is such that $C(0, 0, \nu) = 1$. With these definitions the OSI in the image takes the form

$$\hat{I}_{\rm im}(\mathbf{x}, \nu) = \hat{\Gamma}'(0, \nu) \iint d\mathbf{f}'_1\, d\mathbf{f}'_2\, C(\mathbf{f}'_1, \mathbf{f}'_2, \nu)\, \mathring{t}_{\rm ob}(m\mathbf{f}'_1, \nu)\, \mathring{t}^*_{\rm ob}(m\mathbf{f}'_2, \nu) \times \exp[+i2\pi\mathbf{x} \cdot (\mathbf{f}'_1 - \mathbf{f}'_2)]. \tag{6-86}$$

This is Hopkins' final answer. Comparison of this with the object $\hat{I}_{\rm ob}(\boldsymbol{\xi}, \nu)$ of Eq. (6-77) shows that if $C(\mathbf{f}'_1, \mathbf{f}'_2, \nu) = 1$, the image will be a perfect scaled replica of the object. The function C is the response of the system to the two frequency pairs, (f'_1, g'_1) and (f'_2, g'_2). However, it is not a transfer function in the sense of $T_{\rm pcoh}$ because the problem is not linear in terms of the OSI values $\hat{I}$ of the object and image. We shall call it *Hopkins' frequency response* (HFR) function.

In order to simplify the notation, let us put

$$C_0 = \frac{(ns'/n's)^2}{\pi s_0^2 \hat{\Gamma}'(0, \nu)} \tag{6-87}$$

and study

$$C(\mathbf{f}'_1, \mathbf{f}'_2, \nu) = C_0 \int d\boldsymbol{\alpha}_0\, \hat{I}_{\rm eff}(\boldsymbol{\alpha}_0, \nu)\, \mathfrak{L}\left(-\boldsymbol{\alpha}'_1 - \frac{a'\sigma\boldsymbol{\alpha}_0}{a_0}\right) \mathfrak{L}^*\left(-\boldsymbol{\alpha}'_2 - \frac{a'\sigma\boldsymbol{\alpha}_0}{a_0}\right), \tag{6-88}$$

where $\mathbf{f}' = \boldsymbol{\alpha}'n'/\lambda s'$ with subscripts 1 and 2. It is a double convolution of $\hat{I}_{\rm eff}$ with $\mathfrak{L}$ and $\mathfrak{L}^*$. Since all three functions occupy finite areas in the coordinate plane α_0, β_0, the integration in Eq. (6-88) is restricted to the area common to all three. For a circular effective source (condenser aperture) and the system exit pupil $\mathfrak{L}$, the common area is shown shaded in Fig. 6-6.

In the coherent limit, the effective source is reduced to a delta function for which Eq. (6-88) gives

$$C(\mathbf{f}'_1, \mathbf{f}'_2, \nu) = \mathfrak{L}\left(-\frac{\lambda s'\mathbf{f}'_1}{n'}\right) \mathfrak{L}^*\left(-\frac{\lambda s'\mathbf{f}'_2}{n'}\right), \tag{6-89}$$

where the pupil coordinates have been expressed in terms of the respective spatial frequencies. It is a product of the normalized version of $T_{\rm coh}$ of Eq. (6-42) and its conjugate. The factored form of Eq. (6-89) causes a factorization of the expression of Eq. (6-86). In this way the previous relationships

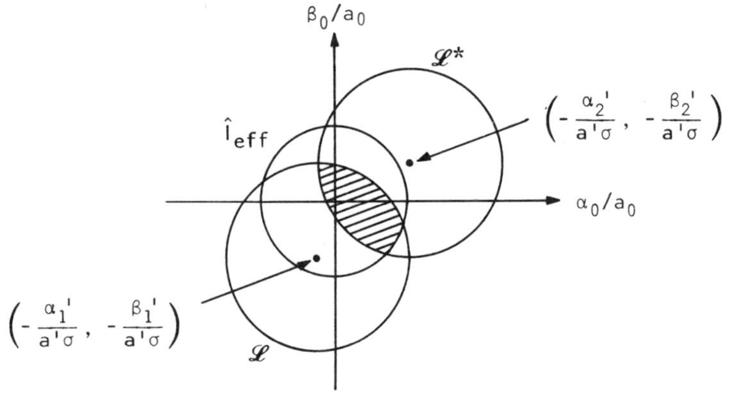

Fig. 6-6. The area of integration (shown shaded) to determine the HFR function, $C(\mathbf{f}'_1, \mathbf{f}'_2, \nu)$. The pupil $\mathfrak{L}^*$ is centered at $\boldsymbol{\alpha}_0/a_0 = -\boldsymbol{\alpha}'_2/a'\sigma$ and $\mathfrak{L}$ is centered at $\boldsymbol{\alpha}_0/a_0 = -\boldsymbol{\alpha}'_1/a'\sigma$. With the coordinates α_0 and β_0 scaled with a_0, the radius of the effective source is unity. At the same time, the coordinates α' and β' of the pupil are scaled with $a'\sigma$. In this description the pupil radius is $1/\sigma$.

for the coherent case may be rederived. For oblique coherent illumination, see Problem 6-6.

In the noncoherent limit the effective source is very large compared to the entrance pupil; more precisely, we are in the limit

$$\sigma = \frac{n_0 s a_0}{n s_0 a} \gg 1. \tag{6-90}$$

Let's suppose that the effective source is uniform,

$$\hat{I}_{\text{eff}}(\boldsymbol{\alpha}_0, \nu) = \hat{I}_{\text{eff}}(0, \nu)\,\text{cyl}\left(\frac{|\boldsymbol{\alpha}_0|}{a_0}\right),$$

where $|\boldsymbol{\alpha}_0| = (\alpha_0^2 + \beta_0^2)^{1/2}$. This form and a change of variable from $\boldsymbol{\alpha}_0$ to $\boldsymbol{\alpha}' = -\boldsymbol{\alpha}'_1 - a'\sigma\boldsymbol{\alpha}_0/a_0$ gives us

$$C(\mathbf{f}'_1, \mathbf{f}'_2, \nu) = \frac{\int d\boldsymbol{\alpha}'\,\text{cyl}(|\boldsymbol{\alpha}' + \boldsymbol{\alpha}'_1|/a'\sigma)\mathfrak{L}(\boldsymbol{\alpha}')\,\mathfrak{L}^*(\boldsymbol{\alpha}' + \boldsymbol{\alpha}'_1 - \boldsymbol{\alpha}'_2)}{\int d\boldsymbol{\alpha}'\,\text{cyl}(|\boldsymbol{\alpha}' + \boldsymbol{\alpha}'_1|/a'\sigma)\,|\mathfrak{L}(\boldsymbol{\alpha}' + \boldsymbol{\alpha}'_1)|^2}.$$

For the noncoherent limit the radius of the source is very large compared to the exit pupil, $a'\sigma \gg a'$. The denominator reduces to the area of $|\mathfrak{L}|^2$, and the numerator may be approximated to the difference-frequency function.

Thus,

$$C(\mathbf{f}'_1 - \mathbf{f}'_2, \nu) = \frac{\int d\boldsymbol{\alpha}'\, \mathcal{L}(\boldsymbol{\alpha}')\, \mathcal{L}^*(\boldsymbol{\alpha}' + \boldsymbol{\alpha}'_1 - \boldsymbol{\alpha}'_2)}{\int d\boldsymbol{\alpha}'\, |\mathcal{L}(\boldsymbol{\alpha}')|^2}. \tag{6-91}$$

for which the common area of $\mathcal{L}$ and $\mathcal{L}^*$ must lie wholly within the effective source. Thus, in the noncoherent limit, the HFR function assumes the form

$$C(\mathbf{f}'_1, \mathbf{f}'_2, \nu) \to C(\mathbf{f}'_1 - \mathbf{f}'_2, \nu),$$

which may be used in Eq. (6-86) to recover the relationships of the noncoherent case (see Problem 6-7). In fact, $C(\mathbf{f}'_1 - \mathbf{f}'_2, \nu)$ is the same as the OTF if we identify $\mathbf{f}'_1 - \mathbf{f}'_2$ as the frequency argument of t_{ncoh} of Eq. (6-58).

A smooth transition to the two limits is possible with the HFR function —a task that is difficult, if not impossible, to perform by use of the transfer function T_{pcoh} of Eq. (6-66).

Cosinusoidal Amplitude Objects

The theory outlined above was used by Hopkins (1953, 1957), who studied the influence of coherence on the images of periodic objects. We shall consider some special cases. The optical system is assumed to be free of aberration and working at perfect focus.

Consider a one-dimensional object whose amplitude transmittance is given by

$$\hat{t}_{\text{ob}}(\xi, \nu) = \tfrac{1}{2}(1 + \cos 2\pi f \xi). \tag{6-92}$$

It is convenient to express the spatial frequency in the object by

$$f = b f_A, \tag{6-93}$$

where b is a unitless parameter and where

$$f_A = \frac{(\text{N.A.})_{\text{ob}}}{\lambda} \tag{6-94}$$

is the Abbe limit of resolution for a microscope in the coherent mode of illumination. The transfer functions for the coherent and noncoherent illumination cases are shown in Fig. 6-7. The frequency is normalized such that the limit of resolution for noncoherent light is 2. On this scale, the

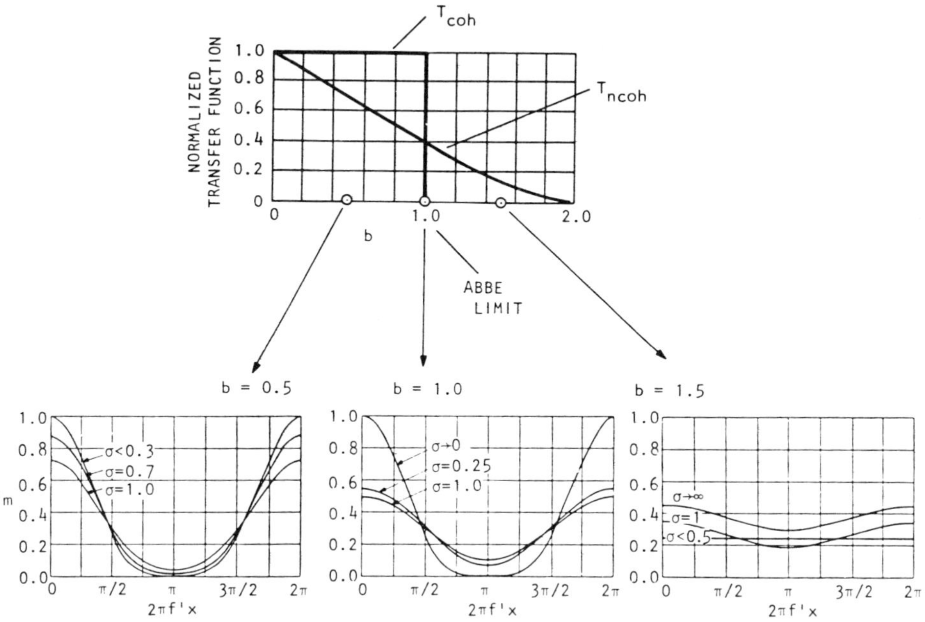

Fig. 6-7. The influence of partial coherence on the image of a cosinusoidal amplitude grating. The plots show the image OSI over one period, $0 \leq x \leq 1/f'$ (from Hopkins, 1957).

Abbe limit is unity and the parameter b itself denotes the spatial frequency of interest. As b approaches 2, the image contrast is only vanishingly small. Of particular interest are the cases $b = 0.5$, 1.0, and 1.5. The OSI distribution, $\hat{I}_{im}$, in the image of the grating, $\hat{t}_{ob}$, is shown for various states of coherence. The state of coherence is varied by adjusting the size of the condenser aperture relative to that of the objective by means of the parameter σ of Eq. (6-82). The coherent limit is approached as $\sigma \to 0$ and the noncoherent limit as $\sigma \to \infty$.

For $b = 0.5$, the grating is rather coarse and is well resolved in coherent light. For $0 < \sigma < 0.5$, the image is the same as in coherent light. As σ is increased, there is a corresponding decrease in contrast, but the effect is not large.

For $b = 1$, the grating is at the Abbe limit, the highest frequency resolved in coherent light. The image shows high contrast at $\sigma = 0$; however, the contrast drops rapidly with increasing σ. With a small increase, $\sigma = 0.25$, the contrast is rather low, and at $\sigma = 1$ it has already reached the noncoherent value of about 4%.

The value $b = 1.5$ corresponds to a fine-structure object. It is well beyond the frequency passed in the coherent mode. The horizontal line representing $\hat{I}_{im}$ exemplifies the inability of the system to resolve it even with some partial coherence, $\sigma < 0.5$. At $\sigma = 1$ the image shows some contrast, and for larger values of σ (as $\sigma \to \infty$) there is further loss of contrast due to the increase of the background level in the image. As $\sigma \to \infty$, the contrast approaches the value (below 2%) for the noncoherent case.

In relation to the three cases discussed, see Problem 6-8 for an alternative interpretation in terms of the coherence width of the illumination.

Effect of Partial Coherence on Edge Objects

Object structures that have "sharp" edges contain a broad band of spatial frequencies. The spectrum of a perfect edge contains all frequencies. The image of an edge is necessarily less sharp, since the imaging system is a lowpass filter. A solution of Problem 6-8 would indicate that light of a certain coherence width that may be regarded as almost noncoherent relative to a coarser frequency may appear to be almost coherent with respect to a very fine-frequency structure. In practice, the sharper the edge the more severe are the requirements to achieve noncoherence. By using the fast Fourier transform routine, a relatively straightforward computer program can show the gross features of the effect of partial coherence on the image of an object in the form of a step function.

Consider an object with amplitude transmittance

$$t_{ob}(\xi) = a + (b - a)s(\xi), \tag{6-95}$$

where a and b are positive constants ($b > a$) and where

$$s(\xi) = \begin{cases} 1, & \xi > 0 \\ 0, & \xi < 0 \end{cases} \tag{6-96}$$

is the step function. The "edge" is at $\xi = 0$. The constants a and b serve to specify the minimum and maximum transmittance levels. The effect of partial coherence on the image of such an object is shown in Fig. 6-8. This sequence of plots, obtained by computer calculations, shows the general trend of the image as the coherence width is varied relative to the impulse response S_{ncoh}, or what is called the *point spread function* (PSF). With coherent light there are fringes parallel to the edge on each side. The crossover point occurs at $I(x) = 0.25$. If the location of the step in the image is defined as the average value, $(a + b)/2$, then the edge appears to be shifted in the positive x direction. As the light is made partially coherent, the fringe structure fades away and the crossover point moves up. It reaches the value 0.32 for the noncoherent case. In this series of plots there are

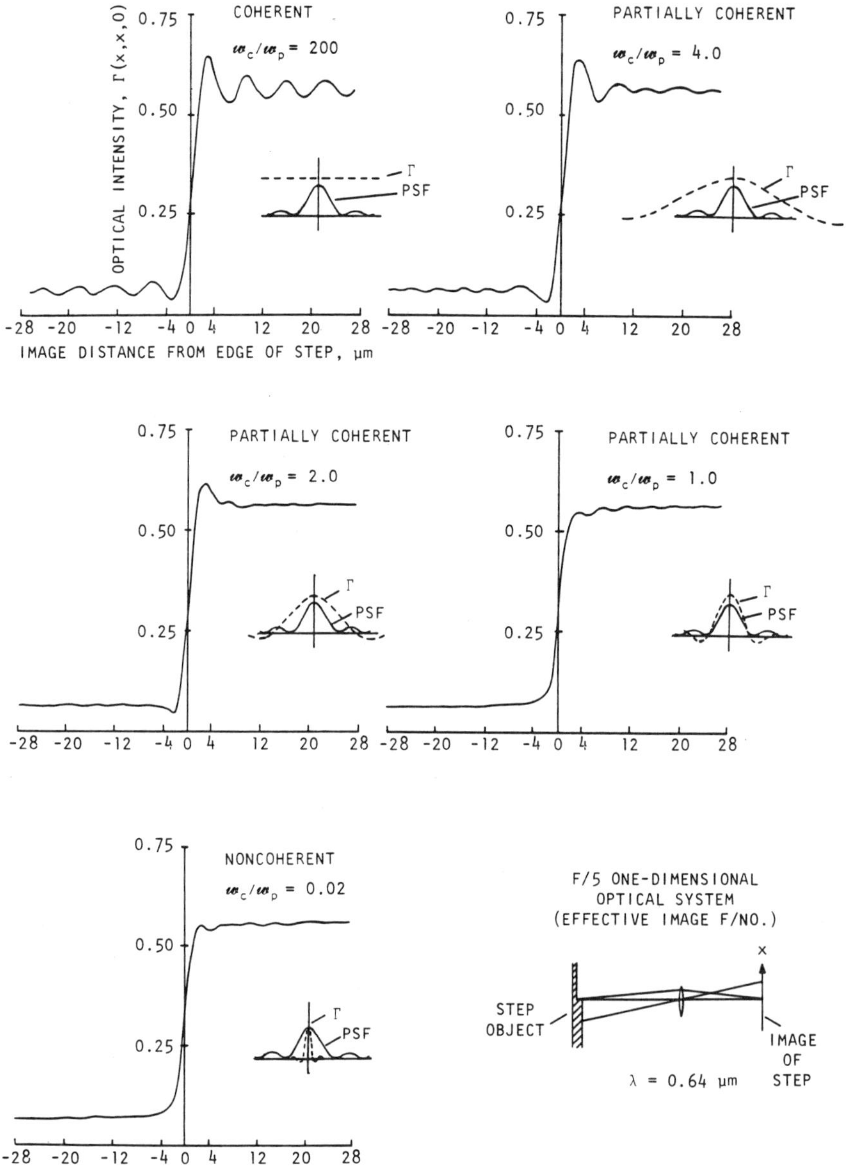

Fig. 6-8. Effect of partial coherence on the image of an object in the form of a step function. (This series of figures is from Lahart et al., 1974.)

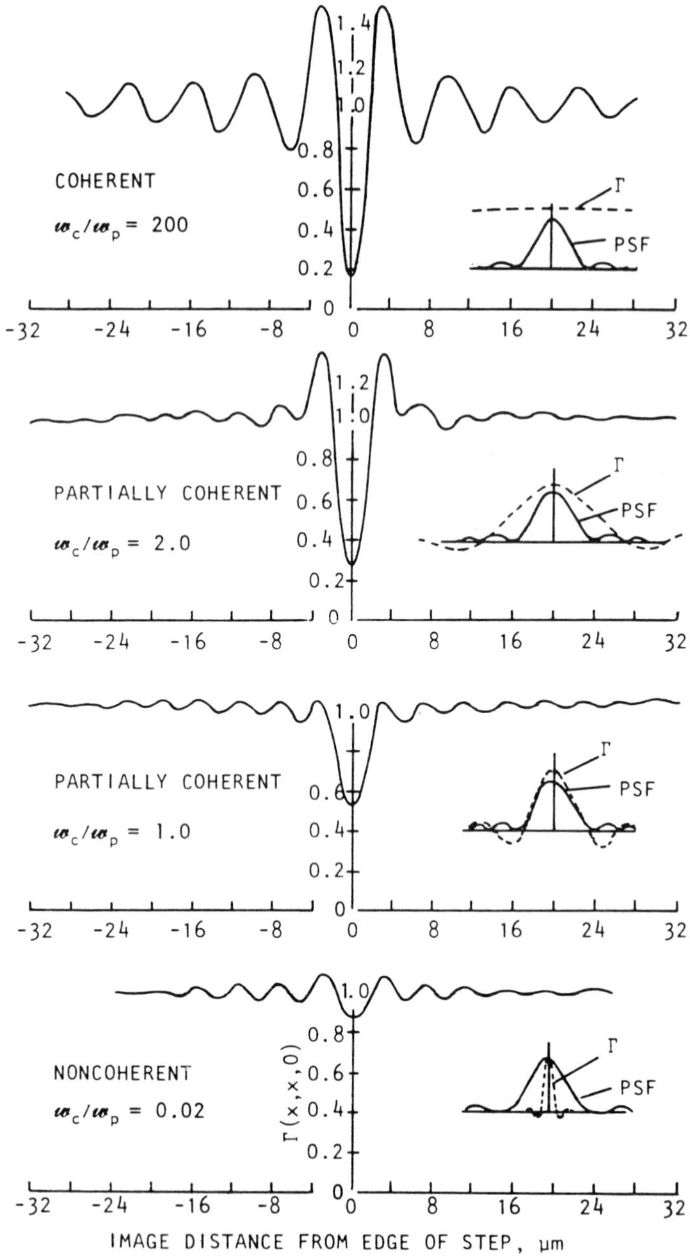

Fig. 6-9. Effect of partial coherence on the image of a phase-step object. (This series of figures is from Lahart et al., 1974.)

inserts showing the form of the coherence function $\Gamma(\xi_1 - \xi_2)$ in comparison with that of the PSF. In the figures, the ratio of the coherence width w_c to that of the width w_p of the PSF ranges from 200 to 0.02, which signify the cases of coherent and noncoherent illumination, respectively.

The requirements of noncoherence become even more severe when a phase-step object is imaged. In Fig. 6-9 the image of the phase step is shown in a sequence of plots, in which the ratio w_c/w_p is varied in the same way as in Fig. 6-8.

For Further Reading

Considine (1966) describes some experiments with partially coherent light. In particular, he shows photographs of three-bar targets in coherent and noncoherent light for the study of resolving power. There is no lack of theoretical papers on the subject. For the effect of nonlinearity on the imaging of sine-wave type objects (amplitude or phase) in partially coherent light, see Becherer and Parrent (1967), Swing and Clay (1967), De and Som (1963), and Som (1967). For further details of imaging amplitude and/or phase objects in partially coherent light, see Ichioka, et al. (1975, 1976) and Ichioka and Suzuki (1976); they have also considered the effect of defocusing error.

Carter (1973) studied the imaging property of a simple lens in terms of the *angular spectrum formalism*. Carpenter and Pask (1977) established an interesting theorem relating the size of the coherence area in the object and image. In their notation the radius R_i of the coherence area in the image is related to the radius R_o in the object by $R_i = MR_o$, provided $MR_o > r_A$, where M is the magnification and r_A is the radius of the Airy disk. However, if $MR_o \leq r_A$, then $R_i = r_A$.

Shore et al. (1966) studied both theoretically and experimentally the Fraunhofer diffraction due to circular apertures with partially coherent light. This work is based on a result due to Schell (1961), which states that the *optical intensity in the Fraunhofer region (far-field) of an aperture is a two-dimensional Fourier transform of the product of the complex autocorrelation function of the aperture and the mutual optical intensity function that describes the spatially stationary coherence across the aperture*. This is a useful result. Using it, the reader is asked to do Problem 6-9. The diffraction of light due to an annular aperture illuminated by partially coherent light is described by Singh and Dhillon (1969). Before closing this subsection we mention two relatively recent references. For the understanding of the influence of partial coherence on a variety of optical instrumentation, the review article by Reynolds and DeVelis (1979) is very helpful. Some additional discussion on Hopkins' effective source and a useful graphical method is found in Goodman (1979, 1981).

IMAGE FORMATION IN THE QUASIMONOCHROMATIC APPROXIMATION

As seen in the QM-field approximation, $\Delta\nu \ll \bar{\nu}$; and if we limit our attention to small time delays τ, the mutual coherence in the object may be expressed as

$$\Gamma_{ob}(\xi_1, \xi_2, \tau) \simeq \Gamma_{ob}(\xi_1, \xi_2, 0)\exp(-i2\pi\bar{\nu}\tau), \qquad |\tau| \ll \frac{1}{\Delta\nu}. \quad (6\text{-}97)$$

We mention again that in the QM-field approximation the mutual coherence function is not known for all τ, and hence taking the Fourier transform of Eq. (6-97) to obtain the mutual power spectral density is out of the question.

The simplification obtained through the use of the QM-field approximation will be evident by starting with Eq. (6-22), namely,

$$\begin{aligned}\hat{\Gamma}_{im}(\mathbf{x}_1, \mathbf{x}_2, \nu) &= \exp\left[\frac{ik}{2s'}(\mathbf{x}_1 \cdot \mathbf{x}_1 - \mathbf{x}_2 \cdot \mathbf{x}_2)\right] \\ &\times \int\int d\xi_1\, d\xi_2\, \hat{\Gamma}_{ob}(\xi_1, \xi_2, \nu)\exp\left[\frac{ik}{2s}(\xi_1 \cdot \xi_1 - \xi_2 \cdot \xi_2)\right] \\ &\times A(\mathbf{x}_1 - m\xi_1)\, A^*(\mathbf{x}_2 - m\xi_2), \end{aligned} \quad (6\text{-}98)$$

and

$$\Gamma_{im}(\mathbf{x}_1, \mathbf{x}_2, \tau) = \int_0^\infty \hat{\Gamma}_{im}(\mathbf{x}_1, \mathbf{x}_2, \nu)\exp(-i2\pi\nu\tau)\, d\nu. \quad (6\text{-}99)$$

In the QM-field approximation the MSDF $\hat{\Gamma}_{im}$ in the image is assumed to be significant only over a narrow range $\Delta\nu$. If we further limit our attention to small enough time delays τ, we find that

$$\Gamma_{im}(\mathbf{x}_1, \mathbf{x}_2, \tau) = \Gamma_{im}(\mathbf{x}_1, \mathbf{x}_2, 0)\exp(-i2\pi\bar{\nu}\tau), \qquad |\tau| \ll \frac{1}{\Delta\nu}, \quad (6\text{-}100)$$

where the mutual optical intensity function $\Gamma_{im}(\mathbf{x}_1, \mathbf{x}_2, 0)$ is given by

$$\Gamma_{im}(\mathbf{x}_1, \mathbf{x}_2, 0) = \int_0^\infty \hat{\Gamma}_{im}(\mathbf{x}_1, \mathbf{x}_2, \nu)\, d\nu. \quad (6\text{-}101)$$

This mutual optical intensity in the image may be connected with the corresponding quantity in the object, provided the frequency-dependent quantities in Eq. (6-98) are *slowly varying compared to* $\hat{\Gamma}_{ob}$. Recall that the function $A(\mathbf{x})$ has an implicit frequency dependence. Under these circumstances we may remove the frequency-dependent quantities from the integral over ν and evaluate them at the mean frequency $\bar{\nu}$. Thus,

$$\begin{aligned}\Gamma_{\text{im}}(\mathbf{x}_1, \mathbf{x}_2, 0) = \exp\left[\frac{i\bar{k}}{2s'}(\mathbf{x}_1 \cdot \mathbf{x}_1 - \mathbf{x}_2 \cdot \mathbf{x}_2)\right] \\ \times \int\int d\xi_1\, d\xi_2 \exp\left[\frac{i\bar{k}}{2s}(\xi_1 \cdot \xi_1 - \xi_2 \cdot \xi_2)\right] \\ \times A_{\bar{\nu}}(\mathbf{x}_1 - m\xi_1)\, A_{\bar{\nu}}^*(\mathbf{x}_2 - m\xi_2) \\ \times \int_0^\infty \hat{\Gamma}_{\text{ob}}(\xi_1, \xi_2, \nu)\, d\nu.\end{aligned}$$

The subscript $\bar{\nu}$ on A indicates that it is evaluated at the mean frequency $\bar{\nu}$, and $\bar{k} = 2\pi/\bar{\lambda} = 2\pi\bar{\nu}/c$. The integral over $\hat{\Gamma}_{\text{ob}}$ is simply the mutual optical intensity in the object. Thus in the QM-field approximation the image-forming equation reads

$$\begin{aligned}\Gamma_{\text{im}}(\mathbf{x}_1, \mathbf{x}_2, 0) = \exp\left[\frac{i\bar{k}}{2s'}(\mathbf{x}_1 \cdot \mathbf{x}_1 - \mathbf{x}_2 \cdot \mathbf{x}_2)\right] \\ \times \int\int d\xi_1\, d\xi_2\, \Gamma_{\text{ob}}(\xi_1, \xi_2, 0) \exp\left[\frac{i\bar{k}}{2s}(\xi_1 \cdot \xi_1 - \xi_2 \cdot \xi_2)\right] \\ \times A_{\bar{\nu}}(\mathbf{x}_1 - m\xi_1) A_{\bar{\nu}}^*(\mathbf{x}_2 - m\xi_2). \qquad (6\text{-}102)\end{aligned}$$

The usefulness of this approximation stems from the fact that the mutual optical intensity in the object and image are all that need to be studied instead of the full mutual coherence function. The equation of image formation is linear in terms of the mutual optical intensity. As such, the appropriate impulse response and the transfer function may be defined. But we observe that the form of Eq. (6-102) is similar to that of Eq. (6-98). The frequency-dependent quantities, except $\hat{\Gamma}_{\text{ob}}$ and $\hat{\Gamma}_{\text{im}}$ of Eq. (6-98), enter into Eq. (6-102) with their values at the mean frequency $\bar{\nu}$. This observation allows us to conclude that all the results of the polychromatic analysis of the previous sections for coherent, noncoherent, and partially coherent cases may be carried over into the QM-field approximation by replacing the frequency ν (in the appropriate places) by the mean frequency $\bar{\nu}$.

If, however, the image is known through the use of the QM-field approximation via Eq. (6-102), it is not possible to infer anything about the polychromatic image. This should be evident, since the QM-field image is in terms of the mutual optical intensity $\Gamma_{\text{im}}(\mathbf{x}_1, \mathbf{x}_2, 0)$, which is obtained by integrating the MSDF $\hat{\Gamma}_{im}(\mathbf{x}_1, \mathbf{x}_2, \nu)$; that is,

$$\Gamma_{im}(\mathbf{x}_1, \mathbf{x}_2, 0) = \int_0^\infty \hat{\Gamma}_{im}(\mathbf{x}_1, \mathbf{x}_2, \nu)\, d\nu.$$

A unique determination of $\hat{\Gamma}_{im}(\mathbf{x}_1, \mathbf{x}_2, \nu)$ is not possible through the knowledge of the mutual optical intensity $\Gamma_{im}(\mathbf{x}_1, \mathbf{x}_2, 0)$ alone. The region of validity of the approximation discussed above is found by an analysis similar to the one carried out in Problem 5-3, related to the propagation of QM fields.

POLYCHROMATIC TRANSFER FUNCTION IN THE NONCOHERENT LIMIT

Very often in practice it is of interest to know what the effective transfer function is over a band of temporal frequencies for a lens system working with noncoherent illumination. As a special case, the effective transfer function for white light is also of interest. In this section, we shall study the conditions under which an effective transfer function for a linear imaging system may be meaningfully defined.

Let us suppose that the image is detected by measuring its irradiance, $\mathsf{C} I_{im}(\mathbf{x})$ [W/m^2], by means of a detector with a constant response over the band of frequencies of interest. With this assumption we need not explicitly introduce the detector response, and we define the optical intensity $I_{im}(\mathbf{x})$ by

$$I_{im}(\mathbf{x}) = \int_0^\infty \hat{I}_{im}(\mathbf{x}, \nu)\, d\nu. \tag{6-103}$$

Although the upper limit of integration is formally taken to be ∞, the range is effectively limited by the band of frequencies contained in $\hat{I}_{im}$, which in turn depends on the frequencies contained in the object, $\hat{I}_{ob}(\boldsymbol{\xi}, \nu)$. In Eq. (6-101) of the previous section, the mutual optical intensity was defined as an integral over $\hat{\Gamma}_{im}$. Thus Eq. (6-103) is a special case of Eq. (6-101), which is obtained by setting $\mathbf{x}_1 = \mathbf{x}_2 = \mathbf{x}$, where

$$\hat{I}_{im}(\mathbf{x}, \nu) \equiv \hat{\Gamma}_{im}(\mathbf{x}, \mathbf{x}, \nu), \tag{6-104}$$

and

$$I_{im}(\mathbf{x}) \equiv \Gamma_{im}(\mathbf{x}, \mathbf{x}, 0). \tag{6-105}$$

We thus seek a linear relationship between $I_{im}(\mathbf{x})$ and $I_{ob}(\boldsymbol{\xi})$. To this end we begin with Eq. (6-22) used in the partially coherent polychromatic case.

In that equation set $\mathbf{x}_1 = \mathbf{x}_2 = \mathbf{x}$ and integrate over the frequency ν to get

$$I_{\mathrm{im}}(\mathbf{x}) = \int_0^\infty \hat{\Gamma}_{\mathrm{im}}(\mathbf{x}, \mathbf{x}, \nu)\, d\nu$$
$$= \int_0^\infty d\nu \int d\boldsymbol{\xi}_1 \int d\boldsymbol{\xi}_2\, \hat{\Gamma}_{\mathrm{ob}}(\boldsymbol{\xi}_1, \boldsymbol{\xi}_2, \nu) h(\mathbf{x}, \boldsymbol{\xi}_1) h^*(\mathbf{x}, \boldsymbol{\xi}_2), \quad (6\text{-}106)$$

where purely for convenience of discussion we have defined a function h of two variables $\mathbf{x}$ and $\boldsymbol{\xi}$ by

$$h(\mathbf{x}, \boldsymbol{\xi}) \equiv \exp\left(\frac{ik}{2s'} \mathbf{x} \cdot \mathbf{x} \right) \exp\left(\frac{ik}{2s} \boldsymbol{\xi} \cdot \boldsymbol{\xi} \right) A(\mathbf{x} - m\boldsymbol{\xi}). \quad (6\text{-}107)$$

At this point we introduce the noncoherence of the polychromatic object by putting

$$\hat{\Gamma}_{\mathrm{ob}}(\boldsymbol{\xi}_1, \boldsymbol{\xi}_2, \nu) = \frac{4\pi}{k^2} \hat{I}_{\mathrm{ob}}(\boldsymbol{\xi}_1, \nu)\, \delta(\boldsymbol{\xi}_2 - \boldsymbol{\xi}_1), \quad (6\text{-}108)$$

which reduces Eq. (6-106) to the form

$$I_{\mathrm{im}}(\mathbf{x}) = \int_0^\infty d\nu \int d\boldsymbol{\xi}\, \hat{I}_{\mathrm{ob}}(\boldsymbol{\xi}, \nu) \frac{c^2}{\pi \nu^2} |A_\nu(\mathbf{x} - m\boldsymbol{\xi})|^2. \quad (6\text{-}109)$$

In this equation a subscript ν is put on the function A to remind us that A is an implicit function of ν.

From Eq. (6-109) it is evident that we cannot achieve our stated goal of obtaining a linear relationship between $I_{\mathrm{im}}(\mathbf{x})$ and $I_{\mathrm{ob}}(\boldsymbol{\xi})$. The specification of the object through the function $\hat{I}_{\mathrm{ob}}(\boldsymbol{\xi}, \nu)$ allows for the possibility of different spectral colors at different locations $\boldsymbol{\xi}$ of the object. Such a situation occurs, for example, when a color transparency is illuminated by white light. For this kind of an object, we observe through Eq. (6-109) that a meaningful polychromatic transfer function cannot be defined!

However, let us consider an object with the same spectral content, $\hat{\gamma}(\nu)$, at every point $\boldsymbol{\xi}$. In this case, the object description factors into the form

$$\hat{I}_{\mathrm{ob}}(\boldsymbol{\xi}, \nu) = I_{\mathrm{ob}}(\boldsymbol{\xi})\, \hat{\gamma}(\nu), \quad (6\text{-}110)$$

where the spectrum is normalized such that

$$\int_0^\infty d\nu\, \hat{\gamma}(\nu) = 1.$$

This normalization allows us to preserve the relationship

$$I_{\text{ob}}(\boldsymbol{\xi}) = \int_0^\infty d\nu\, \hat{I}_{\text{ob}}(\boldsymbol{\xi}, \nu) \qquad (6\text{-}111)$$

for the object. An example of this situation in practice is a black and white transparency (with its gray levels) illuminated by white light or more generally by light with a spectrum $\hat{\gamma}(\nu)$.

For the special object described by Eq. (6-110), the relationship of image formation of Eq. (6-109) takes the desired form

$$I_{\text{im}}(\mathbf{x}) = \int d\boldsymbol{\xi}\, I_{\text{ob}}(\boldsymbol{\xi})\, S_{\text{poly}}(\mathbf{x} - m\boldsymbol{\xi}). \qquad (6\text{-}112)$$

In it the polychromatic impulse response is defined by

$$\begin{aligned} S_{\text{poly}}(\mathbf{x} - m\boldsymbol{\xi}) &= \int_0^\infty \left[\frac{c^2}{\pi \nu^2} | A_\nu(\mathbf{x} - m\boldsymbol{\xi}) |^2 \right] \hat{\gamma}(\nu)\, d\nu \\ &= \int_0^\infty S_{\text{ncoh},\,\nu}(\mathbf{x} - m\boldsymbol{\xi})\, \hat{\gamma}(\nu)\, d\nu, \end{aligned} \qquad (6\text{-}113)$$

where we have modified S_{ncoh} to show that it also depends on ν. Thus the system, Eq. (6-112), is not only linear but also spatially stationary, for which a polychromatic transfer function T_{poly} may be defined in the usual way,

$$T_{\text{poly}}(\mathbf{f}') = \int d\mathbf{x}\, S_{\text{poly}}(\mathbf{x}) \exp(-i2\pi \mathbf{x} \cdot \mathbf{f}'). \qquad (6\text{-}114)$$

However, the matter is not so simple because the spatial frequency $\mathbf{f}'$ was defined as

$$\mathbf{f}' = \frac{\boldsymbol{\alpha}'}{\lambda s'} = \frac{\nu \boldsymbol{\alpha}'}{c s'}$$

and it is not clear which wavelength λ (or frequency ν) should be used. The fact is that the wavelength λ is no longer a good optical number with which to scale the relevant quantities. But this need not stop us from defining a formal Fourier transform relationship as in Eq. (6-114), except that now the spatial frequency variable $\mathbf{f}'$ with units of $(\text{length})^{-1}$ is no longer simply related to the aperture variable $\boldsymbol{\alpha}'$. In terms of $\mathbf{f}'$, the polychromatic transfer function may be related to T_{ncoh},

$$T_{\text{poly}}(\mathbf{f}') = \int_0^\infty T_{\text{ncoh},\,\nu}(\mathbf{f}')\, \hat{\gamma}(\nu)\, d\nu. \qquad (6\text{-}115)$$

The subscript shows that T_{ncoh} also depends on ν. The polychromatic functions are thus expressible as weighted averages with the spectrum $\hat{\gamma}(\nu)$ of the corresponding functions of the noncoherent case. To understand the implication of this, recall that the noncoherent transfer function in terms of the exit pupil function is

$$T_{\text{ncoh}}\left(\frac{\boldsymbol{\alpha}'}{\lambda s'}\right) = \frac{1}{\pi s^2}\int d\boldsymbol{\alpha}_1'\, \mathfrak{L}(\boldsymbol{\alpha}_1')\, \mathfrak{L}^*(\boldsymbol{\alpha}_1' + \boldsymbol{\alpha}'),$$

wherein as far as T_{ncoh} is concerned, the relationship $\mathbf{f}' = \boldsymbol{\alpha}'/\lambda s'$ continues to hold. The complex amplitude transmittance $\mathfrak{L}$ of the system is a function not only of position but also of the wavelength of light. As such, it also includes the effect of dispersion. For example, the fact that all the wavelengths do not come to a common focus would indicate that if the correct Gaussian focus is chosen for say the sodium D light, the corresponding defocus terms will have to be included for the other wavelengths. These considerations do not surface when we are thinking about T_{ncoh} in relation to one wavelength (monochromatic aberrations) but have to be properly accounted for when we are dealing with broadband light. For a discussion about chromatic aberrations a book by Welford (1974) is highly recommended. Now the zero ordinate of T_{poly} is

$$T_{\text{poly}}(0) = \frac{1}{\pi s^2}\int_0^\infty \hat{\gamma}(\nu)\left[\int d\boldsymbol{\alpha}_1'\, |\mathfrak{L}(\boldsymbol{\alpha}_1')|^2\right] d\nu. \tag{6-116}$$

Suppose the wavelength transmittance of the pupil is independent of position $\boldsymbol{\alpha}_1'$. Then we may use $T(\nu)$ as the optical spectral intensity transmittance for the whole pupil. As an example, consider a circular pupil of radius a'. Then the area integral in Eq. (6-116) leads to

$$T_{\text{poly}}(0) = \frac{a'^2}{s^2}\int_0^\infty \hat{\gamma}(\nu)\, T(\nu)\, d\nu.$$

For the calculation of $T_{\text{poly}}(\mathbf{f}')$ as in Eq. (6-115) when $\mathbf{f}' \neq 0$ and/or when the wavelength transmittance of the pupil is position dependent, $\mathfrak{L}$ should be regarded as a function of both $\boldsymbol{\alpha}_1'$ and ν in order to account for the amplitude transmittance. To go further and also include the detector response we refer to Barnes (1971).

Finally, from the definition of OTF of a single frequency ν, the polychromatic OTF is

$$t_{\text{poly}}(\mathbf{f}') = \frac{T_{\text{poly}}(\mathbf{f}')}{T_{\text{poly}}(0)} = \frac{\displaystyle\int_0^\infty \hat{\gamma}(\nu)\, T_{\text{ncoh},\nu}(\mathbf{f}')\, d\nu}{\displaystyle\int_0^\infty \hat{\gamma}(\nu)\, T_{\text{ncoh},\nu}(0)\, d\nu}. \tag{6-117}$$

In conclusion, we observe that the polychromatic noncoherent transfer function may be defined only when the object has the same spectral content at all its points. Although this condition is very restrictive in practice, t_{poly} may be used as a measure of the average performance of the system over a broad band of frequencies. Removal of the restriction compels us to study the OTF in the noncoherent case or the HFR function in the partially coherent case as a function of the temporal frequency ν.

OPTICAL SYSTEMS IN CASCADE

In this section we discuss briefly the problem of image formation for a train of optical systems, $\mathfrak{L}_1, \mathfrak{L}_2, \ldots$, and so on. In this train the image of any one element is the object of the next element. Let us start with the basic linear systems point of view. Consider the input ψ_{in}, related to the output, ψ_{out}, by the convolution integral,

$$\psi_{\text{out}}^{(j)}(\mathbf{x}_j) = \int d\xi_j \, \psi_{\text{in}}^{(j)}(\xi_j)\, S_j(\mathbf{x}_j - m_j \xi_j). \tag{6-118}$$

Then in the Fourier transform domain we have the product relationship for the corresponding transforms,

$$\tilde{\psi}_{\text{out}}^{(j)}(\mathbf{f}'_j) = \tilde{\psi}_{\text{in}}^{(j)}(m_j \mathbf{f}'_j)\, T_j(\mathbf{f}'_j). \tag{6-119}$$

In Eqs. (6-118) and (6-119) the label j is used to signify the jth system.

When the jth and the $(j+1)$th systems are considered together, it can be shown that the effective impulse response is the convolution of the impulse responses of the two systems. Furthermore, it can be shown that in the transform domain the equation of image formation for the two systems together reads

$$\tilde{\psi}_{\text{out}}^{(j+1)}(\mathbf{f}'_{j+1}) = \tilde{\psi}_{\text{in}}^{(j)}(m_j m_{j+1} \mathbf{f}'_{j+1})\left[T_{j+1}(\mathbf{f}'_{j+1})\, T_j(m_{j+1}\mathbf{f}'_{j+1})\right]. \tag{6-120}$$

Thus the combined transfer function is the product of the individual transfer functions with their arguments appropriately scaled with the magnification factors (see Problem 6-10).

We now apply the above considerations to the coherent, partially coherent, and noncoherent imaging problems.

Coherent: The correct expression for the impulse response is given in Eq. (6-28), indicating that although the system is linear it is not stationary. However, if the approximating conditions (C1) through (C3) are fulfilled, the coherent imaging system is spatially stationary. But for a cascade, condition (C3) is not applicable from one element to the next. Therefore, the combined transfer function in the coherent case is not equal to the product of the individual transfer functions.

Partially Coherent: The discussion for the coherent case applies. Hence, here too the combined transfer function is not a product of the individual transfer functions.

Noncoherent: In the noncoherent case the problem of image formation is linear and spatially stationary (up to the quadratic approximation) in terms of the optical spectral intensity $\hat{I}_{ob}(\xi, \nu)$. However, the description of image formation in terms of $\hat{I}_{ob}(\xi, \nu)$ is *not* applicable for a cascade. This is evident by observing that if the object of the first system is noncoherently illuminated the image is partially coherent. Hence for systems $j = 2, 3, \ldots$, and so on, the analysis of the partially coherent image formation must be used. Therefore, expressing the effective transfer function in terms of a product of the individual noncoherent transfer functions is out of the question.

We thus come to the unavoidable conclusion (see Problem 6-11) that there is no case in image formation where the combined transfer function is a product of the individual functions! In practice, however, the condition of spatial stationarity may be approximately justifiable for a cascaded system working with partially coherent illumination. For the details of this discussion the reader may refer to DeVelis and Parrent (1967).

SUMMARY FOR IMAGE FORMATION

Table 6-1 summarizes the results for various conditions of coherence. It is assumed that conditions (C1) through (C3) are fulfilled for the coherent and partially coherent cases. The table also lists the polychromatic transfer function and Hopkins' frequency response (HFR) function.

If and when the QM-field approximation is applicable, the impulse response and the transfer functions of Table 6-1 are evaluated at the mean frequency; that is, ν is replaced by $\bar{\nu}$. The object and image descriptions are then given in terms of the mutual optical intensity. For example, in the partially coherent case, $\hat{\Gamma}_{ob}(\xi_1, \xi_2, \nu)$ is replaced by $\Gamma_{ob}(\xi_1, \xi_2, 0)$.

In this chapter, our principal aim was to formulate a theory of image formation with the object and image described in terms of observables. To maintain linearity we were compelled to use the wave amplitude for the coherent case, the mutual spectral density function for the partially coherent case, and the optical spectral intensity for the noncoherent case. It is

TABLE 6-1. SYSTEM CHARACTERISTICS FOR VARIOUS CONDITIONS OF COHERENCE

State of coherence	Object Description	Impulse Response	Transfer Function
Coherent	Wave amplitude $U_{ob}(\boldsymbol{\xi})$	$S_{coh}(\mathbf{x} - m\boldsymbol{\xi}) = A(\mathbf{x} - m\boldsymbol{\xi})$	$T_{coh}(\mathbf{f}') = \left(\frac{s'}{s}\right) \mathcal{L}(-\lambda s'\mathbf{f}')$
Partially coherent	Mutual spectral density function $\hat{\Gamma}_{ob}(\boldsymbol{\xi}_1, \boldsymbol{\xi}_2, \nu)$	$S_{pcoh}(\mathbf{x}_1 - m\boldsymbol{\xi}_1, \mathbf{x}_2 - m\boldsymbol{\xi}_2, \nu) = A(\mathbf{x}_1 - m\boldsymbol{\xi}_1)\, A^*(\mathbf{x}_2 - m\boldsymbol{\xi}_2)$	$T_{pcoh}(\mathbf{f}'_1, \mathbf{f}'_2, \nu) = \left(\frac{s'}{s}\right)^2 \mathcal{L}(-\lambda s'\mathbf{f}'_1)\, \mathcal{L}^*(-\lambda s'\mathbf{f}'_2)$
Noncoherent	Optical spectral intensity $\hat{I}_{ob}(\boldsymbol{\xi}, \nu)$	$S_{ncoh}(\mathbf{x} - m\boldsymbol{\xi}) = \frac{4\pi}{k^2} \lvert A(\mathbf{x} - m\boldsymbol{\xi}) \rvert^2$	$T_{ncoh}(\mathbf{f}') = T_{ncoh}\left(\frac{\boldsymbol{\alpha}'}{\lambda s'}\right) = \frac{1}{\pi s^2} \int d\boldsymbol{\alpha}'_1\, \mathcal{L}(\boldsymbol{\alpha}'_1)\, \mathcal{L}^*(\boldsymbol{\alpha}'_1 + \boldsymbol{\alpha}')$
Noncoherent (spectral averaging)	Optical intensity $I_{ob}(\boldsymbol{\xi})$	$S_{poly}(\mathbf{x} - m\boldsymbol{\xi}) = \int_0^\infty S_{ncoh,\,\nu}(\mathbf{x} - m\boldsymbol{\xi})\hat{\gamma}(\nu)\, d\nu$	$T_{poly}(\mathbf{f}') = \int_0^\infty T_{ncoh,\,\nu}(\mathbf{f}')\, \hat{\gamma}(\nu)\, d\nu$
Partially coherent (effective source)	Optical spectral intensity $\hat{I}_{ob}(\boldsymbol{\xi}, \nu)$	HFR, $C(\mathbf{f}'_1, \mathbf{f}'_2, \nu) = C\left(\frac{\boldsymbol{\alpha}'_1 n'}{\lambda s'}, \frac{\boldsymbol{\alpha}'_2 n'}{\lambda s'}, \nu\right) = C_0 \int d\boldsymbol{\alpha}_0\, \hat{I}_{eff}(\boldsymbol{\alpha}_0, \nu)\, \mathcal{L}\left(-\boldsymbol{\alpha}'_1 - \frac{a'\sigma\boldsymbol{\alpha}_0}{a_0}\right) \mathcal{L}^*\left(-\boldsymbol{\alpha}'_2 - \frac{a'\sigma\boldsymbol{\alpha}_0}{a_0}\right)$	

customary to measure only the OSI or spectral irradiance of the object and the image. This practice makes the general problem nonlinear, and the concept of a transfer function loses significance. A theory that encompasses the general cases of practical interest and that deals with the observed spectral irradiance is the one based on the effective source. The spatial frequency content of the image is then studied by means of Hopkins' frequency response function.

In the theory of image formation based on Fourier analysis, the object and image are described by a superposition of sine and cosine waves of infinite extent, each with its appropriate amplitude and phase. No particular restriction regarding the physical extent of the object is explicitly imposed. The theory leads to the description of optical systems as low-pass filters. No considerations of "noise" in the measurement were included in the theory, yet the implication is that optical systems inherently have a finite resolution limit. The limit indicates the ability of the system to resolve high-frequency (sine-wave) structure in the object and may be called the sine-wave limit. Similarly, the so-called Rayleigh limit deals with the ability of the system, in the presence of diffraction, to resolve two points in proximity. It is particularly helpful for visual considerations. The capability of the human eye to distinguish between neighboring optical intensity values is pertinent in arriving at this limit with the use of diffraction theory. Toraldo di Francia (1952) was the first to point out that this limit is more or less dictated by practical considerations rather than an in-principle theoretical limit (see also Goodman, 1968).

Consider, for example, an object consisting of two pinholes located at $\boldsymbol{\xi}_{01}$ and $\boldsymbol{\xi}_{02}$, respectively, in an otherwise opaque plane. If they are noncoherently illuminated and have equal strength, Eq. (6-48) shows that the OSI in the image is proportional to

$$\hat{I}_{\mathrm{im}}^{(2)}(\mathbf{x}, \nu) \simeq \frac{4\pi}{k^2}\left[|A(\mathbf{x} - m\boldsymbol{\xi}_{01})|^2 + |A(\mathbf{x} - m\boldsymbol{\xi}_{02})|^2\right],$$

whereas for a single pinhole at say $\boldsymbol{\xi}_{01}$ we get

$$\hat{I}_{\mathrm{im}}^{(1)}(\mathbf{x}, \nu) \simeq \frac{4\pi}{k^2}|A(\mathbf{x} - m\boldsymbol{\xi}_{01})|^2.$$

In the absence of noise in the measurement, the image of two pinholes can always be distinguished from the image of a single pinhole, no matter how close the pinholes are placed. By absence of noise we mean that the process of measurement can distinguish indefinitely between neighboring optical intensity values. How close the pinholes can approach each other has only a structural restriction of how small they are and how close together they may be placed and yet be regarded as two holes.

More generally, the object may extend only over a limited region of the object plane. The image-forming light comes from within this region and not from outside. The object spectrum is then band-limited. It has certain analytic properties (Goodman, 1968) that enable one to reconstruct object detail beyond the above-mentioned classical limits (see Harris, 1964a, b). Rebuilding the object beyond these limits is called *superresolution*. We refer to Frieden (1969) for the general problem of superresolution by capitalizing on the finite size of the object.

RESOLUTION CRITERIA IN TERMS OF THE COHERENCE FUNCTION

For the study of resolution in optics there are several criteria customarily used. The question of resolution is always accompanied by considerations of noise in the measurement, be it subjective or objective. We shall assume that the quantity measured is optical intensity or irradiance.

It is found empirically that the Rayleigh criterion correlates well with visual (subjective) observation. It deals with resolving two closely spaced, mutually noncoherent point sources. The individual sources are assumed to be unresolved. That is, their shape and/or optical intensity distribution is not detectable; they appear as points. Assuming them to be of equal optical intensity, they are considered as just resolved in the aberration-free case if the first zero of the diffraction pattern (PSF or Airy disk) of one source coincides with the central maximum of the other. The just-resolved point separation δ' in the image space is found to be

$$\delta' = \frac{0.61\,\bar{\lambda} s'}{n'a'} \tag{6-121}$$

in the notation of Fig. 6-5, where $\bar{\lambda}$ is the mean wavelength of the quasimonochromatic (QM) field. For the image with magnification m, the point separation $\delta = \delta'/m$ in the object is given by

$$\delta = \frac{0.61\,\bar{\lambda} s}{na} = \frac{0.61\,\bar{\lambda}}{(\mathrm{N.A.})_o}, \tag{6-122}$$

where $(\mathrm{N.A.})_o$ is the numerical aperture of the objective.

The Sparrow criterion is also commonly used. According to it, the two point sources are considered as just resolved if there is no minimum in the total (resulting) diffraction pattern of the two point sources. It is derived by using a second derivative of the optical intensity distribution of the resulting pattern. It leads to

$$\delta = \frac{0.5\,\bar{\lambda}}{(\text{N.A.})_o}. \tag{6-123}$$

The numerical factor is largely arbitrary; it depends on what is regarded as "just resolved." The functional dependence on the wavelength, the object distance, and the size of the entrance pupil is important.

In the study of the spatial frequency response of optical systems the sine-wave resolution is specified. In the noncoherent case the value of the transfer function T_{ncoh} at the cutoff frequency $f'_{\max}$ of Eq. (6-59) is zero. That is, the sine wave at this frequency is not visible. The value of $T_{\text{ncoh}}\,(\frac{3}{4}f'_{\max})$ is slightly less than 0.2 in the aberration-free case. If we suppose that this spatial frequency structure is "just resolved" and denote it by f'_R we find that

$$f'_R = \tfrac{3}{4}f'_{\max} = \frac{3}{2}\,\frac{a'n'}{\bar{\lambda}s'}, \tag{6-124}$$

where $\bar{\lambda}$ is the mean wavelength in vacuum of the QM field and n' is the refractive index of image space as used in Fig. 6-5. The corresponding period p'_R of the frequency just resolved is

$$p'_R = \frac{0.67\,\bar{\lambda}s'}{a'n'}, \tag{6-125}$$

which in object space translates to

$$p_R = \frac{0.67\,\bar{\lambda}s}{an}. \tag{6-126}$$

Objects in the form of step functions or three-bar targets (Cobb charts or Air Force resolution targets) are also commonly employed for evaluating optical systems. Elaborate mathematical formulas are available to relate three-bar information to the sine-wave response.

Resolution criteria deal with aberration-free optical systems. They are used as a kind of standard for the limiting performance of an optical system. This is followed by the study of image quality criteria and aberration tolerances.

In practice, objects of interest are rarely if ever two points, sine waves, or three bars, although object scenes commonly may contain detail that looks like an "edge." Therefore it is not obvious how to translate information from these various resolution criteria to decide whether a particular kind of object detail is visible or resolvable.

A more general approach to the above topic is to use the coherence function. We continue to assume that the basic quantity that is measured in the image is optical intensity or irradiance. A resolution criterion formulated in terms of the coherence function is more general because the two-point and sine-wave criteria may be shown to be its special cases. Furthermore, it will indicate whether a particular kind of object detail is visible or resolvable.

Consider a noncoherently illuminated object. If its coherence function (mutual optical intensity, MOI) is "narrower" than the aperture diameter of the optical system, then as seen before (e.g., in the discussion of the Michelson stellar interferometer) the object is resolved by the optical system. On the other hand, if a small enough source is used as an object, then the PSF of the system is all that is produced in the image plane. The source object is thus regarded as unresolved by the optical system. In this case the size, shape, and optical intensity distribution across the source area are not visible or detectable. By the van Cittert–Zernike theorem, the optical intensity distribution across the aperture is uniform and the coherence function due to such a source is very broad compared to the aperture diameter.

A resolution criterion to be formulated in what is to follow compares the size of the coherence function to the aperture diameter. If the former is narrow enough compared to the latter the object is resolved; otherwise it is unresolved.

Two-Point Object

Consider an object consisting of two mutually noncoherent point sources, as given by

$$I_{\text{ob}}(\xi, 0) = I_0\left[\mathcal{A}_1\delta(\xi - \tfrac{1}{2}\delta)\,\delta(\eta) + \mathcal{A}_2\delta(\xi + \tfrac{1}{2}\delta)\,\delta(\eta)\right], \quad (6\text{-}127)$$

where $\mathcal{A}_1$ and $\mathcal{A}_2$ are real and positive and have units of area. The sources are separated by a distance δ along the ξ axis. The coherence function at the entrance pupil of the optical system is

$$\begin{aligned}\Gamma_{\text{ent}}(\boldsymbol{\alpha}_1, \boldsymbol{\alpha}_2, 0) = \frac{I_0}{\pi s^2} \exp&\left[\frac{i\bar{k}n}{2s}(\boldsymbol{\alpha}_1 \cdot \boldsymbol{\alpha}_1 - \boldsymbol{\alpha}_2 \cdot \boldsymbol{\alpha}_2)\right] \\ &\times\left[(\mathcal{A}_1 + \mathcal{A}_2)\cos\frac{\pi n(\alpha_1 - \alpha_2)\delta}{\bar{\lambda}s}\right. \\ &\left.-i(\mathcal{A}_1 - \mathcal{A}_2)\sin\frac{\pi n(\alpha_1 - \alpha_2)\delta}{\bar{\lambda}s}\right]. \qquad (6\text{-}128)\end{aligned}$$

For simplicity of argument, we may consider $\mathscr{A}_1 = \mathscr{A}_2$; that is, the sources have equal strength. The absolute value of the coherence function is proportional to that of the cosine term. The variables α_1 and α_2 are restricted to lie within the finite area of the pupil. For a circular entrance pupil of radius a, $|\alpha_1 - \alpha_2| \le 2a$. In order to include significant variation of Γ_{ent} within the pupil we shall arbitrarily set the condition that its first zero, which occurs at $\bar{\lambda}s/2n\delta$, must fulfill

$$\frac{\bar{\lambda}s}{2n\delta} = \varepsilon 2a, \tag{6-129}$$

where ε is a parameter less than unity. The point sources separated by

$$\delta = \frac{\bar{\lambda}s}{4\varepsilon na} \tag{6-130}$$

may be regarded as just resolved. Comparison of this condition with Eq. (6-122) of the Rayleigh criterion suggests that $\varepsilon = 0.41$.

Cosinusoidal Object

Now consider the object to be a noncoherent cosine structure,

$$\Gamma_{\text{ob}}(\xi_1, \xi_2, 0) = \frac{\bar{\lambda}^2}{\pi n^2} I_0 \left(1 + \cos\frac{2\pi\xi_1}{p_{\text{ob}}}\right) \delta(\xi_1 - \xi_2), \tag{6-131}$$

where p_{ob} stands for its period. By the van Cittert–Zernike theorem, the coherence function in the aperture plane is

$$\begin{aligned}\Gamma_{\text{ent}}(\boldsymbol{\alpha}_1, \boldsymbol{\alpha}_2, 0) = {} & \frac{\bar{\lambda}^2}{\pi n^2} I_0 \exp\left[\frac{i\bar{k}n}{2s}(\boldsymbol{\alpha}_1 \cdot \boldsymbol{\alpha}_1 - \boldsymbol{\alpha}_2 \cdot \boldsymbol{\alpha}_2)\right] \delta(\beta_1 - \beta_2) \\ & \times \left[\delta(\alpha_1 - \alpha_2) + \tfrac{1}{2}\delta\left(\alpha_1 - \alpha_2 - \frac{\bar{\lambda}s}{np_{\text{ob}}}\right)\right. \\ & \left. + \tfrac{1}{2}\delta\left(\alpha_1 - \alpha_2 + \frac{\bar{\lambda}s}{np_{\text{ob}}}\right)\right],\end{aligned} \tag{6-132}$$

where δ stands for the Dirac delta function. For a one-dimensional structure along the ξ axis, the spectra are along the α axis at $\alpha_1 - \alpha_2 = 0$ and $\pm\bar{\lambda}s/np_{\text{ob}}$. To assure that the coherence function is narrow compared to the aperture, we set

$$\frac{\bar{\lambda}s}{np_{\text{ob}}} = \varepsilon 2a,$$

where ε is a parameter less than unity. Thus the cosine structure may be regarded as just resolved if its period fulfills

$$p_{\text{ob}} = \frac{\bar{\lambda}s}{2\varepsilon na}. \tag{6-133}$$

In particular, if we put $\varepsilon = 0.75$, the above condition becomes identical to that of Eq. (6-126) obtained from the noncoherent transfer function.

General Object Detail

These two seemingly unrelated examples show that a resolution criterion based on the coherence function may be profitable. By using it we may state a criterion of visibility or detectability for any noncoherently illuminated isolated object. Light from any such object produces a uniform optical intensity distribution in the aperture plane, and the coherence function Γ_{ent} in that plane is dictated by the van Cittert–Zernike theorem. Now the absolute value $|\Gamma_{\text{ent}}|$ is a function of $\boldsymbol{\alpha}_1 - \boldsymbol{\alpha}_2$. In this space let w_{12} be the width that includes significant variations of $|\Gamma_{\text{ent}}|$ in a given azimuth, and let D_{apert} be the extent of the aperture in that azimuth. If the condition

$$w_{12} = \varepsilon D_{\text{apert}}, \qquad \varepsilon < 1, \tag{6-134}$$

is fulfilled for all azimuths, then the object will be visible. If it is violated for any azimuth, then the object detail in that azimuth would not be visible. (If the aperture is not circular, then D_{apert} will vary with azimuth.) Formulation of the resolution criterion as given above has the advantage that it makes contact with known criteria and, in addition, indicates whether a particular kind of object detail is visible.

PROBLEMS

6-1. Verify the paraxial optical relationships in Eqs. (6-5)–(6-7). For this purpose let l and l' be the object and image distances, respectively, from the principal planes and let l_p and l'_p be the respective distances of the entrance and exit pupils from the principal planes. The symbols s and s' are

$$s = l - l_p \quad \text{and} \quad s' = l' - l'_p.$$

6-2. Verify that the right-hand side of Eq. (6-23) is real and nonnegative. *Hint*: Show that $\Gamma_{im}(\mathbf{x}, \mathbf{x}, 0)$ is a spatial Fourier transform of an autocorrelation function.

6-3. Specify appropriate criteria by which the relevant phase factors may be approximated to unity and derive mathematical inequalities to verify conditions (C1) and (C2) and the inequalities in Eq. (6-37).

6-4. (a) For an unaberrated lens system with a clear circular aperture of radius a', show that the normalized transfer function for the noncoherent illumination is

$$t_{ncoh}(\sigma) = \frac{1}{\pi}(2b - \sin 2b)$$

where $b = \cos^{-1}|\sigma/2|$, and σ is the normalized spatial frequency in the azimuth in which the response is to be determined. The spatial frequencies f' and g' are related to σ by

$$\frac{f'}{f'_{max}} = \sigma \cos \psi, \qquad \frac{g'}{g'_{max}} = \sigma \sin \psi$$

where ψ is the azimuth shown in Fig. 6-4 and $f'_{max} = a'/\lambda s'$ as defined in the coherent case.

(b) Consider two objects with amplitude transmittance (1) $t_{ob} = \cos 2\pi f_{ob}\xi$ and (2) $t_{ob} = |\cos 2\pi f_{ob}\xi|$. Calculate the value of T_{coh} and the contrast value t_{ncoh} for the two objects when $f'_{max} = 1$ and $f_{ob} = 0.9$. (*Answer*: $T_{coh} = 1$ and 0, and $t_{ncoh} = 0.45$ and 0.04 for the respective objects.)

6-5. A Ronchi ruling with opaque strips the same size as the transparent strips with period p is used as an object. Consider the cases of coherent and noncoherent illumination of wavelength λ. How large must the radius of the exit pupil of the lens system be for it to just resolve the object in the two cases?

6-6. Find the transfer function for oblique coherent illumination as obtained by an off-axis point source. Show that it is possible to double the resolution limit in coherent light by a proper inclination of the illuminating plane wave. Calculate the visibility of a cosinusoidal grating image and show that the visibility in this case is superior to that of the same grating illuminated with noncoherent light (see Hopkins, 1953).

6-7. In the noncoherent limit the HFR function assumes the form

$$C(\mathbf{f}'_1, \mathbf{f}'_2, \nu) \rightarrow C(\mathbf{f}'_1 - \mathbf{f}'_2, \nu).$$

Use this form in Eq. (6-86) and show that it reduces to Eq. (6-52), namely,

$$\hat{I}_{\text{im}}(\mathbf{x}, \nu) = \int d\xi\, \hat{I}_{\text{ob}}(\xi, \nu)\, S_{\text{ncoh}}(\mathbf{x} - m\xi)$$

provided we define

$$S_{\text{ncoh}}(\mathbf{x} - m\xi) = \frac{\hat{\Gamma}'(0, \nu)}{m^2 \hat{\Gamma}(0, \nu)} \int d\mathbf{f}'_{12}\, C(\mathbf{f}'_{12}, \nu) \exp[i2\pi(\mathbf{x} - m\xi) \cdot \mathbf{f}'_{12}]$$

where $\mathbf{f}'_{12} = \mathbf{f}'_1 - \mathbf{f}'_2$.

6-8. Give an interpretation of Hopkins' plots of Fig. 6-7 by comparing the coherence width of the illumination with the grating period. Estimate the value of the *visibility* v of the grating image in each case and plot it on the graph that shows the coherent and noncoherent transfer functions in that figure.

6-9. Using Schell's result, show that the optical intensity in the Fraunhofer region of an aperture illuminated by spatially stationary partially coherent light is a two-dimensional convolution of the optical intensity in that region due to the same aperture with *coherent* illumination and Hopkins' effective source. Use your result to study the two limiting cases of coherence and noncoherence.

6-10. Verify that the combined transfer function of a system made up of several components is a product of the transfer function of the individual components provided that (a) the output of one is the input to the next component and (b) the impulse response is spatially stationary for each component. This verification requires establishing Eq. (6-120).

6-11. Consider a system made up of two components, each described by a spatially nonstationary impulse response:

$$\psi^{(1)}_{\text{out}}(\xi_2) = \int d\xi_1\, \psi^{(1)}_{\text{in}}(\xi_1)\, S_1(\xi_2, \xi_1),$$

$$\psi^{(2)}_{\text{out}}(\xi_3) = \int d\xi_2\, \psi^{(2)}_{\text{in}}(\xi_2)\, S_2(\xi_3, \xi_2).$$

The output of one is the input for the next, $\psi_{\text{in}}^{(2)}(\xi_2) = \psi_{\text{out}}^{(1)}(\xi_2)$. Show that the impulse response of the combined system is

$$S(\xi_3, \xi_1) = \int d\xi_2\, S_2(\xi_3, \xi_2)\, S_1(\xi_2, \xi_1).$$

If the combined transfer function is defined by

$$T(\mathbf{f}_3, \mathbf{f}_1) = \int S(\xi_3, \xi_1) \exp[-i2\pi(\xi_3 \cdot \mathbf{f}_3 + \xi_1 \cdot \mathbf{f}_1)]\, d\xi_3\, d\xi_1,$$

then show that

$$T(\mathbf{f}_3, \mathbf{f}_1) = \int d\mathbf{f}_2\, T_2(\mathbf{f}_3, \mathbf{f}_2)\, T_1(-\mathbf{f}_2, \mathbf{f}_1),$$

where T_2 and T_1 are the respective functions defined for S_2 and S_1.

REFERENCES

Baker, L. R. (1955). An interferometer for measuring the spatial frequency response of a lens system, *Proc. Phys. Soc. Lond. Sec. B* **68** (11):871–880.

Barnes, K. R. (1971). *The Optical Transfer Function*, American Elsevier, New York, 76 pp.

Becherer, R. J., and G. B. Parrent, Jr. (1967). Nonlinearity in optical imaging systems, *J. Opt. Soc. Am.* **57** (12):1479–1486.

Born, M., and E. Wolf (1970). *Principles of Optics*, 4th ed., Pergamon Press, New York, Art. 10.5.2.

Bromilow, N. S. (1958). Geometrical–optical calculation of frequency response for systems with spherical aberration, *Proc. Phys. Soc. Lond.* **71** (2):231–237.

Carpenter, D. J., and C. Pask (1977). Coherence properties in the image of a partially coherent object, *J. Opt. Soc. Am.* **67** (1):115–117.

Carter, W. H. (1973). Wave theory for a simple lens, *Opt. Acta* **20** (10):805–826.

Considine, P. S. (1966). Effects of coherence on imaging systems, *J. Opt. Soc. Am.* **56** (8):1001–1009.

De, M. (1955). The influence of astigmatism on the response function of an optical system, *Proc. R. Soc. Lond. Ser. A* **233**:91–104.

De, M., and S. C. Som (1963). Diffraction images of circular phase objects in partially coherent light, *J. Opt. Soc. Am.* **53** (7):779–787.

DeVelis, J. B., and G. B. Parrent, Jr. (1967). Transfer function for cascaded optical systems, *J. Opt. Soc. Am.* **57** (12):1486–1490.

Frieden, B. R. (1969). On arbitrarily perfect imagery with a finite aperture, *Opt. Acta* **16** (6):795–807.

Gaskill, J. D. (1978). *Linear Systems, Fourier Transforms and Optics*, Wiley, New York, 554 pp.

Goodbody, A. M. (1958). The influence of spherical aberration on the response function of an optical system, *Proc. Phys. Soc. Lond.* **72** (3):411–424.

Goodman, D. S. (1979). *Stationary Optical Projectors*, Ph.D. dissertation, University of Arizona.

Goodman, D. S. (1981). Graphical method for image irradiance determination (Abstr.), *J. Opt. Soc. Am.* **71** (12), 1626.

Goodman, J. W. (1968). *Introduction to Fourier Optics*, McGraw-Hill, San Francisco, 287 pp.

Harris, J. L. (1964a). Resolving power and decision theory, *J. Opt. Soc. Am.* **54** (5):606–611.

Harris, J. L. (1964b). Diffraction and resolving power, *J. Opt. Soc. Am.* **54** (7):931–936.

Hopkins, H. H. (1953). On the diffraction theory of optical images, *Proc. R. Soc. Lond. Ser. A* **217**:408–432.

Hopkins, H. H. (1955a). The frequency response of defocused optical systems, *Proc. R. Soc. Lond. Ser. A* **231**:91–103.

Hopkins, H. H. (1955b). Interferometric methods for the study of diffraction images, *Opt. Acta* **2** (1):23–29.

Hopkins, H. H. (1957). Applications of coherence theory in microscopy and interferometry, *J. Opt. Soc. Am.* **47** (6):508–526.

Ichioka, Y., and T. Suzuki (1976). Image of a periodic complex object in an optical system under partially coherent illumination, *J. Opt. Soc. Am.* **66** (9):921–932.

Ichioka, Y., K. Yamamoto, and T. Suzuki (1975). Image of a sinusoidal complex object in a partially coherent optical system, *J. Opt. Soc. Am.* **65** (8):892–902.

Ichioka, Y., K. Yamamoto, and T. Suzuki (1976). Defocused image of a periodic complex object in an optical system under partially coherent illumination, *J. Opt. Soc. Am.* **66** (9):932–938.

Jenkins, F. A., and H. E. White (1976). *Fundamentals of Optics*, 4th ed., McGraw-Hill, New York, 746 pp.

Lahart, M. J., A. S. Marathay, and R. E. Wagner (1974). Unpublished material developed in a course at Optical Sciences Center, University of Arizona.

Marathay, A. S. (1959). Geometrical optical calculation of frequency response for systems with coma, *Proc. Phys. Soc. Lond.* **74** (6):721–730.

O'Neill, E. L. (1956). Transfer function for an annular aperture, *J. Opt. Soc. Am.* **46** (4):285–288.

O'Neill, E. L. (1963). *Introduction to Statistical Optics*, Addison-Wesley, Reading, MA, 179 pp.

Reynolds, G. O., and J. B. DeVelis (1979). Review of optical coherence effects in instrument design, Session 1, pp. 2–33, *Applications of Optical Coherence*, W. H. Carter, Ed., Proceedings of the Society of the Photo-Optical Instrumentation Engineers, Vol. 194.

Schell, A. C. (1961). *Multiple Plate Antenna*, Ph.D. thesis, Massachusetts Institute of Technology.

Shore, R. A., B. J. Thompson, and R. E. Whitney (1966). Diffraction by apertures illuminated with partially coherent light, *J. Opt. Soc. Am.* **56** (6):733–738.

Singh, K., and H. S. Dhillon (1969). Diffraction of partially coherent light by an aberration-free annular aperture, *J. Opt. Soc. Am.* **59** (4):395–401.

Som, S. C. (1967). Diffraction images of annular and disklike objects under partially coherent illumination, *J. Opt. Soc. Am.* **57** (12):1499–1509.

Swing, R. E., and J. R. Clay (1967). Ambiguity of the transfer function with partially coherent illumination, *J. Opt. Soc. Am.* **57** (10):1180–1189.

Toraldo di Francia, G. (1952). Super-gain antennas and optical resolving power, *Nuovo Cimento Suppl.* **9** (3):426–435.

Welford, W. T. (1974). *Aberrations of the Symmetrical Optical System*, Academic Press, New York, 240 pp.

Zernike, F. (1938). The concept of degree of coherence and its application to optical problems, *Physica* (*The Hague*) **5** (8):785–795.

7

Radiometry

Radiometry is a science of detection of electromagnetic radiation. Its beginning may be marked by the significant contributions of Bouguer and Lambert in the 1750s (Bouguer, 1760; Walsh, 1965). It has been successfully used since then and is widely used today. The ideas and concepts of this venerable science are based on geometrical or ray optics. It largely ignores the wave nature of light, its diffraction effects, and its states of partial coherence. However, it is well known that diffraction does affect radiant power distribution. It is also generally believed that radiometry, as it stands, in some sense applies to noncoherent sources or fields. But this has never been proved. It is therefore highly desirable to derive the laws of radiometry from the basic wave theory of light, as only then may we find the conditions under which radiometry is applicable. Furthermore, we will then have a formalism to deal with the measurement of light in any state of coherence and to account for diffraction effects in a unified way.

We shall refer to radiometry as it stands as *conventional* radiometry (c-radiometry). It is to be distinguished from the radiometry formulated by using wave optics, which we shall call *generalized* radiometry (g-radiometry). We shall begin with a brief discussion of the concepts and definitions of c-radiometry.

CONVENTIONAL RADIOMETRY: DEFINITIONS AND SYMBOLS

Conventional radiometry is largely empirical. Since its inception, several workers having to deal with the detection of light under different experimental conditions have found it necessary to define and use quantities applicable to their particular cases. To unify these various concepts and definitions and to develop a scheme applicable to a majority of practical situations is indeed a monumental task. A notable attempt at unification was made by Jones (1963). An authoritative and most complete exposition of this topic is found in a series of self-study manuals edited by Nicodemus (1976). The reader may also refer to the *Infrared Handbook* edited by Wolfe and Zissis (1978) and the *Handbook of Military Infrared Technology* edited by Wolfe (1965). At the textbook level there are several sources, the more recent ones being Grum and Becherer (1979), Wyatt (1978), and Klein (1970). We shall follow the SI system of units (see Mechtly, 1969, and MacAdam, 1967, and for a general reference see *USA Standard Nomenclature and Definitions for Illuminating Engineering*, Illuminating Engineering Society, 1968).

In this brief section we cannot do justice to the broad topic of radiometry. Our purpose is mainly to show the interrelationships among the basic radiometric quantities in common use. We have been using some of the terminology throughout the book. Now we shall pull it together and complete the picture. We begin with the (*total*) *radiant power* $\mathbf{\Phi}$ with units of watts [W]. The *spectral radiant power* is generally expressed as a function of wavelength λ or frequency ν with the respective subscript. However, anticipating a temporal Fourier transform relationship, we shall denote the spectral radiant power by $\hat{\mathbf{\Phi}}$, with units of W Hz^{-1} and regard it as a function of ν. Its integral over all frequencies gives the total radiant power,

$$\underset{[\mathrm{W}]}{\mathbf{\Phi}} = \underset{[\mathrm{W\,Hz^{-1}}]}{\int_0^\infty \hat{\mathbf{\Phi}}(\nu)\, d\nu.} \tag{7-1}$$

In relation to the radiation emanating from an optical source we speak of the *radiant exitance* M with units of W m^{-2}. It is a function of position $\mathbf{r}$ on the source. As in Eq. (7-1),

$$\underset{[\mathrm{W\,m^{-2}}]}{\mathsf{M}(\mathbf{r})} = \underset{[\mathrm{W\,m^{-2}\,Hz^{-1}}]}{\int_0^\infty \hat{\mathsf{M}}(\mathbf{r}, \nu)\, d\nu,} \tag{7-2}$$

where $\hat{\mathsf{M}}$ is the *spectral radiant exitance*. For a plane source of shape $\mathcal{A}(x, y)$, area $\mathcal{A}$, and in the plane at $z =$ constant, the total radiant power is defined by the area integral

$$\begin{aligned} \mathbf{\Phi}(z) &= \iint_{\mathcal{A}} \mathsf{M}(x, y, z)\, dx\, dy \\ &\equiv \iint \mathcal{A}(x, y)\, \mathsf{M}(x, y, z)\, dx\, dy. \end{aligned} \tag{7-3}$$

The coordinate z labels the location of the source. When light from the source falls on a receiving surface, the radiation on it is described by *radiant incidence* or *irradiance* E with units of W m^{-2}. It too is a function of position $\mathbf{r}$. The *spectral irradiance* $\hat{\mathsf{E}}(\mathbf{r}, \nu)$ has units of W m^{-2} Hz^{-1}. These quantities obey relationships similar to Eqs. (7-2) and (7-3). Reconsider the plane source of area $\mathcal{A}$. Its radiant exitance M is a function of position $\mathbf{r} = \hat{\mathbf{i}}\, x + \hat{\mathbf{j}}\, y + \hat{\mathbf{k}}\, z$, and from every position (x, y) on the source it may channel different amounts of light in different directions $\hat{\mathbf{n}}$, with components

$$\begin{aligned} \hat{\mathbf{n}} &= \hat{\mathbf{i}}\, p + \hat{\mathbf{j}} q + \hat{\mathbf{k}}\, m \\ &= \hat{\mathbf{i}}(\sin\theta \cos\phi) + \hat{\mathbf{j}}(\sin\theta \sin\phi) + \hat{\mathbf{k}} \cos\theta, \end{aligned} \tag{7-4}$$

where θ is the polar angle made with the z axis and ϕ is the azimuthal angle. The element of solid angle $d\Omega$, in steradians [sr], about this direction is

$$d\Omega = \frac{dp\,dq}{m} = \sin\theta\,d\theta\,d\phi. \tag{7-5}$$

The first equality is established by using the Jacobian of the change of variables from (p, q) to (θ, ϕ).

To account for the variation of the source property as a function of position and direction, a quantity called *radiance* $\mathsf{L}(\mathbf{r}, \hat{\mathbf{n}})$ is used, with units of W m^{-2} sr^{-1}. From the point of view of an observer looking at the source in the direction $\hat{\mathbf{n}}$, the projection $(d\mathcal{A})_\perp = \cos\theta\,d\mathcal{A}$, shown in Fig. 7-1 is effective. The total radiant power in terms of L is given by

$$\underset{[\mathrm{W}]}{\boldsymbol{\Phi}(z)} = \int_{\mathcal{A}}\int_{1/2} \underset{[\mathrm{W\,m^{-2}\,sr^{-1}}]}{\mathsf{L}(\mathbf{r}, \hat{\mathbf{n}})} \cos\theta\,d\Omega\,d\mathcal{A}, \tag{7-6}$$

where the z value indicates the source location. The symbol $\mathcal{A}$ under the integrals stands for the integration over $\mathcal{A}(x, y)$, and the $\frac{1}{2}$ under the integral implies that the angle integration is limited to the right half-space, $0 \leq \theta \leq \pi/2$ and $0 \leq \phi \leq 2\pi$. The product $d\Omega \cos\theta$ is often called the *projected solid angle* element. The spectral radiant power $\hat{\boldsymbol{\Phi}}(z, \nu)$ is found by using the *spectral radiance* $\hat{\mathsf{L}}(\mathbf{r}, \hat{\mathbf{n}}, \nu)$ with units of W m^{-2} sr^{-1} Hz^{-1}.

Now as the observer moves away from the source, its details as a function of position (x, y) assume less importance. Far enough away, the source appears as a point source and its characteristic angular distribution of radiation is all that is evident. To describe this situation we define the term *radiant intensity* $\mathsf{I}(z, \hat{\mathbf{n}})$ with units of W sr^{-1}. Its solid angle integral over the

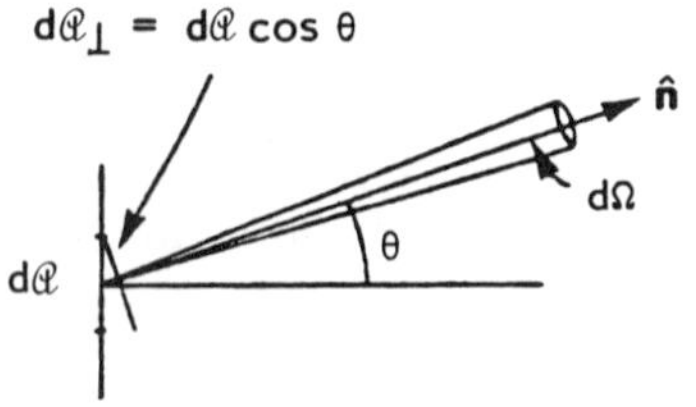

Fig. 7-1. Diagram showing the projection of an elemental area $d\mathcal{A}$ of the source. The projection $d\mathcal{A}_\perp$ is taken perpendicular to the direction $\hat{\mathbf{n}}$ of observation.

TABLE 7-1. INTERRELATIONSHIPS AMONG THE SPECTRAL FUNCTIONS OF CONVENTIONAL RADIOMETRY[a]

$$\hat{\mathsf{L}} \quad [\mathrm{W\,m^{-2}\,sr^{-1}\,Hz^{-1}}]$$

$$\underset{[\mathrm{W\,Hz^{-1}}]}{\hat{\boldsymbol{\Phi}}} = \int_{\mathcal{A}}\int_{1/2} \hat{\mathsf{L}} \cos\theta \, d\Omega \, d\mathcal{A}$$

$$\underset{[\mathrm{W\,m^{-2}\,Hz^{-1}}]}{\hat{\mathsf{M}}} = \int_{1/2} \hat{\mathsf{L}} \cos\theta \, d\Omega, \qquad \underset{[\mathrm{W\,sr^{-1}}\,Hz^{-1}]}{\hat{\mathsf{I}}} = \cos\theta \int_{\mathcal{A}} \hat{\mathsf{L}} \, d\mathcal{A}$$

$$\hat{\boldsymbol{\Phi}} = \int_{\mathcal{A}} \hat{\mathsf{M}} \, d\mathcal{A}, \qquad \hat{\boldsymbol{\Phi}} = \int_{1/2} \hat{\mathsf{I}} \, d\Omega$$

[a]Functional arguments are omitted.

right half-space yields the total radiant power,

$$\underset{[\mathrm{W}]}{\boldsymbol{\Phi}(z)} = \int_{1/2} \underset{[\mathrm{W\,sr^{-1}}]}{\mathsf{I}(z, \hat{\mathbf{n}})} \, d\Omega. \tag{7-7}$$

The solid angle integral over the *spectral radiant intensity* $\hat{\mathsf{I}}(z, \hat{\mathbf{n}}, \nu)$ with units of W $\mathrm{sr^{-1}}$ $\mathrm{Hz^{-1}}$ leads to the spectral radiant power.

Among the various functions defined above, the basic one is the radiance L. In terms of it, the other functions may be derived by regrouping and performing the appropriate integral. The interrelationships among the spectral functions are displayed in Table 7-1. The analogous relationships among the total functions are found by integrating over all frequencies $0 \leq \nu \leq \infty$.

Lambertian Source

Although sources in general do have position-dependent and/or angle-dependent properties, it is often advantageous to study the limiting case of a source whose properties are independent of position and direction. This is an idealization, never realized in practice in the strict sense but approached to a good approximation.

Consider a source whose spectral radiance is independent of both $\mathbf{r}$ and $\hat{\mathbf{n}}$, for which we put

$$\hat{\mathsf{L}}(\mathbf{r}, \hat{\mathbf{n}}, \nu) = \hat{\mathsf{L}}_0(\nu). \tag{7-8}$$

For such a source it follows from the relationships shown in Table 7-1 that

$$\begin{aligned}\hat{\mathsf{\Phi}} &= \pi \mathscr{A} \hat{\mathsf{L}}_0,\\ \hat{\mathsf{M}} &= \pi \hat{\mathsf{L}}_0,\\ \hat{\mathsf{I}} &= \cos\theta\, \mathscr{A} \hat{\mathsf{L}}_0.\end{aligned} \tag{7-9}$$

A source described by Eq. (7-8) and having the properties listed in Eq. (7-9) is called a *Lambertian* source. In the first two relationships the factor π [sr] is the value of the hemispherical solid angle with proper account of the $\cos\theta$ weighting. The spectral radiant intensity relationship correctly accounts for the projection ($\mathscr{A}\cos\theta$) along the viewing direction of the source area $\mathscr{A}$. Such a source would appear uniformly bright to an observer viewing it from different directions.

Outlook

With this brief discussion of c-radiometry we proceed to discuss some results that may be derived by applying the Rayleigh–Sommerfeld diffraction theory of light to the special case of a noncoherent source. We shall then formulate a generalized radiometry (g-radiometry) to account for the wave nature of light and its states of partial coherence.

NONCOHERENT SOURCE

In Chapter 4 the noncoherent source was described as the one whose coherence function is so sharply peaked that its spatial frequency spectrum is "white" or constant over the real waves of its Fourier decomposition. The physical interpretation is that such a source delivers the same amount of power (on the average) in each direction dictated by each spatial frequency in the right half-space. Consistent with this interpretation, we found in Chapter 5 that the beam from such a source spreads more than the beam from a partially coherent source with the same size aperture.

In this section we begin with the basic propagation law of the mutual spectral density function (MSDF) $\hat{\Gamma}_{12}(\nu)$, given in Eq. (5-1). It is derived by using the Rayleigh–Sommerfeld (RS) diffraction theory. We shall apply it to the special case of a noncoherent source in a plane and study the following:

1. The spectral irradiance in a plane parallel to the source plane.
2. The spectral irradiance on a hemisphere covering the source aperture.
3. The spectral radiant intensity distribution far enough away from the source.

For this purpose we may use the δ-function representation to describe noncoherence of the source in the $z = 0$ plane,

$$\hat{\Gamma}_s(x_{s1}, y_{s1}, x_{s2}, y_{s2}, 0, \nu) = \frac{\lambda^2}{\pi}\hat{\Gamma}_s(x_{s1}, y_{s1}, x_{s1}, y_{s1}, 0, \nu)\,\delta(x_{s1} - x_{s2})\,\delta(y_{s1} - y_{s2}), \tag{7-10}$$

although the other functions listed in Eq. (4-47) are just as good to describe the situation. We identify the spectral radiant exitance with the MSDF with equal arguments,

$$\hat{\mathsf{M}}(x, y, 0, \nu) = \mathsf{C}\hat{\Gamma}(x, y, x, y, 0, \nu). \tag{7-11}$$

The aforementioned calculations are in terms of this quantity; the radiant exitance itself is the integral over the frequency ν.

Spectral Irradiance on a Plane

The planar noncoherent source is in the plane $z = 0$. We wish to obtain the spectral irradiance $\hat{\mathsf{E}}(x, y, z, \nu)$ in the plane $z =$ constant. Use of Eq. (5-1) along with noncoherence of Eq. (7-10) yields

$$\hat{\mathsf{E}}(x, y, z, \nu) = \frac{\lambda^2}{4\pi^3}\int\!\!\int \hat{\mathsf{M}}(x_s, y_s, 0, \nu)\cos^2\theta\left(\frac{1 + k^2\rho^2}{\rho^4}\right)dx_s\,dy_s, \tag{7-12}$$

where, as before,

$$\rho = |\mathbf{r} - \mathbf{s}| = \left[(x - x_s)^2 + (y - y_s)^2 + z^2\right]^{1/2},$$
$$\cos\theta = \frac{z}{\rho}. \tag{7-13}$$

In the radiation zone, since $k\rho \gg 1$, we have to a good approximation

$$\hat{\mathsf{E}}(x, y, z, \nu) \simeq \frac{1}{\pi}\int\!\!\int \hat{\mathsf{M}}(x_s, y_s, 0, \nu)\frac{\cos^2\theta}{\rho^2}\,dx_s\,dy_s. \tag{7-14}$$

Consider an elemental area $(dx_s\,dy_s)$ at x_s, y_s in the source aperture (Fig. 7-2). The differential spectral irradiance contributed by it to the observation

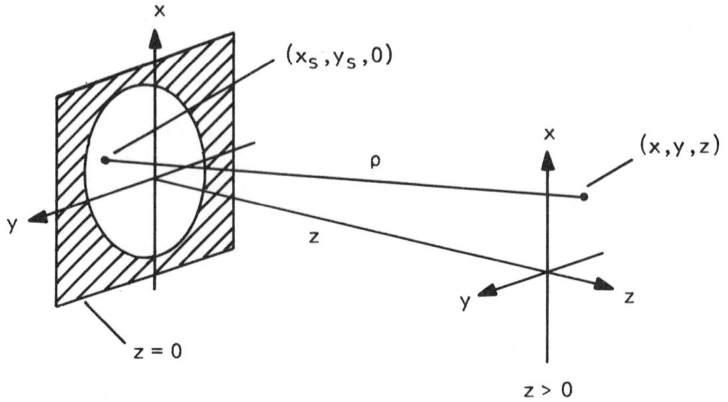

Fig. 7-2. Diagram to calculate the irradiance in a plane parallel to the source plane. The source is in the plane $z = 0$.

plane has the angular dependence given by

$$d\hat{\mathsf{E}} = \frac{1}{\pi z^2}\hat{\mathsf{M}}(x_s, y_s, 0, \nu)\cos^4\theta\, dx_s\, dy_s. \tag{7-15}$$

It has the characteristic z^{-2} falloff and exhibits the $\cos^4\theta$ behavior of a Lambertian source (see Grum and Becherer, 1979).

Spectral Irradiance on a Hemisphere

The *x-z* section of the hemispherical surface of observation is shown in Fig. 7-3. The center of the hemisphere is at the aperture point $(x_{s0}, y_{s0}, 0)$ and its radius is R. Also shown is a plane parallel to the aperture plane intersecting the hemisphere at a z value greater than zero.

Recall the review of the RS theory in Appendix 5.1 and the application of this theory in Chapter 5. Observe that the spatial Fourier transforms of the relevant functions are on the x, y coordinates for a fixed value of $z > 0$. Furthermore, the product relationships, Eqs. (A5-8) and (5-5), indicate that the spectrum of the diffracted field and/or the MCF is found for constant z values when it is known for the aperture at $z = 0$. That is, the RS theory provides the diffracted amplitude and/or the MSDF in *planes* parallel to the aperture plane. More generally, the RS theory provides the diffracted field on a surface dictated by the surface over which the Green's function is made to vanish (see Marathay, 1975; Marathay and Prasad, 1980; and Appendix 7.1).

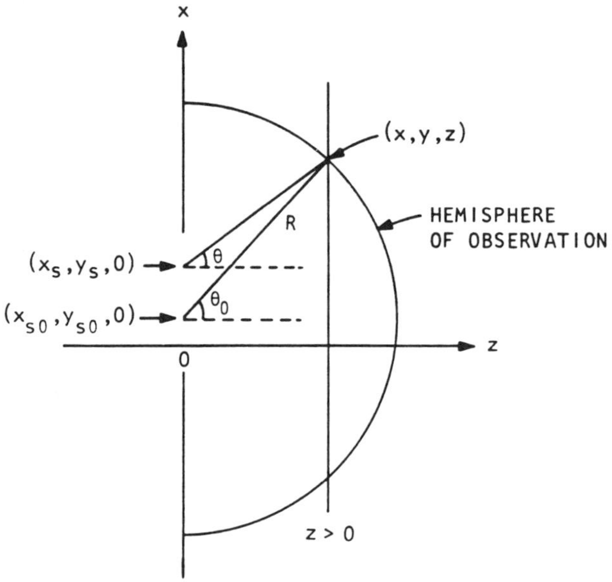

Fig. 7-3. Geometry for the diffraction problem and the hemisphere of observation.

Thus, the spectral irradiance $\hat{\mathsf{E}}_R$ on the hemisphere of radius R may be found by taking a projection of $\hat{\mathsf{E}}$ on the plane $z > 0$ given by Eq. (7-12). Let $d\sigma_P = dx\,dy$ be the area element on the plane $z > 0$. The spectral radiant power is $\hat{\mathsf{E}}\,d\sigma_P$. The corresponding element $d\sigma_R$ at the intersection of the hemisphere and the plane is $d\sigma_R = d\sigma_P \cos\theta_0$. The spectral radiant power on the hemisphere is $\hat{\mathsf{E}}_R\,d\sigma_R = \hat{\mathsf{E}}_R\,d\sigma_P \cos\theta_0$. We therefore conclude that (up to a constant involving a ratio of spectral powers)

$$\hat{\mathsf{E}}_R(x, y, \nu) = \frac{\hat{\mathsf{E}}(x, y, z, \nu)}{\cos\theta_0}, \tag{7-16}$$

where the z value is fixed by the relationship

$$z = +\left(R^2 - x^2 - y^2\right)^{1/2}.$$

The combination of Eqs. (7-14) and (7-16) yields

$$\hat{\mathsf{E}}_R(x, y, \nu) \simeq \frac{1}{\pi\cos\theta_0}\int\!\!\int \hat{\mathsf{M}}(x_s, y_s, 0, \nu)\frac{\cos^2\theta}{\rho^2}\,dx_s\,dy_s. \tag{7-17}$$

This is the spectral irradiance on the hemisphere. The differential contribution $d\hat{\mathsf{E}}_R$ due to an area element $dx_s\,dy_s$ is

$$d\hat{\mathsf{E}}_R \simeq \frac{1}{\pi\rho^2}\frac{\cos^2\theta}{\cos\theta_0}\hat{\mathsf{M}}(x_s, y_s, 0, \nu)\,dx_s\,dy_s. \tag{7-18}$$

If the area element is chosen at $(x_{s0}, y_{s0}, 0)$, which is the center of the hemisphere, then $\rho = R$ and $\theta = \theta_0$. The spectral irradiance $d\hat{\mathsf{E}}_R$ is proportional to $\cos\theta_0$, a characteristic of a Lambertian source. In general, if the radius R is large enough compared to the aperture dimensions, $\rho \simeq R$, the variation of $\cos\theta$ from the value $\cos\theta_0$ may be negligible over the aperture. To this approximation, any area element in the aperture will exhibit a $\cos\theta_0$ variation of the spectral irradiance on the hemisphere.

Spectral Radiant Intensity

The solid angle integration of the spectral radiant intensity over the right half-space yields the spectral radiant power. Therefore, the spectral radiant intensity may be calculated by using

$$\hat{\mathsf{I}} = \frac{\partial\hat{\mathbf{\Phi}}}{\partial\Omega}. \tag{7-19}$$

The spectral radiant power in some plane $z > 0$, analogous to Eq. (7-3) or Eqs. (5-24) and (5-25), is

$$\hat{\mathbf{\Phi}}(z, \nu) = \int\!\!\int \hat{\mathsf{E}}(x, y, z, \nu)\,dx\,dy$$

$$\equiv \mathsf{C}\int\!\!\int \hat{\Gamma}(x, y, x, y, z, \nu)\,dx\,dy, \tag{7-20}$$

which in the spatial frequency domain has the form

$$\hat{\mathbf{\Phi}}(z, \nu) = \mathsf{C}\int\!\!\int_{-\infty}^{\infty} \mathring{\Gamma}(\kappa p, \kappa q, \kappa p, \kappa q, z, \nu)\,d(\kappa p)\,d(\kappa q). \tag{7-21}$$

The concept of $\hat{\mathsf{I}}$ applies to distances far away from the source, where we may neglect the evanescent waves contained in $p^2 + q^2 > 1$. In terms of the spherical polar angles of Eq. (7-5), we may approximate Eq. (7-21) to

$$\hat{\mathbf{\Phi}}(z, \nu) = \mathsf{C}\iint_{1/2} \mathring{\Gamma}(\kappa p, \kappa q, \kappa p, \kappa q, z, \nu)\kappa^2 m\,d\Omega,$$

wherein the arguments of $\mathring{\Gamma}$ are still written in terms of p, q to save space. The spectral radiant intensity is then

$$\hat{\mathrm{I}}(z, \hat{\mathbf{n}}, \nu) = \frac{\partial \hat{\mathbf{\Phi}}(z, \nu)}{\partial \Omega} = \kappa^2 m \mathrm{C} \mathring{\Gamma}(\kappa p, \kappa q, \kappa p, \kappa q, z, \nu). \qquad (7\text{-}22)$$

Due to the product relationship of Eq. (5-5) and after neglecting the evanescent waves, we get

$$\hat{\mathrm{I}}(z, \hat{\mathbf{n}}, \nu) = \kappa^2 m \mathrm{C} \mathring{\Gamma}(\kappa p, \kappa q, \kappa p, \kappa q, 0, \nu).$$

For the case of a completely noncoherent source of Eq. (7-10), whose spectral radiant exitance is uniform or otherwise, the above expression reduces to

$$\hat{\mathrm{I}}(z, \hat{\mathbf{n}}, \nu) = \frac{m}{\pi} \hat{\mathbf{\Phi}}(0, \nu), \qquad (7\text{-}23)$$

where $\hat{\mathbf{\Phi}}(0, \nu)$ is the spectral radiant power of the source. The angular dependence of $\hat{\mathrm{I}}$ is $m = \cos\theta$, and the above relationship is consistent with those given in Eq. (7-9) for a Lambertian source.

The three properties of the noncoherent source discussed in these last three subsections show the likeness of the noncoherent source to the Lambertian source of c-radiometry. We shall return to this comparison after developing the formulation of g-radiometry.

GENERALIZED RADIOMETRY

In g-radiometry we wish to formulate a scheme of functions to account for the wave nature, propagation, partial coherence, and detection of light. The wave amplitude propagates, meaning that if it is known, say on a plane or an aperture, it can be determined on a parallel plane downstream by means of diffraction theory. However, it is not accessible to measurement. On the other hand, the optical intensity, which is real and nonnegative, is directly related to quantities that are detectable. However, it cannot propagate; that is, the optical intensity in a plane cannot be determined by knowledge of the optical intensity *alone* in a previous plane.

The attributes of propagation and accessibility to measurement are indeed rolled into one function, namely, the mutual coherence function. It is most suited for constructing the functions of generalized radiometry.

The MCF $\Gamma_{12}(\tau)$ is a function of two space points and the time delay τ. The g-radiometric functions defined in terms of it are then also functions of two points. Furthermore, the angle dependence will also enter in pairs. There is no reason to presuppose that the functions of g-radiometry and their interrelationships look like and have the same form as those of c-radiometry. Under the appropriate conditions and/or approximations, the scheme of g-radiometry should reduce to that of c-radiometry discussed in the beginning of this chapter. We shall continue to use the same symbol scheme for c-radiometry to define the g-radiometric functions, except that their respective list of arguments will be longer. To save space we shall use the two-dimensional vectors used in Chapter 6. Thus, for the x, y coordinates in a plane labeled by $z =$ constant, we shall write

$$\mathbf{x} = \hat{\mathbf{i}}\, x + \hat{\mathbf{j}}\, y.$$

For the pair of spatial frequencies, use

$$\mathbf{f} = \hat{\mathbf{i}}\, f + \hat{\mathbf{j}}\, g.$$

And for the pair of direction cosines, use

$$\mathbf{p} = \hat{\mathbf{i}}\, p + \hat{\mathbf{j}}\, q,$$

for which the third direction cosine is

$$m = \begin{cases} (1 - \mathbf{p} \cdot \mathbf{p})^{1/2}, & p^2 + q^2 = \mathbf{p} \cdot \mathbf{p} \leq 1 \\ +i(\mathbf{p} \cdot \mathbf{p} - 1)^{1/2}, & p^2 + q^2 = \mathbf{p} \cdot \mathbf{p} > 1. \end{cases}$$

The area element is denoted by

$$d\mathbf{x} \equiv dx\, dy,$$

and similarly, $d\mathbf{f} \equiv df\, dg$ and $d\mathbf{p} \equiv dp\, dq$. The caret (ˆ) will be used over a function symbol to denote the temporal Fourier transform, the tilde (˜) to denote the two-dimensional spatial Fourier transform, and the degree sign (°) to denote the spatial and temporal Fourier transform.

Let us begin with the already familiar spectral radiant power $\hat{\mathbf{\Phi}}(z, \nu)$ with units of W Hz^{-1} for the plane z. It is real and nonnegative. The *generalized* (total) *radiant power* may be defined through the Fourier relationship

$$\underset{[\text{W}]}{\mathbf{\Phi}(z, \tau)} = \int_0^\infty \underset{[\text{W Hz}^{-1}]}{\hat{\mathbf{\Phi}}(z, \nu)} \exp(-i 2\pi\nu\tau)\, d\nu. \tag{7-24}$$

In general, $\boldsymbol{\Phi}(z, \tau)$ is complex-valued. Its zero ordinate,

$$\boldsymbol{\Phi}(z,0) = \int_0^{\infty} \hat{\boldsymbol{\Phi}}(z, \nu)\, d\nu, \tag{7-25}$$

is real and positive and it corresponds to the total radiant power of c-radiometry.

If the source coherence function $\Gamma_{12}(\tau)$ is known, then the *generalized radiant exitance* may be defined by

$$\mathsf{M}(\mathbf{x}_1, \mathbf{x}_2, z, \tau) = \mathsf{C}\Gamma(\mathbf{x}_1, \mathbf{x}_2, z, \tau), \tag{7-26}$$

and the *generalized spectral radiant exitance* by

$$\hat{\mathsf{M}}(\mathbf{x}_1, \mathbf{x}_2, z, \nu) = \mathsf{C}\hat{\Gamma}(\mathbf{x}_1, \mathbf{x}_2, z, \nu), \tag{7-27}$$

with the Fourier relationship

$$\underset{[\mathrm{W\,m^{-2}}]}{\mathsf{M}(\mathbf{x}_1, \mathbf{x}_2, z, \tau)} = \int_0^{\infty} \underset{[\mathrm{W\,m^{-2}\,Hz^{-1}}]}{\hat{\mathsf{M}}(\mathbf{x}_1, \mathbf{x}_2, z, \nu)} \exp(-i2\pi\nu\tau)\, d\nu \tag{7-28}$$

Although M is, in general, a complex cross correlation, its zero ordinate ($\mathbf{x}_1 = \mathbf{x}_2$ and $\tau = 0$) is real and positive,

$$\mathsf{M}(\mathbf{x}, \mathbf{x}, z, 0) = \int_0^{\infty} \hat{\mathsf{M}}(\mathbf{x}, \mathbf{x}, z, \nu)\, d\nu, \tag{7-29}$$

where $\hat{\mathsf{M}}(\mathbf{x}, \mathbf{x}, z, \nu)$ is real and nonnegative. These last two functions correspond to the c-radiometric functions.

The relationship to the generalized radiant power is through the area integral,

$$\boldsymbol{\Phi}(z, \tau) = \int_{\mathscr{A}} \mathsf{M}(\mathbf{x}, \mathbf{x}, z, \tau)\, d\mathbf{x}, \tag{7-30}$$

wherein the limit $\tau = 0$ corresponds to the c-radiometric functions. Its spectral version is

$$\hat{\boldsymbol{\Phi}}(z, \nu) = \int_{\mathscr{A}} \hat{\mathsf{M}}(\mathbf{x}, \mathbf{x}, z, \nu)\, d\mathbf{x}. \tag{7-31}$$

When necessary, the area integral in the above relationships may be shown explicitly by using $\mathscr{A}(x, y)$ as in Eq. (7-3).

For the radiation incident on a plane z, one uses the symbol $\mathsf{E}(\mathbf{x}_1, \mathbf{x}_2, z, \tau)$ for the *generalized irradiance* in place of M in Eqs. (7-26) and (7-30). Thus,

$$\boldsymbol{\Phi}(z, \tau) = \int_{\mathscr{A}} \mathsf{E}(\mathbf{x}, \mathbf{x}, z, \tau)\, d\mathbf{x}. \tag{7-32}$$

The spectral version of this is obtained by using the symbol $\hat{\mathsf{E}}(\mathbf{x}_1, \mathbf{x}_2, z, \nu)$ for the *generalized spectral irradiance* in place of $\hat{\mathsf{M}}$ in Eqs. (7-27) and (7-31).

Generalized Radiance

The definition of generalized radiance and its reduction to the c-radiometric function is not as straightforward as in the above discussion. For one thing, a mixed function of both the space coordinate and the Fourier conjugate angle argument is not acceptable. By performing experiments, such a joint function cannot be accurately determined for both of its arguments simultaneously. The situation is different for a *known* function, for example, the Fourier kernel itself, $\exp(i2\pi\mathbf{x} \cdot \mathbf{p})$, which depends on both $\mathbf{x}$ and $\mathbf{p}$.

We begin with the spatial and temporal Fourier transform of $\Gamma_{12}(\tau)$, namely,

$$\Gamma(\mathbf{x}_1, \mathbf{x}_2, z, \tau) = \int_0^{\infty} d\nu \iint_{-\infty}^{\infty} d(\kappa\mathbf{p}_1)\, d(\kappa\mathbf{p}_2)\, \mathring{\Gamma}(\kappa\mathbf{p}_1, \kappa\mathbf{p}_2, z, \nu) \times \exp\{-i2\pi[\nu\tau - \kappa(\mathbf{p}_1 \cdot \mathbf{x}_1 - \mathbf{p}_2 \cdot \mathbf{x}_2)]\}, \tag{7-33}$$

and the inverse

$$\mathring{\Gamma}(\kappa\mathbf{p}_1, \kappa\mathbf{p}_2, z, \nu) = \int_{-\infty}^{\infty} d\tau \iint_{-\infty}^{\infty} d\mathbf{x}_1\, d\mathbf{x}_2\, \Gamma(\mathbf{x}_1, \mathbf{x}_2, z, \tau) \times \exp\{+i2\pi[\nu\tau - \kappa(\mathbf{p}_1 \cdot \mathbf{x}_1 - \mathbf{p}_2 \cdot \mathbf{x}_2)]\}. \tag{7-34}$$

We convert the first of these relationships to integrations over the angles by using Eq. (7-5),

$$\Gamma(\mathbf{x}_1, \mathbf{x}_2, z, \tau) = \int_0^{\infty} d\nu \exp(-i2\pi\nu\tau)\, \kappa^4 \iint_{-\infty}^{\infty} m_1\, d\Omega_1\, m_2\, d\Omega_2 \times \mathring{\Gamma}(\kappa\mathbf{p}_1, \kappa\mathbf{p}_2, z, \nu) \exp[i2\pi\kappa(\mathbf{p}_1 \cdot \mathbf{x}_1 - \mathbf{p}_2 \cdot \mathbf{x}_2)],$$

wherein $d\Omega$ with its subscripts may be interpreted as the solid angle element for real waves.

Now let us define a function $\hat{F}$ by

$$\hat{F}(\mathbf{x}_1, \mathbf{x}_2, z; \mathbf{p}_1, \mathbf{p}_2; \nu) \equiv \kappa^4 \mathring{\Gamma}(\kappa \mathbf{p}_1, \kappa \mathbf{p}_2, z, \nu) \exp[i2\pi\kappa(\mathbf{p}_1 \cdot \mathbf{x}_1 - \mathbf{p}_2 \cdot \mathbf{x}_2)] \tag{7-35}$$

and its temporal Fourier transform by

$$F(\mathbf{x}_1, \mathbf{x}_2, z; \mathbf{p}_1, \mathbf{p}_2; \tau) = \int_0^\infty \hat{F}(\mathbf{x}_1, \mathbf{x}_2, z; \mathbf{p}_1, \mathbf{p}_2; \nu) \exp(-i2\pi\nu\tau)\, d\nu. \tag{7-36}$$

In terms of F the MCF may be reexpressed as

$$\Gamma(\mathbf{x}_1, \mathbf{x}_2, z, \tau) = \int\int_{-\infty}^{\infty} F(\mathbf{x}_1, \mathbf{x}_2, z; \mathbf{p}_1, \mathbf{p}_2; \tau) m_1\, d\Omega_1\, m_2\, d\Omega_2. \tag{7-37}$$

Ordinarily, a mixed function of both $\mathbf{x}$ and $\kappa\mathbf{p}$ is unacceptable since they are Fourier conjugate variables. However, there is no problem in the definition of $\hat{F}$ in Eq. (7-35) because its $\mathbf{x}$ and $\mathbf{p}$ dependence comes solely through a *known* function, namely, the exponential, which in turn is multiplied by $\mathring{\Gamma}$, which depends on $\mathbf{p}$ and is independent of $\mathbf{x}$. Furthermore, $\mathring{\Gamma}$ is indirectly accessible by measurement through the MCF $\Gamma_{12}(\tau)$.

Now $\Gamma_{12}(\tau)$ is a complex cross correlation of the wave amplitude from the point $\mathbf{x}_1$ in the z plane and the amplitude from $\mathbf{x}_2$ also in the z plane and with a time delay τ. Similarly, the function F is a complex cross correlation of the wave amplitude from the point $\mathbf{x}_1$ in the z plane in the direction $\mathbf{p}_1$ and from $\mathbf{x}_2$ also in the z plane but in the direction $\mathbf{p}_2$ with a time delay τ. For the ensemble average definition of F, refer to Marathay (1976a).

Let the *generalized radiance* be defined in terms of F by

$$\mathsf{L}(\mathbf{x}_1, \mathbf{x}_2, z; \mathbf{p}_1, \mathbf{p}_2; \tau) \equiv \mathsf{C} F(\mathbf{x}_1, \mathbf{x}_2, z; \mathbf{p}_1, \mathbf{p}_2; \tau) \tag{7-38}$$

and the *generalized spectral radiance* by

$$\hat{\mathsf{L}}(\mathbf{x}_1, \mathbf{x}_2, z; \mathbf{p}_1, \mathbf{p}_2; \nu) \equiv \mathsf{C} \hat{F}(\mathbf{x}_1, \mathbf{x}_2, z; \mathbf{p}_1, \mathbf{p}_2; \nu). \tag{7-39}$$

With these proposed definitions, Eq. (7-37) may be rewritten as

$$\underset{[\mathrm{W\,m^{-2}}]}{\mathsf{M}(\mathbf{x}_1, \mathbf{x}_2, z, \tau)} = \int\int_{-\infty}^{\infty} \underset{[\mathrm{W\,m^{-2}\,sr^{-2}}]}{\mathsf{L}(\mathbf{x}_1, \mathbf{x}_2, z; \mathbf{p}_1, \mathbf{p}_2; \tau)} m_1\, d\Omega_1\, m_2\, d\Omega_2 \tag{7-40}$$

and its spectral version

$$\hat{\mathsf{M}}(\mathbf{x}_1, \mathbf{x}_2, z, \nu) = \iint_{-\infty}^{\infty} \hat{\mathsf{L}}(\mathbf{x}_1, \mathbf{x}_2, z; \mathbf{p}_1, \mathbf{p}_2; \nu) m_1\, d\Omega_1\, m_2\, d\Omega_2.$$

$$[\mathrm{W\ m^{-2}\ Hz^{-1}}] \qquad [\mathrm{W\ m^{-2}\ sr^{-2}\ Hz^{-1}}] \qquad (7\text{-}41)$$

The units of L and $\hat{\mathsf{L}}$ shown in the above equations contain sr^{-2} corresponding to the two directions labeled by $\mathbf{p}_1$ and $\mathbf{p}_2$. Observe that this is quite different from the units of (spectral) radiance which depends on a single direction in c-radiometry. Furthermore, the complex function L (or $\hat{\mathsf{L}}$) does *not* reduce to its counterpart in c-radiometry when we put $\mathbf{x}_1 = \mathbf{x}_2$, $\mathbf{p}_1 = \mathbf{p}_2$, and $\tau = 0$, and it contains both real and evanescent waves. We shall take up the question of contact with c-radiometry after the discussion about intensity.

Generalized Radiant Intensity

We begin with the generalized radiant power of Eq. (7-30) and express it in terms of L,

$$\begin{aligned} \boldsymbol{\Phi}(z, \tau) &= \int_{\mathscr{A}} \mathsf{M}(\mathbf{x}, \mathbf{x}, z, \tau)\, d\mathbf{x} \\ &= \int_{\mathscr{A}} \iint_{-\infty}^{\infty} \mathsf{L}(\mathbf{x}, \mathbf{x}, z; \mathbf{p}_1, \mathbf{p}_2; \tau) m_1\, d\Omega_1\, m_2\, d\Omega_2\, d\mathbf{x}. \qquad (7\text{-}42) \end{aligned}$$

A rearrangement of this expression allows us to define the *generalized radiant intensity* I as that function whose double solid angle integral yields the generalized radiant power. Thus,

$$\mathsf{I}(z; \mathbf{p}_1, \mathbf{p}_2; \tau) = m_1 m_2 \int_{\mathscr{A}} \mathsf{L}(\mathbf{x}, \mathbf{x}, z; \mathbf{p}_1, \mathbf{p}_2; \tau)\, d\mathbf{x} \qquad (7\text{-}43)$$

so that

$$\boldsymbol{\Phi}(z, \tau) = \iint_{-\infty}^{\infty} \mathsf{I}(z; \mathbf{p}_1, \mathbf{p}_2; \tau)\, d\Omega_1\, d\Omega_2.$$

$$[\mathrm{W}] \qquad [\mathrm{W\ sr^{-2}}] \qquad (7\text{-}44)$$

The function $\mathsf{I}\, d\Omega_1\, d\Omega_2$ describes the angular correlation of the total radiation from the source in the direction $\mathbf{p}_1$ within solid angle $d\Omega_1$ and from the direction $\mathbf{p}_2$ within $d\Omega_2$ with a time delay τ.

The spectral version of Eq. (7-44) reads

$$\hat{\Phi}(z,\nu) = \iint_{-\infty}^{\infty} \hat{\mathsf{I}}(z;\mathbf{p}_1,\mathbf{p}_2;\nu)\,d\Omega_1\,d\Omega_2, \qquad (7\text{-}45)$$

$$[\mathrm{W\,Hz^{-1}}] \qquad\qquad [\mathrm{W\,sr^{-2}\,Hz^{-1}}]$$

where $\hat{\mathsf{I}}$ is the *generalized spectral radiant intensity.* The function $\hat{\mathsf{I}}$ is the temporal Fourier transform of I; therefore we also have

$$\hat{\mathsf{I}}(z;\mathbf{p}_1,\mathbf{p}_2;\nu) = m_1 m_2 \int_{\mathscr{A}} \hat{\mathsf{L}}(\mathbf{x},\mathbf{x},z;\mathbf{p}_1,\mathbf{p}_2;\nu)\,d\mathbf{x}. \qquad (7\text{-}46)$$

Table 7-2 lists the spectral functions and their relationships in g-radiometry. The temporal Fourier transform leads to the total quantities of g-radiometry as functions of time delay τ.

The expression of $\hat{\mathsf{I}}$ of Eq. (7-46) includes an area integration, and as in Eq. (7-3) it is over the source shape $\mathscr{A}(\mathbf{x})$, or the region of interest in the receiving plane. A substitution of the radiance $\hat{\mathsf{L}}$ in terms of $\hat{F}$ from Eqs. (7-35) and (7-39) shows that

$$\hat{\mathsf{I}}(z;\mathbf{p}_1,\mathbf{p}_2;\nu) = \mathsf{C}\kappa^4 m_1 m_2 \mathring{\Gamma}(\kappa\mathbf{p}_1,\kappa\mathbf{p}_2,z,\nu)\,\tilde{\mathscr{A}}^*(\kappa\mathbf{p}_1-\kappa\mathbf{p}_2), \qquad (7\text{-}47)$$

where the asterisk denotes the complex conjugate and we have used the

TABLE 7-2. INTERRELATIONSHIPS AMONG THE SPECTRAL FUNCTIONS OF GENERALIZED RADIOMETRY

$\hat{\Phi}(z,\nu)$ [W Hz^{-1}]	$=$	$\int_{\mathscr{A}}\iint_{-\infty}^{\infty} \hat{\mathsf{L}}(\mathbf{x},\mathbf{x},z;\mathbf{p}_1,\mathbf{p}_2;\nu)m_1\,d\Omega_1\,m_2\,d\Omega_2\,d\mathbf{x}$ [W m^{-2} sr^{-2} Hz^{-1}]
$\hat{\mathsf{M}}(\mathbf{x}_1,\mathbf{x}_2,z,\nu)$ [W m^{-2} Hz^{-1}]	$=$	$\iint_{-\infty}^{\infty} \hat{\mathsf{L}}(\mathbf{x}_1,\mathbf{x}_2,z;\mathbf{p}_1,\mathbf{p}_2;\nu)m_1\,d\Omega_1\,m_2\,d\Omega_2$
$\hat{\Phi}(z,\nu)$	$=$	$\int_{\mathscr{A}} \hat{\mathsf{M}}(\mathbf{x},\mathbf{x},z,\nu)\,d\mathbf{x}$
$\hat{\mathsf{I}}(z;\mathbf{p}_1,\mathbf{p}_2;\nu)$ [W sr^{-2} Hz^{-1}]	$=$	$m_1 m_2 \int_{\mathscr{A}} \hat{\mathsf{L}}(\mathbf{x},\mathbf{x},z;\mathbf{p}_1,\mathbf{p}_2;\nu)\,d\mathbf{x}$
$\hat{\Phi}(z,\nu)$	$=$	$\iint_{-\infty}^{\infty} \hat{\mathsf{I}}(z;\mathbf{p}_1,\mathbf{p}_2;\nu)\,d\Omega_1\,d\Omega_2$

Fourier relationship,

$$\tilde{\mathscr{A}}(\kappa \mathbf{p}_1 - \kappa \mathbf{p}_2) = \int \mathscr{A}(\mathbf{x}) \exp[-i2\pi\kappa(\mathbf{p}_1 - \mathbf{p}_2) \cdot \mathbf{x}]\, d\mathbf{x}. \qquad (7\text{-}48)$$

For $\mathbf{p}_1 = \mathbf{p}_2$, $\tilde{\mathscr{A}}(0)$ equals $\mathscr{A}$, which is the area considered. As with the spectral radiance $\hat{\mathsf{L}}$ (or with L), we repeat the caution that the spectral radiant intensity $\hat{\mathsf{I}}$ does *not* reduce to its counterpart in c-radiometry if we simply put $\mathbf{p}_1$ equal to $\mathbf{p}_2$ and that its units contain sr^{-2} instead of sr^{-1}. For the ensemble-averaged definition of generalized radiant intensity, refer to Marathay (1976b).

In the formulation of g-radiometry, the position coordinates as well as the direction cosines come in pairs: $\mathbf{x}_1, \mathbf{x}_2$ and $\mathbf{p}_1, \mathbf{p}_2$. It is frequently advantageous to work with the average and difference variables. On several occasions we have used such variables with $\mathbf{x}_1$ and $\mathbf{x}_2$; the same scheme may be used with $\mathbf{p}_1$ and $\mathbf{p}_2$. Thus, put

$$\mathbf{p} = \tfrac{1}{2}(\mathbf{p}_1 + \mathbf{p}_2) \quad \text{and} \quad \mathbf{p}_{12} = \mathbf{p}_1 - \mathbf{p}_2. \qquad (7\text{-}49)$$

The spherical polar angles θ_1, ϕ_1 and θ_2, ϕ_2 are used, respectively, with $\mathbf{p}_1$ and $\mathbf{p}_2$. Likewise, we introduce

$$\begin{aligned} p &= \sin\theta\cos\phi, & p_{12} &= \sin\theta_{12}\cos\phi_{12}, \\ q &= \sin\theta\sin\phi, & q_{12} &= \sin\theta_{12}\sin\phi_{12}. \end{aligned} \qquad (7\text{-}50)$$

The respective third direction cosines, namely, m and m_{12}, are constrained by the conditions

$$\begin{aligned} p^2 + q^2 + m^2 &= 1, \\ p_{12}^2 + q_{12}^2 + m_{12}^2 &= 1. \end{aligned} \qquad (7\text{-}51)$$

The relationship of m and m_{12} with m_1 and m_2 is complicated, but that does not in any way reduce the utility of these variables in dealing with multiple integrals. With the differential area elements we found that $d\mathbf{x}_1\, d\mathbf{x}_2 = d\mathbf{x}\, d\mathbf{x}_{12}$; likewise with the appropriate Jacobian for the change of variables, the following relationships are established:

$$\begin{aligned} dp_1\, dq_1\, dp_2\, dq_2 &= dp\, dq\, dp_{12}\, dq_{12}, \\ dp_1\, dq_1\, dp_2\, dq_2 &= \cos\theta\, d\Omega \cos\theta_{12}\, d\Omega_{12} \end{aligned} \qquad (7\text{-}52)$$

where $d\Omega$ and $d\Omega_{12}$ are the corresponding solid angle elements for the average and difference variables. The abbreviated notation,

$$d(\kappa\mathbf{p}) = \kappa^2\, d\mathbf{p} = \kappa^2\, dp\, dq,$$
$$d(\kappa\mathbf{p}_{12}) = \kappa^2\, d\mathbf{p}_{12} = \kappa^2\, dp_{12}\, dq_{12}, \tag{7-53}$$

will be used. Finally, the combination of variables occurring in the Fourier kernel has the form

$$\kappa(\mathbf{p}_1 \cdot \mathbf{x}_1 - \mathbf{p}_2 \cdot \mathbf{x}_2) = \kappa(\mathbf{p} \cdot \mathbf{x}_{12} + \mathbf{p}_{12} \cdot \mathbf{x}). \tag{7-54}$$

From this relationship it is clear that the Fourier conjugate variables are $\kappa\mathbf{p}$, $\mathbf{x}_{12}$ and $\kappa\mathbf{p}_{12}, \mathbf{x}$.

We shall now apply the formalism of g-radiometry to the following three cases: (i) uniform noncoherent source, (ii) partially coherent source, and (iii) a Carter–Wolf source. In each case we shall study the conditions under which contact with c-radiometry may be established. We will then also treat the cases of the completely coherent source and the blackbody source.

Uniform Noncoherent Source and Conventional Radiometry

Consider a noncoherent source whose mutual spectral density function (MSDF) is given by

$$\hat{\Gamma}_s(\mathbf{x}_1, \mathbf{x}_2, 0, \nu) = \hat{I}_s(0, \nu)\, \mathcal{Q}(\mathbf{x}_1) \frac{2J_1(kr_{12})}{kr_{12}}, \tag{7-55}$$

where $r_{12} = (x_{12}^2 + y_{12}^2)^{1/2}$ and the subscript s from the source coordinates has been dropped for notational simplicity. The optical spectral intensity (OSI) of the uniform source is $\hat{I}_s(0, \nu)$, and the shape of the source in the $z = 0$ plane is $\mathcal{Q}(\mathbf{x}_1)$. The chosen form of the MSDF is a good approximation for the noncoherent source, whose coherence, described by $2J_1(kr_{12})/kr_{12}$, is very narrow compared to any dimension of the source area $\mathcal{Q}$. This approximation makes it simpler to reveal the salient features of the transition to c-radiometry.

The Besinc form used above is one of a class of functions discussed in Chapter 4 whose spatial frequency spectrum (SFS) is constant (or nearly so) over the real plane waves. In particular, the above coherence function has the property

$$\int_{-\infty}^{\infty} \frac{2J_1(kr_{12})}{kr_{12}} \exp(-i2\pi\kappa\mathbf{p} \cdot \mathbf{x}_{12})\, d\mathbf{x}_{12} = \frac{\lambda^2}{\pi} \operatorname{cyl}\left[\frac{(p^2 + q^2)^{1/2}}{1}\right], \tag{7-56}$$

where the variable $\mathbf{p}$ is the average—$\mathbf{p} = \frac{1}{2}(\mathbf{p}_1 + \mathbf{p}_2)$—and the integral is zero for evanescent waves—$p^2 + q^2 > 1$.

To study the g-radiometry of such a source we need the double Fourier transform over both points $\mathbf{x}_1$ and $\mathbf{x}_2$. Thus,

$$\mathring{\Gamma}(\kappa\mathbf{p}_1, \kappa\mathbf{p}_2, 0, \nu) = \frac{\lambda^2}{\pi}\hat{I}_s(0, \nu)\,\tilde{\mathscr{A}}(\kappa\mathbf{p}_1 - \kappa\mathbf{p}_2)\,\mathrm{cyl}\left[\frac{(p^2+q^2)^{1/2}}{1}\right]. \tag{7-57}$$

Expressions for the g-radiometric functions are found by starting with the generalized spectral radiant exitance for the source,

$$\hat{\mathsf{M}}(\mathbf{x}_1, \mathbf{x}_2, 0, \nu) = \mathsf{C}\hat{I}_s(0, \nu)\,\mathscr{A}(\mathbf{x}_1)\frac{2J_1(kr_{12})}{kr_{12}}, \tag{7-58}$$

which for the generalized spectral radiant power gives

$$\begin{aligned}\hat{\boldsymbol{\Phi}}(0, \nu) &= \int_{\mathscr{A}} \hat{\mathsf{M}}(\mathbf{x}, \mathbf{x}, 0, \nu)\, d\mathbf{x} \\ &= \mathsf{C}\mathscr{A}\hat{I}_s(0, \nu).\end{aligned} \tag{7-59}$$

Use of Eqs. (7-39), (7-35), and (7-57) gives the generalized spectral radiance,

$$\begin{aligned}\hat{\mathsf{L}}(\mathbf{x}_1, \mathbf{x}_2, 0; \mathbf{p}_1, \mathbf{p}_2; \nu) = {}& \frac{\kappa^2}{\pi}\mathsf{C}\hat{I}_s(0, \nu)\,\tilde{\mathscr{A}}(\kappa\mathbf{p}_1 - \kappa\mathbf{p}_2) \\ & \times \mathrm{cyl}\left[\frac{(p^2+q^2)^{1/2}}{1}\right]\exp[i2\pi\kappa(\mathbf{p}_1 \cdot \mathbf{x}_1 - \mathbf{p}_2 \cdot \mathbf{x}_2)].\end{aligned} \tag{7-60}$$

The generalized spectral radiant intensity is found by starting with Eq. (7-47) to give

$$\hat{\mathsf{I}}(0; \mathbf{p}_1, \mathbf{p}_2; \nu) = \frac{\kappa^2}{\pi}\mathsf{C}m_1 m_2 \hat{I}_s(0, \nu)\,|\tilde{\mathscr{A}}(\kappa\mathbf{p}_1 - \kappa\mathbf{p}_2)|^2 \mathrm{cyl}\left[\frac{(p^2+q^2)^{1/2}}{1}\right] \tag{7-61}$$

The chosen coherence function depends on the coordinate difference $\mathbf{x}_{12}$, which in turn leads to the restriction of constant SFS over the average variable, $\mathbf{p} = \frac{1}{2}(\mathbf{p}_1 + \mathbf{p}_2)$. The condition on the difference variable, $\mathbf{p}_{12} = \mathbf{p}_1$

$-\mathbf{p}_2$, is imposed by the nature of the transform $\tilde{\mathscr{A}}$ of the shape of the uniform source. With the expressions given above, the interrelationships among the spectral functions as given in Table 7-2 are easily verified for self-consistency by using the average and difference variables.

To establish contact with c-radiometry, let us study the transform $\tilde{\mathscr{A}}$ of the source shape function $\mathscr{A}(\mathbf{x})$. As an example, let the uniform noncoherent source be rectangular in shape with sides $2a_0$ and $2b_0$. We find that

$$\tilde{\mathscr{A}}(\kappa\mathbf{p}_1 - \kappa\mathbf{p}_2) = \{2a_0 \operatorname{Sinc}[2\pi a_0 \kappa(p_1 - p_2)]\}\{2b_0 \operatorname{Sinc}[2\pi b_0 \kappa(q_1 - q_2)]\}. \tag{7-62}$$

Now consider two points $\mathbf{x}_1'$ and $\mathbf{x}_2'$ belonging to a plane $z =$ constant, parallel to the source plane so that $p_1 - p_2 \simeq (x_1' - x_2')/z$ and $q_1 - q_2 \simeq (y_1' - y_2')/z$. Thus the expression in Eq. (7-62) is precisely what one would obtain for the spatial coherence in the plane z by applying the van Cittert–Zernike theorem to the uniform, noncoherent, rectangular source. Observe that, for values of $|p_1 - p_2| \geq \lambda/2a_0$, the Sinc function either is zero or attains negligible values compared to unity. For a source of side $2a_0 = 1$ mm and $\lambda = 500$ nm, the ratio $\lambda/2a_0$ equals 0.5×10^{-3}. Thus, in the field there is significant correlation when the directions are separated by less than $\lambda/2a_0$, which is on the order of a milliradian. For a large enough source ($2a_0 \gg \lambda$) or if the size of λ is neglected (ray optical approximation), the Sinc function may be approximated by a delta function.

This example suggests that, in the general case of a uniform noncoherent source of shape $\mathscr{A}(\mathbf{x})$ with dimensions large enough compared to λ, the approximation

$$\tilde{\mathscr{A}}(\kappa\mathbf{p}_1 - \kappa\mathbf{p}_2) \simeq \delta(\kappa\mathbf{p}_1 - \kappa\mathbf{p}_2) \tag{7-63}$$

may be used. It amounts to *neglecting the coherence in the field*. We no longer distinguish between the directions $\mathbf{p}_1$ and $\mathbf{p}_2$.

To make contact with c-radiometry, we shall set

(i) $\mathbf{x}_1 = \mathbf{x}_2 = \mathbf{x}$;
(ii) $\tau = 0$ (when dealing with total functions);

limit our attention to:

(iii) Real waves: $p^2 + q^2 \leq 1$, that is, $0 \leq \theta \leq \pi/2$ and $0 \leq \phi \leq 2\pi$.
(iv) Neglect coherence in the field.

We proceed as follows: First we approximate the generalized spectral radiance function of Eq. (7-60) by

$$\hat{\mathsf{L}}(\mathbf{x}, \mathbf{x}, 0; \mathbf{p}_1, \mathbf{p}_2; \nu) \simeq \hat{\mathsf{L}}_0(\mathbf{x}, 0; \mathbf{p}_1; \nu)\, \delta(\mathbf{p}_1 - \mathbf{p}_2), \tag{7-64}$$

where the *reduced spectral radiance* function $\hat{\mathsf{L}}_0$ is defined by

$$\hat{\mathsf{L}}_0(0, 0; 0; \nu) \equiv \frac{1}{\pi} \mathsf{C} \hat{\mathcal{I}}_s(0, \nu). \tag{7-65}$$

This reduced function is a special case of the one to be defined in a later section in relation to the partially coherent source. In the present case it is independent of $\mathbf{x}$ as well as of $\mathbf{p}_1$, and we bear in mind that it applies only to real waves, $p_1^2 + q_1^2 \leq 1$. By use of the delta function we lost the band-limited nature of $\tilde{\mathcal{Q}}$. Therefore in Eqs. (7-64) and (7-65) we stipulate that $\hat{\mathsf{L}}_0$ is zero outside the source domain.

The presence of the delta function reduces the relationship

$$\hat{\mathsf{M}}(\mathbf{x}_1, \mathbf{x}_2, 0, \nu) = \int\int_{-\infty}^{\infty} \hat{\mathsf{L}}(\mathbf{x}_1, \mathbf{x}_2, 0; \mathbf{p}_1, \mathbf{p}_2; \nu) m_1\, d\Omega_1\, m_2\, d\Omega_2$$

to

$$\hat{\mathsf{M}}(\mathbf{x}, \mathbf{x}, 0, \nu) = \int_{1/2} \hat{\mathsf{L}}_0(\mathbf{x}, 0; \mathbf{p}_1; \nu) m_1\, d\Omega_1, \tag{7-66}$$

where we have put $\mathbf{x}_1 = \mathbf{x}_2$. In the example considered, both sides are independent of $\mathbf{x}$, and $\hat{\mathsf{L}}_0$ is independent of $\mathbf{p}_1$ as well. Thus, the angle integral is

$$\int_{1/2} m_1\, d\Omega_1 = \int_0^{2\pi} \int_0^{\pi/2} \cos\theta_1 \sin\theta_1\, d\theta_1\, d\phi_1 = \pi.$$

Therefore Eq. (7-66) leads to the Lambertian relationship of c-radiometry,

$$\hat{\mathsf{M}}(0, 0, 0, \nu) = \pi \hat{\mathsf{L}}_0(0, 0; 0; \nu). \tag{7-67}$$

In this relationship both $\hat{\mathsf{M}}$ and $\hat{\mathsf{L}}_0$ are real and nonnegative with units of W m^{-2} Hz^{-1} and W m^{-2} sr^{-1} Hz^{-1}, respectively. Observe that the units of $\hat{\mathsf{L}}_0$ are those of the spectral radiance of c-radiometry, which came about from neglecting field coherence. The functions of wave theory appearing in Eq. (7-67) are to be identified with those of c-radiometry. Furthermore, if the expression for $\hat{\mathsf{L}}_0$ is substituted from Eq. (7-65), the above relationship reduces to an identity.

Now recall the expression of spectral radiant power $\hat{\boldsymbol{\Phi}}$ of Table 7-2 and use in it the special form of $\hat{\mathsf{L}}$ of Eq. (7-64) to give

$$\hat{\boldsymbol{\Phi}}(0,\nu) = \iint_{1/2} \mathscr{C}(\mathbf{x})\,\hat{\mathsf{L}}_0(\mathbf{x},0;\mathbf{p}_1;\nu)m_1\,d\Omega_1\,d\mathbf{x}. \tag{7-68}$$

We may now define the *reduced spectral radiant intensity* $\hat{\mathsf{I}}_0$ as that function whose single solid angle integral yields the spectral radiant power. Thus, we define

$$\hat{\mathsf{I}}_0(0;\mathbf{p}_1;\nu) \equiv m_1\int \mathscr{C}(\mathbf{x})\,\hat{\mathsf{L}}_0(\mathbf{x},0;\mathbf{p}_1;\nu)\,d\mathbf{x} \tag{7-69}$$

so that

$$\hat{\boldsymbol{\Phi}}(0,\nu) = \int_{1/2} \hat{\mathsf{I}}_0(0;\mathbf{p}_1;\nu)\,d\Omega_1. \tag{7-70}$$

The units of $\hat{\mathsf{I}}_0$ are W sr^{-1} Hz^{-1}. For the case considered, we have

$$\hat{\mathsf{I}}_0(0;\mathbf{p}_1;\nu) = \mathsf{C}\,\frac{m_1}{\pi}\,\hat{I}_s(0,\nu)\,\mathscr{C}, \tag{7-71}$$

and therefore

$$\hat{\boldsymbol{\Phi}}(0,\nu) = \mathsf{C}\mathscr{C}\hat{I}_s(0,\nu). \tag{7-72}$$

The wave theoretic function $\hat{\mathsf{I}}_0$ is real and nonnegative; it is to be identified

TABLE 7-3. RADIOMETRY OF A NONCOHERENT SOURCE

$\hat{\mathsf{M}}(0,0,0,\nu)$	$=\pi\hat{\mathsf{L}}_0(0,0;0;\nu)$
[W m^{-2} Hz^{-1}]	[sr] [W m^{-2} sr^{-1} Hz^{-1}]
$\hat{\mathsf{I}}_0(0;\mathbf{p}_1;\nu)$	$=m_1\mathscr{C}\hat{\mathsf{L}}_0(0,0;0;\nu)$
[W sr^{-1} Hz^{-1}]	
$\hat{\boldsymbol{\Phi}}(0,\nu)$	$=\mathsf{C}\mathscr{C}\hat{I}_s(0,\nu)$
[W Hz^{-1}]	
where	
$\hat{\mathsf{L}}_0(0,0;0;\nu)$	$=\dfrac{1}{\pi}\,\mathsf{C}\hat{I}_s(0,\nu)$

with that of c-radiometry. Observe that for the noncoherent source the angle dependence of $\hat{\mathrm{I}}_0$ comes solely from $m_1 = \cos\theta_1$. The $\cos\theta_1$ dependence of spectral radiant intensity is characteristic of a Lambertian source.

In Table 7-3 we summarize the results obtained for a noncoherent source.

Review of Noncoherent Source Properties

We began our discussion of noncoherence in Chapter 4. The noncoherence was described by a sufficiently sharply peaked function of the coordinate difference $\mathbf{x}_1 - \mathbf{x}_2$. The sharply peaked function was one of a class of functions all having the property that their spatial frequency spectrum (SFS) is a constant over the real plane waves. Regardless of the detailed mathematical form, any one of the class of functions would have been just as good to describe noncoherence as any other. They all lead to the same result in the radiation zone, that is, for distances much larger than the wavelength.

The constant SFS is interpreted to mean that a uniform noncoherent source delivers the same amount of radiant power in every direction of travel of the real plane waves in the right half-space. The property of the uniformity of power distribution is also shared by the Lambertian source of c-radiometry.

The irradiance on the plane of observation calculated by means of the van Cittert–Zernike theorem for a noncoherent source shows the characteristic $1/z^2$ dependence for distances $z \gg \lambda$. The expression for the irradiance, Eq. (5-128), in the plane of observation in the radiation zone is independent of λ, suggesting a ray optical behavior in the field like that of a Lambertian source. As the coherence function of a partially coherent source is made progressively narrower it approaches the conditions of a noncoherent source. The plots in Fig. 5-18 showed how the beam power spreads as the radiation leaves the source. It was evident from these graphs that the beam from a noncoherent source spreads uniformly into the right half-space as soon as it leaves the source, as would be expected of a Lambertian source.

In the present chapter, we have studied the noncoherent source in the framework of the Rayleigh–Sommerfeld diffraction theory and derived three results. The first showed that the spectral irradiance in a plane parallel to the noncoherent source plane has a $\cos^4\theta$ dependence [see Eq. (7-15)]. The second showed that the spectral irradiance on a hemisphere has a $\cos\theta$ dependence [see Eq. (7-18)]. The third result showed that the spectral radiant intensity has a $\cos\theta$ dependence [see Eq. (7-23)]. These three results are also shared by the Lambertian source of c-radiometry. A moment's reflection will indicate that these results and the plots in Fig. 5-18 are entirely consistent with the interpretation of the constant SFS of the coherence function describing the uniform noncoherent source.

After formulating g-radiometry we established that, if we neglect field coherence for the case of a noncoherent source and set $\mathbf{x}_1 = \mathbf{x}_2$, the functions of g-radiometry reduce to those of c-radiometry. The results summarized in Table 7-3 are characteristic of a Lambertian source.

All this evidence gathered together points to the one important connection that *the mathematical idealization of a noncoherent source of the theory of partial coherence is the same as the idealization of the Lambertian source of c-radiometry*. Both these concepts are limiting forms and are only approachable at best to a good approximation by real sources.

Partially Coherent Source and Conventional Radiometry

Reduced Functions for Generalized Radiometry

Generalized radiometry as applied to a source of any state of coherence was discussed before and summarized in Table 7-2. The generalized functions are defined in terms of cross-correlation functions of the field variable or in terms of their spatial Fourier transforms. In general, the functions of generalized radiometry are complex and accessible to measurement to the same extent that the field correlation itself is an observable. The purpose of this section is to formulate a suitable set of reduced functions for g-radiometry so that its structure will look much like that of c-radiometry.

We assume that the MSDF $\hat{\Gamma}(\mathbf{x}_1, \mathbf{x}_2, 0, \nu)$, and hence the generalized spectral radiant exitance $\hat{\mathsf{M}}(\mathbf{x}_1, \mathbf{x}_2, 0, \nu)$, of a partially coherent source in the plane $z = 0$ is known. We begin with the relationship

$$\hat{\mathsf{M}}(\mathbf{x}_1, \mathbf{x}_2, 0, \nu) = \int\int_{-\infty}^{\infty} \hat{\mathsf{L}}(\mathbf{x}_1, \mathbf{x}_2, 0; \mathbf{p}_1, \mathbf{p}_2; \nu) m_1 \, d\Omega_1 \, m_2 \, d\Omega_2, \quad (7\text{-}73)$$

which, in terms of the average and difference variables of Eqs. (7-49) and (7-52), takes the form

$$\hat{\mathsf{M}}(\mathbf{x}_1, \mathbf{x}_2, 0, \nu) = \int\int_{-\infty}^{\infty} \hat{\mathsf{L}}(\mathbf{x}_1, \mathbf{x}_2, 0; \mathbf{p} + \tfrac{1}{2}\mathbf{p}_{12}, \mathbf{p} - \tfrac{1}{2}\mathbf{p}_{12}; \nu) m \, d\Omega \, m_{12} \, d\Omega_{12}. \quad (7\text{-}74)$$

To make contact with c-radiometry we consider coincident space points, but instead of neglecting field coherence we define the *reduced spectral radiance* by

$$\hat{\mathsf{L}}_0(\mathbf{x}, 0; \mathbf{p}; \nu) \equiv \int_{-\infty}^{\infty} \hat{\mathsf{L}}(\mathbf{x}, \mathbf{x}, 0; \mathbf{p} + \tfrac{1}{2}\mathbf{p}_{12}, \mathbf{p} - \tfrac{1}{2}\mathbf{p}_{12}; \nu) m_{12} \, d\Omega_{12}. \quad (7\text{-}75)$$

We bear in mind that, for contact with c-radiometry, $\hat{\mathsf{L}}_0$ is defined for real waves; and we observe that

$$\hat{\mathsf{M}}(\mathbf{x},\mathbf{x},0,\nu) = \int_{1/2} \hat{\mathsf{L}}_0(\mathbf{x},0;\mathbf{p};\nu)\, m\, d\Omega. \tag{7-76}$$

The units of $\hat{\mathsf{L}}_0$ are W m^{-2} sr^{-1} Hz^{-1}. The left-hand side is real and nonnegative and so is the right-hand side. However, the function $\hat{\mathsf{L}}_0$ itself is *not* in general nonnegative. To see this property we express it in terms of the total Fourier transform $\mathring{\Gamma}$ by putting Eqs. (7-39) and (7-35) into Eq. (7-75). The relationship is

$$\hat{\mathsf{L}}_0(\mathbf{x},0;\mathbf{p};\nu) = \mathsf{C}\kappa^2 \int_{-\infty}^{\infty} \mathring{\Gamma}(\kappa\mathbf{p} + \tfrac{1}{2}\kappa\mathbf{p}_{12}, \kappa\mathbf{p} - \tfrac{1}{2}\kappa\mathbf{p}_{12}, 0, \nu)$$
$$\times \exp(i2\pi\kappa\mathbf{p}_{12}\cdot\mathbf{x})\, d(\kappa\mathbf{p}_{12}), \tag{7-77}$$

where we expressed the solid angle integration $d\Omega_{12}$ of Eq. (7-75) in terms of the difference vector $d(\mathbf{p}_{12})$ as in Eq. (7-53). Now, since $\Gamma_{12}(\tau) = \Gamma_{21}^*(-\tau)$, we have $\hat{\Gamma}_{12}(\nu) = \hat{\Gamma}_{21}^*(\nu)$, which in turn demands that

$$\mathring{\Gamma}(\kappa\mathbf{p}_1, \kappa\mathbf{p}_2, z, \nu) = \mathring{\Gamma}^*(\kappa\mathbf{p}_2, \kappa\mathbf{p}_1, z, \nu). \tag{7-78}$$

Use of this property shows that $\hat{\mathsf{L}}_0$ is real but the integral in Eq. (7-77) need not be nonnegative in general. For the general partially coherent source, the reduced function $\hat{\mathsf{L}}_0$ is the best we can do to establish contact with c-radiometry. This function is neither in the g-radiometric nor in the c-radiometric framework. It depends on the position coordinate $\mathbf{x}$, which is *not* Fourier conjugate to the direction $\mathbf{p}$ [see Eq. (7-54)], which makes it an acceptable function in wave theory.

For the partially coherent source the spectral radiant power is

$$\hat{\mathbf{\Phi}}(0,\nu) = \int \mathcal{Q}(\mathbf{x})\, \hat{\mathsf{M}}(\mathbf{x},\mathbf{x},0,\nu)\, d\mathbf{x},$$

which in terms of the reduced spectral radiance is

$$\hat{\mathbf{\Phi}}(0,\nu) = \int \mathcal{Q}(\mathbf{x}) \int_{1/2} \hat{\mathsf{L}}_0(\mathbf{x},0;\mathbf{p};\nu)\, m\, d\Omega\, d\mathbf{x}, \tag{7-79}$$

where the angle integral is limited to real waves. A reduced spectral radiant intensity $\hat{\mathsf{I}}_0$ may be defined by

$$\hat{\mathsf{I}}_0(0;\mathbf{p};\nu) \equiv m \int \mathcal{Q}(\mathbf{x})\, \hat{\mathsf{L}}_0(\mathbf{x},0;\mathbf{p};\nu)\, d\mathbf{x}, \tag{7-80}$$

so that its single solid angle integral gives

$$\hat{\mathbf{\Phi}}(0, \nu) = \int_{1/2} \hat{\mathsf{I}}_0(0; \mathbf{p}; \nu)\, d\Omega. \tag{7-81}$$

The units of $\hat{\mathsf{I}}_0$ are W sr^{-1} Hz^{-1}. To understand its properties we express it in terms of $\mathring{\Gamma}$ by putting Eq. (7-77) into Eq. (7-80). Thus,

$$\hat{\mathsf{I}}_0(0; \mathbf{p}; \nu) = \mathsf{C} m\kappa^2 \int \mathring{\Gamma}\left(\kappa\mathbf{p} + \tfrac{1}{2}\kappa\mathbf{p}_{12}, \kappa\mathbf{p} - \tfrac{1}{2}\kappa\mathbf{p}_{12}, 0, \nu\right) \times \tilde{\mathcal{Q}}^*(\kappa\mathbf{p}_{12})\, d(\kappa\mathbf{p}_{12}), \tag{7-82}$$

where the asterisk denotes complex conjugate, and the Fourier relationship of Eq. (7-48) is used. The function $\tilde{\mathcal{Q}}$ itself is the coherence in the field contributed by a *uniform*, *noncoherent* source of the same size and shape as the partially coherent source. If this coherence is neglected, as in Eq. (7-63), then

$$\hat{\mathsf{I}}_0(0; \mathbf{p}; \nu) = \mathsf{C} m\kappa^2 \mathring{\Gamma}(\kappa\mathbf{p}, \kappa\mathbf{p}, 0, \nu). \tag{7-83}$$

It is not only real but also nonnegative. This function from wave theory is to be identified with the spectral radiant intensity of c-radiometry. The results of this section are gathered together in Table 7-4.

TABLE 7-4. RADIOMETRY OF A PARTIALLY COHERENT SOURCE IN THE PLANE $z = 0$

$\hat{\mathbf{\Phi}}(0, \nu)$	$= \int_{\mathcal{Q}} \int_{1/2} \hat{\mathsf{L}}_0(\mathbf{x}, 0; \mathbf{p}; \nu)\, m\, d\Omega\, d\mathbf{x}$
[W Hz^{-1}]	[W m^{-2} sr^{-1} Hz^{-1}]
$\hat{\mathsf{M}}(\mathbf{x}, \mathbf{x}, 0, \nu)$	$= \int_{1/2} \hat{\mathsf{L}}_0(\mathbf{x}, 0; \mathbf{p}; \nu)\, m\, d\Omega$
[W m^{-2} Hz^{-1}]	
$\hat{\mathsf{I}}_0(0; \mathbf{p}; \nu)$	$= \mathsf{C} m\kappa^2 \mathring{\Gamma}(\kappa\mathbf{p}, \kappa\mathbf{p}, 0, \nu)$
[W sr^{-1} Hz^{-1}]	
where	
$\hat{\mathsf{L}}_0(\mathbf{x}, 0; \mathbf{p}; \nu)$	$= \mathsf{C}\kappa^2 \int \mathring{\Gamma}(\kappa\mathbf{p} + \tfrac{1}{2}\kappa\mathbf{p}_{12}, \kappa\mathbf{p} - \tfrac{1}{2}\kappa\mathbf{p}_{12}, 0, \nu) \times \exp(i2\pi\kappa\mathbf{p}_{12} \cdot \mathbf{x})\, d(\kappa\mathbf{p}_{12})$

In the limit of a uniform, noncoherent source, that is, $\mathring{\Gamma}(\kappa\mathbf{p}, \kappa\mathbf{p}, 0, \nu) \to$ constant, the reduced spectral radiant intensity exhibits a $\cos\theta$ dependence as in Table 7-3. In this way the rest of the relationships of Table 7-3 may also be reproduced by starting with Table 7-4.

Finally, we mention that the reduced spectral radiance $\hat{\mathsf{L}}_0$, defined in Eqs. (7-64) and (7-65) for the noncoherent source, is a special case of the one defined in Eq. (7-75).

Carter–Wolf Source and Conventional Radiometry

The MSDF of the Carter–Wolf source (Carter and Wolf, 1977) has the factored form

$$\begin{aligned}\hat{\Gamma}_s(\mathbf{x}_1, \mathbf{x}_2, 0, \nu) &= \hat{I}_s[\tfrac{1}{2}(\mathbf{x}_1 + \mathbf{x}_2), \nu]\, \hat{g}_s(\mathbf{x}_1 - \mathbf{x}_2, \nu)\\ &= \hat{I}_s(\mathbf{x}, \nu)\, \hat{g}_s(\mathbf{x}_{12}, \nu).\end{aligned} \tag{7-84}$$

The subscript s signifies source-related functions in the $z = 0$ plane. Its OSI is a function of the average variable $\mathbf{x}$, and its coherence is described by $\hat{g}_s$ (see Chapter 5). Now, $\hat{I}_s$ is real and nonnegative, and it is very broad compared to $\hat{g}_s$, which is a suitably chosen autocorrelation function with the properties

$$\begin{aligned}\hat{g}_s(0, \nu) &= 1,\\ \hat{g}_s(\mathbf{x}_{12}, \nu) &= \hat{g}_s^*(\mathbf{x}_{21}, \nu).\end{aligned} \tag{7-85}$$

The fortuitous separation of variables of Eq. (7-84) leads to some special properties in regard to the radiometry of this source.

First we need the spatial Fourier transform of the MSDF of Eq. (7-84),

$$\begin{aligned}\mathring{\Gamma}(\kappa\mathbf{p}_1, \kappa\mathbf{p}_2, 0, \nu) = \iint_{-\infty}^{\infty} &\hat{I}_s[\tfrac{1}{2}(\mathbf{x}_1 + \mathbf{x}_2), \nu]\, \hat{g}_s(\mathbf{x}_1 - \mathbf{x}_2, \nu)\\ &\times \exp[-i2\pi\kappa(\mathbf{p}_1 \cdot \mathbf{x}_1 - \mathbf{p}_2 \cdot \mathbf{x}_2)]\, d\mathbf{x}_1\, d\mathbf{x}_2.\end{aligned}$$

Use of the average and difference variables of Eqs. (7-49) to (7-54) directly gives

$$\mathring{\Gamma}\left(\kappa\mathbf{p} + \tfrac{1}{2}\kappa\mathbf{p}_{12}, \kappa\mathbf{p} - \tfrac{1}{2}\kappa\mathbf{p}_{12}, 0, \nu\right) = \mathring{I}_s(\kappa\mathbf{p}_{12}, \nu)\, \mathring{g}_s(\kappa\mathbf{p}, \nu). \tag{7-86}$$

Since the Carter–Wolf source is partially coherent, the spatial Fourier transform given above does *not* describe the coherence properties in the field as it did in the case of the completely noncoherent source. For a partially coherent source one has to use the generalized van Cittert–Zernike result of Eq. (5-1) or the angular spectrum formulation of Eq. (5-5) to study the spatial coherence. (We had studied this aspect of the problem in Chapter 5, but it is repeated here for ease of reference.) This second approach leads to the result

$$\begin{aligned}\hat{\Gamma}(\mathbf{x}+\tfrac{1}{2}\mathbf{x}_{12},\mathbf{x}-\tfrac{1}{2}\mathbf{x}_{12},z,\nu)\\ &=\int\int \mathring{I}_s(\kappa\mathbf{p}_{12},\nu)\,\mathring{g}_s(\kappa\mathbf{p},\nu)\exp[ik(m_1-m_2^*)z]\\ &\quad\times\exp[i2\pi\kappa(\mathbf{p}\cdot\mathbf{x}_{12}+\mathbf{p}_{12}\cdot\mathbf{x})]\,d(\kappa\mathbf{p})\,d(\kappa\mathbf{p}_{12}).\end{aligned}$$

For the forward direction (small angles), the quadratic approximation of the combination $m_1 - m_2^*$ gives

$$m_1 - m_2^* \simeq -\mathbf{p}\cdot\mathbf{p}_{12}.$$

In this approximation the coherence in the plane $z=$ constant may be expressed in one of two equivalent ways:

$$\begin{aligned}\hat{\Gamma}(\mathbf{x}+\tfrac{1}{2}\mathbf{x}_{12},\mathbf{x}-\tfrac{1}{2}\mathbf{x}_{12},z,\nu)\\ &=\int \mathring{I}_s(\kappa\mathbf{p}_{12},\nu)\,\hat{g}_s(\mathbf{x}_{12}-\mathbf{p}_{12}z,\nu)\exp[i2\pi\kappa\mathbf{p}_{12}\cdot\mathbf{x}]\,d(\kappa\mathbf{p}_{12})\\ &=\int \hat{I}_s(\mathbf{x}-\mathbf{p}z,\nu)\,\mathring{g}_s(\kappa\mathbf{p},\nu)\exp[i2\pi\kappa\mathbf{p}\cdot\mathbf{x}_{12}]\,d(\kappa\mathbf{p}).\end{aligned} \tag{7-87}$$

In the first expression, both $\mathring{I}_s$ and $\hat{g}_s$ are very narrow. In the second, $\hat{I}_s$ and $\mathring{g}_s$ are both very broad functions. Instead of making any further approximations in this direction, let us look at the functions of g-radiometry.

The generalized spectral radiance as defined in Eq. (7-39) is

$$\begin{aligned}\hat{\mathsf{L}}(\mathbf{x}+\tfrac{1}{2}\mathbf{x}_{12},\mathbf{x}-\tfrac{1}{2}\mathbf{x}_{12},0;\mathbf{p}+\tfrac{1}{2}\mathbf{p}_{12},\mathbf{p}-\tfrac{1}{2}\mathbf{p}_{12};\nu)\\ &=\mathsf{C}\kappa^4\mathring{I}_s(\kappa\mathbf{p}_{12},\nu)\,\mathring{g}_s(\kappa\mathbf{p},\nu)\exp[+i2\pi\kappa(\mathbf{p}\cdot\mathbf{x}_{12}+\mathbf{p}_{12}\cdot\mathbf{x})].\end{aligned} \tag{7-88}$$

According to the definition of the reduced spectral radiance of Eqs. (7-75) and (7-77) we find that

$$\hat{\mathsf{L}}_0(\mathbf{x},0;\mathbf{p};\nu)=\mathsf{C}\kappa^2\hat{I}_s(\mathbf{x},\nu)\,\mathring{g}_s(\kappa\mathbf{p},\nu). \tag{7-89}$$

It is interesting to note that this function is real and nonnegative! The function $\hat{I}_s$ is nonnegative, and $\mathring{g}_s$, being a Fourier transform (spatial) of an autocorrelation function, is also nonnegative.

TABLE 7-5. RADIOMETRY OF THE CARTER–WOLF SOURCE

$\hat{\mathsf{M}}(\mathbf{x}, \mathbf{x}, 0, \nu) = \mathsf{C}\hat{I}_s(\mathbf{x}, \nu)$
$\hat{\mathsf{L}}_0(\mathbf{x}, 0; \mathbf{p}; \nu) = \mathsf{C}\kappa^2\hat{I}_s(\mathbf{x}, \nu)\, \mathring{g}_s(\kappa\mathbf{p}, \nu)$
$\hat{\mathsf{I}}_0(0; \mathbf{p}; \nu) = \mathsf{C}m\kappa^2\mathring{g}_s(\kappa\mathbf{p}, \nu)\, \mathring{I}_s(0, \nu)$
$\hat{\mathbf{\Phi}}(0, \nu) = \mathsf{C}\mathring{I}_s(0, \nu)$

The reduced spectral radiant intensity as defined in Eq. (7-80) is

$$\hat{\mathsf{I}}_0(0; \mathbf{p}; \nu) = \mathsf{C}m\kappa^2\mathring{g}_s(\kappa\mathbf{p}, \nu) \int \mathcal{Q}(\mathbf{x})\, \hat{I}_s(\mathbf{x}, \nu)\, d\mathbf{x}.$$

For the Carter–Wolf source the effective area of the source is dictated by the OSI function $\hat{I}_s$. The limits of integration are from $-\infty$ to ∞, and the source area function $\mathcal{Q}(\mathbf{x})$ may be omitted from the integrand; the integral equals $\mathring{I}_s(0, \nu)$. The reduced spectral radiant intensity thus becomes

$$\hat{\mathsf{I}}_0(0; \mathbf{p}; \nu) = \mathsf{C}m\kappa^2\mathring{g}_s(\kappa\mathbf{p}, \nu)\, \mathring{I}_s(0, \nu). \tag{7-90}$$

It is real and nonnegative. Both $\hat{\mathsf{I}}_0$ and $\hat{\mathsf{L}}_0$ may be identified with the respective functions of c-radiometry.

The spectral radiant power $\hat{\mathbf{\Phi}}$ is found through the solid angle integral of $\hat{\mathsf{I}}_0$ over the real waves as in Eq. (7-81). If the spectrum of $\hat{g}_s$ is essentially over real waves, then to a good approximation we have

$$\hat{\mathbf{\Phi}}(0, \nu) = \mathsf{C}\mathring{I}_s(0, \nu) \int_{-\infty}^{\infty} \mathring{g}_s(\kappa\mathbf{p}, \nu)\, d(\kappa\mathbf{p}).$$

Since the integral is the zero ordinate of $\hat{g}_s(\mathbf{x}_{12}, \nu)$, it equals unity; thus we get

$$\hat{\mathbf{\Phi}}(0, \nu) = \mathsf{C}\mathring{I}_s(0, \nu). \tag{7-91}$$

The radiometric functions for the Carter–Wolf source obtained by wave theory are listed in Table 7-5. Since these functions are real and nonnegative, this partially coherent source behaves most like a c-radiometric source.

Coherent Source

Without discussing a specific example we shall discuss the functions necessary to study the radiometry of coherent sources. This case has no analog in c-radiometry.

The coherent source is a deterministic case; the field from such a source is also coherent. The MSDF of such a source has the factored form

$$\hat{\Gamma}(\mathbf{x}_1, \mathbf{x}_2, z, \nu) = U_{\nu_0}(\mathbf{x}_1, z)\, U^*_{\nu_0}(\mathbf{x}_2, z)\, \delta(\nu - \nu_0). \tag{7-92}$$

The temporal frequency of the coherent light is ν_0. The space variation is described by U_{ν_0}, where the subscript is only to remind us of the field frequency. The related MCF Γ is time harmonic, with the temporal dependence given by $\exp(-i2\pi\nu_0\tau)$.

The spatial Fourier transform of the MSDF is

$$\mathring{\Gamma}(\kappa\mathbf{p}_1, \kappa\mathbf{p}_2, z, \nu) = \tilde{U}_{\nu_0}(\kappa_0\mathbf{p}_1, z)\, \tilde{U}^*_{\nu_0}(\kappa_0\mathbf{p}_2, z)\, \delta(\nu - \nu_0). \tag{7-93}$$

The function $\hat{F}$ of Eq. (7-35) takes the form

$$\hat{F}(\mathbf{x}_1, \mathbf{x}_2, z; \mathbf{p}_1, \mathbf{p}_2; \nu) = \hat{F}_c(\mathbf{x}_1, z; \mathbf{p}_1; \nu_0)\, \hat{F}^*_c(\mathbf{x}_2, z; \mathbf{p}_2; \nu_0)\, \delta(\nu - \nu_0), \tag{7-94}$$

where

$$\hat{F}_c(\mathbf{x}, z; \mathbf{p}; \nu_0) \equiv \kappa_0^2 \tilde{U}_{\nu_0}(\kappa_0\mathbf{p}, z) \exp(i2\pi\kappa_0\mathbf{p}\cdot\mathbf{x}) \tag{7-95}$$

with subscripts 1 or 2 on $\mathbf{x}$ and $\mathbf{p}$. The relationship between $\hat{\Gamma}$ and $\hat{F}$ also factors, which leads to

$$U_{\nu_0}(\mathbf{x}, z) = \int \hat{F}_c(\mathbf{x}, z; \mathbf{p}; \nu_0)\, m\, d\Omega. \tag{7-96}$$

Thus, corresponding to the amplitude U_{ν_0}, the function $\hat{F}_c$ plays the role of "amplitude radiance."

The definition of the spectral radiant power does not factor:

$$\hat{\mathbf{\Phi}}(z, \nu_0) = \mathsf{C}\int \mathcal{Q}(\mathbf{x})\, |\, U_{\nu 0}(\mathbf{x}, z)\,|^2\, d\mathbf{x}. \tag{7-97}$$

From Eq. (7-46) the spectral radiant intensity is

$$\hat{\mathsf{I}}(z; \mathbf{p}_1, \mathbf{p}_2; \nu_0) = m_1 m_2 \mathsf{C}\int \mathcal{Q}(\mathbf{x})\, \hat{F}_c(\mathbf{x}, z; \mathbf{p}_1; \nu_0) \times \hat{F}^*_c(\mathbf{x}, z; \mathbf{p}_2; \nu_0)\, d\mathbf{x}. \tag{7-98}$$

It is such that the spectral power relationship of Eq. (7-45) remains unchanged.

From Generalized Radiometry to Conventional Radiometry

In the last few sections we discussed the formalism of g-radiometry. Starting from it we recreated the structure of c-radiometry:

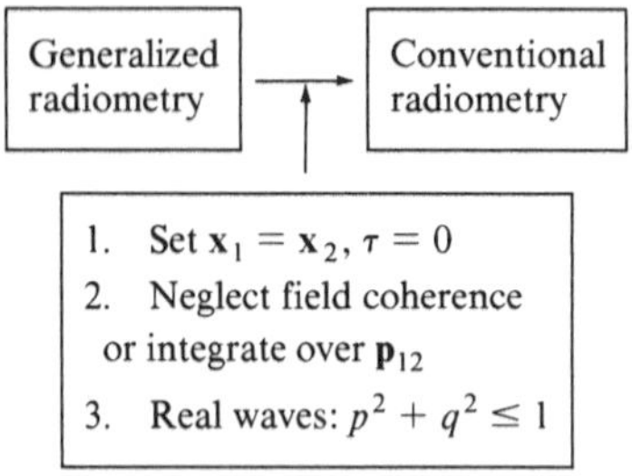

Thus we arrived at the expected conclusion that c-radiometry constitutes a useful approximation. It is valid for noncoherent sources and when dealing with fields whose spatial coherence may be neglected. In this way we found that:

1. The limit of a noncoherent source of coherence theory is the same as the limit of a Lambertian source of c-radiometry.
2. For partially coherent sources, contact with c-radiometry is established by use of reduced functions that in general depend on the position $\mathbf{x}$ and the direction $\mathbf{p}$.

We shall turn our attention to applying the formalism of g-radiometry to study blackbody radiation. Before we do that, we digress a bit to discuss its properties in the section that follows.

BLACKBODY SOURCE

It is characteristic of material bodies that they emit and absorb radiation at any temperature above absolute zero. The rate of emission or total radiant power from the body increases with temperature. Part of the radiation incident on the body is absorbed. This process is described by the term *absorptivity* α, defined by

$$\alpha = \frac{\text{Radiant incidence absorbed [W]}}{\text{Radiant incidence [W]}} = \frac{\mathsf{E}\text{ absorbed}}{\mathsf{E}}. \tag{7-99}$$

Consider an ideal isothermal enclosure containing one or more material bodies. Like the bodies, the walls of the enclosure also emit and absorb radiation. The walls and the material bodies in the radiation environment eventually reach a state of thermal equilibrium, at temperature T. At equilibrium the radiant exitance M must equal the amount of radiation absorbed, which is just α times the radiant incidence E. In the enclosure, E is common to all the bodies involved. Therefore, regardless of the substance of the bodies, we have the equality of the ratios,

$$\frac{\mathsf{M}_1}{\alpha_1} = \frac{\mathsf{M}_2}{\alpha_2} = \cdots = \frac{\mathsf{M}_j}{\alpha_j}, \tag{7-100}$$

where the subscripts $1, 2, \ldots, j$ are body labels. The absorptivity of course depends on the body substance. Its limiting value unity suggests a body that absorbs *all* the radiation incident on it. Such a body is called a *blackbody* (BB), for which $\alpha_{\mathrm{BB}} = 1$. Due to the equality of the ratios of Eq. (7-100), the radiant exitance M_{BB} of a blackbody that is also included in the enclosure will be larger than that from any other body at equilibrium temperature T, that is,

$$\mathsf{M}_{\mathrm{BB}} > \mathsf{M}, \qquad \mathsf{M}_{\mathrm{BB}} = \frac{\mathsf{M}}{\alpha}. \tag{7-101}$$

The right-hand sides of the above refer to any other body in the enclosure, the subscripts having been dropped.

The blackbody is an idealization never achieved but only approached in practice. It is frequently taken as a standard, and the term *emissivity* ε for a material body is defined by the ratio of the radiant exitance of the body to that of a blackbody at the same temperature T, namely,

$$\varepsilon = \frac{\mathsf{M}}{\mathsf{M}_{\mathrm{BB}}}. \tag{7-102}$$

Thus, at thermal equilibrium we have

$$\varepsilon = \alpha. \tag{7-103a}$$

That is, the emissivity of a body equals its absorptivity regardless of the substance. The above equality holds only at thermal equilibrium and is known as *Kirchhoff's radiation law*. As a corollary,

$$\varepsilon_{\mathrm{BB}} = \alpha_{\mathrm{BB}} = 1 \tag{7-103b}$$

for a blackbody at equilibrium temperature T.

The emissivities and absorptivities are frequently defined as functions of coordinate, angle, and/or wavelength (see, e.g., Grum and Becherer, 1979, or Wolfe, 1965). But for a blackbody at equilibrium temperature T, these terms are unity by definition, which makes the blackbody a Lambertian source. With these considerations it is reasonable to expect that the coherence function of a blackbody will be such that its SFS is constant over the real waves (for all angles), and, furthermore, that the spectral functions (or the total functions) and their relationships given in Table 7-3 would also apply for the case of the blackbody. However, one should properly begin with the coherence function of the blackbody and show explicitly that the functions of g-radiometry reduce to those in Table 7-3. Unfortunately, no derivation of the coherence function from first principles for a material blackbody is to be found in the literature. As we shall see in what is to follow, this situation can be remedied.

Although no material body or surface absorbs all the radiation falling on it, the next best thing is to consider *blackbody simulators*. A simulator is an attempt at a practical realization of a blackbody to a good approximation. The approach taken for this purpose is to construct a large enough cavity held at a fixed temperature T with a small enough hole in one of its walls. The walls of the cavity emit and absorb radiation, and ideally thermal equilibrium is attained. The presence of the hole is assumed to cause only a slight perturbation, and it allows a small fraction of the radiation to escape. This is referred to as cavity radiation. The radiation incident on the hole enters the cavity and very little of it escapes; that is, we have $\alpha \to 1$. The hole in the cavity wall is the approximation sought to the ideal blackbody at temperature T. Thus, cavity radiation is used synonymously with blackbody radiation.

The actual making of a blackbody is a difficult engineering problem (see Bartell and Wolfe, 1973, 1976, and also Palmer, 1979, where further references may be found). The emissivities achieved in practice are $\varepsilon \simeq 0.99$.

Cavity Radiation

The spectral distribution of cavity radiation is given by the well known Planck's law. It is customary to express the spectral radiant exitance of the cavity radiation as a function of wavelength,

$$\hat{\mathsf{M}}(\lambda, T) = \frac{2\pi c^2 h}{\lambda^5} \frac{1}{\exp(hc/\lambda k_B T) - 1}; \tag{7-104}$$

its units are W m^{-2} μm^{-1}. In this formula, Planck's constant h equals 6.625×10^{-34} J s; the speed of light c is 2.998×10^{8} m s^{-1}; the Boltzmann constant k_B is 1.380×10^{-23} J K^{-1}; and T is the temperature in kelvins. Blackbody simulators closely approach this spectral distribution. What they fail to do is achieve the uniform angular distribution of radiation to any reasonable degree of accuracy over the full range. Typically, the uniformity of angular distribution available in practice is about $\pm 5°$ from the normal to the hole.

The foundations of the theoretical calculation were laid by Planck. Since the introduction of Planck's quantum hypothesis and the subsequent birth of the quantum theory, various aspects of the interaction of radiation and matter and the theory of radiation have been investigated in great detail. To continue our discussion, we need the results from the theory of radiation in thermal equilibrium with the walls of the cavity. The correlation tensors of the electric and magnetic fields were fully developed for the quantized field by Glauber (1963). This opened a new research field called *quantum coherence theory*. For an in-depth study, see the book by Klauder and Sudarshan (1968). The temporal and spatial coherence properties of blackbody radiation are discussed in a series of articles by Kano and Wolf (1962), Mehta (1963), and Mehta and Wolf (1964a, b); see also the review article by Mandel and Wolf (1965).

We shall mostly follow the Mehta and Wolf notation in presenting some of their results pertaining to our discussion. The calculation is vectorial and begins with the analytic signal representation of the electric field **E**, defined by

$$\mathbf{E}(\mathbf{r}, t) = \mathbf{E}^{(r)}(\mathbf{r}, t) + i\mathbf{E}^{(i)}(\mathbf{r}, t), \tag{7-105}$$

where $\mathbf{E}^{(r)}$ is the real physical field and $\mathbf{E}^{(i)}$ is its Hilbert transform. The transform operation is applied to each of the Cartesian components $E_i^{(r)}$ (where $i = x, y, z$) of the field. The above definition differs by a multiplicative factor $(1/\sqrt{2})$ from the one given in Eq. (2-12) for the scalar field. We assume that the field is statistically stationary in time. The cross correlation of the field from points $\mathbf{r}_1$ and $\mathbf{r}_2$ with a time delay $\tau = t_1 - t_2$ generates a nine-component second-rank tensor denoted by $\mathcal{E}$ and may be displayed as a 3×3 matrix:

$$\mathcal{E} = \begin{bmatrix} \mathcal{E}_{xx} & \mathcal{E}_{xy} & \mathcal{E}_{xz} \\ \mathcal{E}_{yx} & \mathcal{E}_{yy} & \mathcal{E}_{yz} \\ \mathcal{E}_{zx} & \mathcal{E}_{zy} & \mathcal{E}_{zz} \end{bmatrix}. \tag{7-106}$$

Its elements are defined with the time average:

$$\mathcal{E}_{ij}(\mathbf{r}_1, \mathbf{r}_2, \tau) = \langle E_i(\mathbf{r}_1, t + \tau)\, E_j^*(\mathbf{r}_2, t)\rangle. \tag{7-107}$$

The quantity $\mathcal{E}(\mathbf{r}_1, \mathbf{r}_2, \tau)$, formed by using the above elements $\mathcal{E}_{ij}$, is called the correlation tensor of the electric field.

For cavity radiation we find that $\mathcal{E}$ is not a function of $\mathbf{r}_1$ and $\mathbf{r}_2$ separately, but depends on the difference vector $\mathbf{r}_{12} = \mathbf{r}_1 - \mathbf{r}_2$, consistent with the expectation of spatial homogeneity. Furthermore, for $\mathbf{r}_{12} = 0$ and for any value of τ, $\mathcal{E}$ is found to be a multiple of a 3×3 unit matrix, consistent with the expected isotropic nature of the radiation in the cavity. Examination of the spatial coherence, $\mathbf{r}_{12} \neq 0$, breaks the isotropic symmetry by the introduction of a direction $\mathbf{r}_{12}$ in the cavity. For this case the tensor is not diagonal.

Of special interest to us is the spectral version of the tensor. Mehta and Wolf define the *cross-spectral tensor* $W_{ij}^{(e)}$ as the temporal Fourier transform of the corresponding element of $\mathcal{E}$,

$$W_{ij}^{(e)}(\mathbf{r}_1, \mathbf{r}_2, \nu) = \int_{-\infty}^{\infty} \mathcal{E}_{ij}(\mathbf{r}_1, \mathbf{r}_2, \tau) \exp(+i2\pi\nu\tau)\, d\tau. \qquad \text{(7-108)}$$

Note that we used the term "mutual spectral density function" (MSDF) $\hat{\Gamma}_{12}(\nu)$ for the temporal Fourier transform of the MCF $\Gamma_{12}(\tau)$ for the scalar field. The superscript (e) is used to indicate that the electric field is studied.

Like $\mathcal{E}$, $W^{(e)}$ also depends on the point separation $\mathbf{r}_{12}$ instead of on $\mathbf{r}_1$ and $\mathbf{r}_2$ separately. It is convenient to define a unitless vector $\mathbf{R}$ by

$$\mathbf{R} = k\mathbf{r}_{12} = \hat{\mathbf{i}}(kx_{12}) + \hat{\mathbf{j}}(ky_{12}) + \hat{\mathbf{k}}(kz_{12}). \qquad \text{(7-109)}$$

For our discussion we need only the diagonal elements of $W^{(e)}$, which may be expressed as

$$W_{xx}^{(e)}(\mathbf{R}) = \pi A\left\{\frac{\sin R}{R}\left[1 - \frac{1}{R^2} + \frac{3R_x^2}{R^4} - \frac{R_x^2}{R^2}\right] + \cos R\left[\frac{1}{R^2} - \frac{3R_x^2}{R^4}\right]\right\}$$

$$W_{yy}^{(e)}(\mathbf{R}) = \pi A\left\{\frac{\sin R}{R}\left[1 - \frac{1}{R^2} + \frac{3R_y^2}{R^4} - \frac{R_y^2}{R^2}\right] + \cos R\left[\frac{1}{R^2} - \frac{3R_y^2}{R^4}\right]\right\}$$

$$W_{zz}^{(e)}(\mathbf{R}) = \pi A\left\{\frac{\sin R}{R}\left[1 - \frac{1}{R^2} + \frac{3R_z^2}{R^4} - \frac{R_z^2}{R^2}\right] + \cos R\left[\frac{1}{R^2} - \frac{3R_z^2}{R^4}\right]\right\},$$

(7-110)

wherein the symbol A defined by Mehta and Wolf is

$$A \equiv A(\nu, T) = \frac{8\pi h\nu^3}{c^3}\left[\frac{1}{\exp(h\nu/k_B T) - 1}\right]. \qquad \text{(7-111)}$$

This is in fact the frequency ν version of the spectral density given in Eq. (7-104) with λ as the argument.

Now by a proper grouping of the terms and by use of a limiting procedure it can be shown that the diagonal terms approach the value

$$W_{ii}^{(e)}(\mathbf{R}) \to \tfrac{2}{3}\pi A, \qquad i = x, y, z, \tag{7-112}$$

as the point separation R tends toward zero. In this way, the off-diagonal terms may be shown to go to zero in the same limit.

There is no unique way for a transition from a vector theory to a scalar theory. We need to isolate a scalar quantity in the vector formalism that will play the role of the MSDF $\hat{\Gamma}_{12}(\nu)$ in the scalar theory. If a single polarization component is present, it may be taken as the desired scalar. For cavity radiation a convenient scalar is the trace of the electric cross-spectral tensor (see, e.g., Marathay and Parrent, 1970).

The radiation emanating from the hole in the wall of the cavity is identified with that from a blackbody. If the z axis is chosen perpendicular to the plane of the hole, the trace of the 2×2 matrix in the left top corner in Eq. (7-106) may be used with $\mathbf{R}$ chosen in the x-y plane. Alternatively, as done in the literature (see, e.g., Walther, 1968), we may choose $\mathbf{R}$ in any direction and take the trace of the entire tensor; thus,

$$\operatorname{tr} W^{(e)} \equiv W_{xx}^{(e)} + W_{yy}^{(e)} + W_{zz}^{(e)}.$$

The expressions for the diagonal elements in Eq. (7-110) show that

$$\operatorname{tr} W^{(e)} = 2\pi A \frac{\sin R}{R}. \tag{7-113}$$

After normalization, it is this quantity that we shall identify with the scalar version of the spatial coherence of the blackbody radiation.

Spatial Coherence of Blackbody Radiation

Following the Carter–Wolf model for the MSDF, as in Eq. (7-84), we shall use

$$\hat{\Gamma}_{\mathrm{BB}}(\mathbf{x}_1, \mathbf{x}_2, 0, \nu) = \hat{I}_{\mathrm{BB}}(\mathbf{x}, \nu)\, \hat{g}_{\mathrm{BB}}(\mathbf{x}_{12}, \nu) \tag{7-114}$$

for the radiation emanating from the hole in the cavity wall. The factor $\hat{I}_{\mathrm{BB}}$ is a nonzero constant for points inside the hole and zero for points outside the hole. As suggested by Eq. (7-113), the spatial coherence factor is taken to be

$$\hat{g}_{BB}(\mathbf{x}_{12}, \nu) = \frac{\sin kr_{12}}{kr_{12}} \tag{7-115}$$

where

$$r_{12} = \left[(x_1 - x_2)^2 + (y_1 - y_2)^2\right]^{1/2} \equiv \left(x_{12}^2 + y_{12}^2\right)^{1/2}.$$

We assume that the dimensions of the hole are very small compared to the cavity dimension and very large compared to the width over which $\hat{g}_{BB}$ is significant. This form of $\hat{g}_{BB}$ was derived by using the vector theory of the previous section. It may also be obtained by means of the scalar theory by starting from the pair of Helmholtz equations satisfied by $\hat{\Gamma}_{BB}$. To achieve this one must change to average and difference variables for all three coordinates and assume that the solution has the factored form of Eq. (7-114) to bring about a separation of variables. After the condition of isotropy is imposed, it is assumed that $\hat{I}_{BB}$ is a constant for all r. Finally, with the subsidiary conditions of Eqs. (7-85) imposed on $\hat{g}_{BB}$, it follows that it must have the form given in Eq. (7-115).

The g-radiometry of this source is studied by applying the results summarized in Table 7-5. For this purpose we need the two-dimensional spatial Fourier transform of $\hat{\Gamma}_{BB}$. Since $\hat{g}_{BB}$ depends on the radial distance r_{12}, the transform,

$$\mathring{g}_{BB}(\kappa\mathbf{p}, \nu) = \int \hat{g}_{BB}(\mathbf{x}_{12}, \nu) \exp(-i2\pi\kappa\mathbf{p} \cdot \mathbf{x}_{12})\, d\mathbf{x}_{12}, \tag{7-116}$$

is more easily done in polar coordinates:

$$x_{12} = r_{12}\cos\phi, \qquad y_{12} = r_{12}\sin\phi$$
$$p = \rho\cos\psi, \qquad q = \rho\sin\psi,$$

with $d\mathbf{x}_{12} = dx_{12}\, dy_{12} = r_{12}\, dr_{12}\, d\phi$. The integral over the angle ϕ may be performed to give

$$\mathring{g}_{BB}(\kappa\mathbf{p}, \nu) = 2\pi \int_0^\infty \left(\frac{\sin kr_{12}}{kr_{12}}\right) J_0(kr_{12}\rho)\, r_{12}\, dr_{12},$$

where J_0 is the zero-order Bessel function. A table of integrals, such as given by Gradshteyn and Ryzhik (1965), shows that the above integral reduces to

$$\mathring{g}_{BB}(\kappa \mathbf{p}, \nu) = \frac{2\pi}{k^2} \frac{\text{cyl}\left[\left(p^2+q^2\right)^{1/2}/1\right]}{\left[1-\left(p^2+q^2\right)\right]^{1/2}}, \qquad (7\text{-}117)$$

indicating that $\mathring{g}_{BB}$ is zero for the evanescent waves, $p^2 + q^2 > 1$. By using the spherical polar angles, θ and ϕ of Eq. (7-4), we find that

$$\mathring{g}_{BB}(\kappa \mathbf{p}, \nu) = \frac{\lambda^2}{2\pi} \frac{1}{\cos\theta}, \qquad 0 \le \theta < \pi/2. \qquad (7\text{-}118)$$

The angle θ is the polar angle with the z axis of the direction of propagation of the Fourier plane wave.

Clearly, the plane waves traveling at large angles (close to $\pi/2$) dominate over those traveling in the forward direction ($\theta \simeq 0$) with respect to the hole in the cavity wall. Thus, the blackbody SFS is not "white;" that is, it is not uniform or constant over the real plane waves. Therefore, based on the discussion of noncoherence in Chapter 4, the blackbody source is not of the noncoherent type although its interval of spatial coherence is on the order of λ. The individual polarization components have about 2λ or 3λ coherence interval as seen with the graphical plots of Mehta and Wolf (1964a, b). The blackbody has been referred to as a partially coherent source.

From what has been said up to now, the blackbody source is *not* Lambertian. Even worse, the reduced spectral radiance of Eq. (7-89) acquires a $1/\cos\theta$ dependence, and the reduced spectral radiant intensity of Eq. (7-90) becomes independent of θ!

The question of what sort of a *planar* noncoherent source will furnish a spatially stationary coherence function of the type $\hat{g}_{BB}$ is answered with the help of the van Cittert–Zernike theorem. The required OSI distribution of such a planar source is finite at the center and approaches infinity at the edges. There is no physical *planar* source with such attributes; of course, blackbody simulators do exist.

The formalism of g-radiometry leads to nonphysical radiometric properties of the radiation from a blackbody. This situation is unacceptable. However, as we shall soon see, there is no such difficulty in the formulation of radiometry as originally presented by Walther (1968) and subsequently developed by Marchand and Wolf (1974). The formalism of g-radiometry of this chapter differs from the one due to Walther, Marchand, and Wolf (WMW) in the choice of the field quantity that is relevant to radiometric measurement. As we shall see, this difference is reflected in the definition of irradiance (spectral) and its relation to radiant power (spectral). In order to understand this, it is necessary to discuss the flow vector of scalar field theory. The actual discussion of the WMW formalism will be taken up in a later section.

FLOW VECTOR

The flow vector **P** of scalar field theory is analogous to the Poynting vector **S** of electromagnetic theory. We begin by discussing the Poynting vector.

In studying the classical interaction of radiation and matter or charges and currents, a basic conservation law in the form of an equation of continuity is established. For our discussion related to a source-free region it reads

$$\frac{\partial \mathsf{w}}{\partial t} + \nabla \cdot \mathbf{S} = 0, \tag{7-119}$$

where w stands for the energy density, with units of J m^{-3}. It is made up of the sum of electric and magnetic energy densities. By integrating the above equation throughout the volume $\mathcal{V}$ we may convert the divergence of **S** to the integral of **S** itself over the surface $\mathcal{S}$ enclosing the volume; thus,

$$\frac{\partial \mathsf{Q}}{\partial t} + \iint_{\mathcal{S}} \mathbf{S} \cdot \hat{\mathbf{n}}\, d\mathcal{S} = 0, \tag{7-120}$$

where Q is the total energy [J] within the volume $\mathcal{V}$ and $\hat{\mathbf{n}}$ is a unit normal to the surface element $d\mathcal{S}$ pointing out of the volume. The physical interpretation of this equation is that the time rate of decrease (increase) in energy within the volume equals the rate of energy flow out of (into) the volume across the surface.

The expressions for the energy density and the Poynting vector in Gaussian units are

$$\mathsf{w} = \frac{1}{16\pi}(\mathbf{E} \cdot \mathbf{D}^* + \mathbf{B} \cdot \mathbf{H}^*),$$

$$\mathbf{S} = \frac{c}{8\pi}\mathrm{Re}(\mathbf{E} \times \mathbf{H}^*). \tag{7-121}$$

The units of **S** are W m^{-2}. With the complex notation the above formulas directly yield the time-average values for purely time-harmonic fields. An additional time-average operation is called for when one deals with more general fields.

Since the divergence of **S** appears in Eq. (7-119) and its integral over a closed surface is in Eq. (7-120), there is some degree of arbitrariness and ambiguity in both the definition and interpretation of the Poynting vector. This topic has been discussed in detail by several authors: Eyges (1972), Jackson (1963), Born and Wolf (1970), Feynman et al. (1964), and Longhurst (1967). Nevertheless, **S** itself is often interpreted as describing the magnitude and direction of energy flow across an isolated surface element $d\mathbb{S}$. Furthermore, it is customary in optics to identify the optical intensity with the absolute value of the average Poynting vector,

$$I = |\langle \mathbf{S} \rangle| \tag{7-122}$$

(see, e.g., Born and Wolf, 1970, Sections 1.1.4 and 8.4). The Poynting vector is basic to both geometrical and physical optics. In the limit $\lambda \to 0$, it is shown (see Born and Wolf, 1970, Section 3.1.2) that the average energy density propagates with the speed of light in the medium. Also, in the same limit the magnitude of the average Poynting vector [see Eq. (7-122)] equals the average energy density times the speed of light in the medium, and the direction of the vector is along the normal to the geometrical wavefront. The terms reflectivity and transmissivity are defined in terms of the Poynting vector to assure conservation of energy. The need for the definitions in terms of the Poynting vector is particularly evident when one studies the reflection and refraction of polarized light at a plane interface separating two media of different indexes (see Born and Wolf, 1970, Section 1.5.3). These topics are discussed in several other texts, for example, Hecht and Zajac (1976), Klein (1970), and Longhurst (1967, Section 19.5). We shall now discuss the flow vector for a scalar field.

We are interested in a scalar field V that obeys the wave equation,

$$\nabla^2 V - \frac{1}{c^2}\frac{\partial^2 V}{\partial t^2} = 0.$$

For such a scalar field the equation of continuity is

$$\frac{\partial \mathsf{w}}{\partial t} + \nabla \cdot \mathbf{P} = 0, \tag{7-123}$$

where the energy density is

$$\mathsf{w} = c^2 \nabla V \cdot \nabla V^* + \left(\frac{\partial V}{\partial t}\right)\left(\frac{\partial V^*}{\partial t}\right). \tag{7-124}$$

and the flow vector has the expression

$$\mathbf{P} = -c^2\left(\frac{\partial V}{\partial t}\nabla V^* + \frac{\partial V^*}{\partial t}\nabla V\right). \tag{7-125}$$

For a brief discussion of this topic we refer to Skinner (1965). For the derivation of Eq. (7-123), refer to Aharoni (1959). The procedure is based on the calculus of variation as described, for example, by Roman (1965). In following this procedure, one first constructs a suitable Lagrangian density such that the resulting Euler–Lagrange equation is the same as the scalar wave equation. Next, one defines the generalized momentum and using it a Hamiltonian density is constructed. The first partial derivative with respect to time of the Hamiltonian density leads to Eq. (7-123).

The flow vector for a monochromatic field,

$$V \to \hat{V}\exp(-i\omega t),$$

takes the form

$$\hat{\mathbf{P}} = i\omega c^2(\hat{V}\nabla\hat{V}^* - \hat{V}^*\nabla\hat{V}), \tag{7-126}$$

where $\hat{V}$ obeys the Helmholtz equation and $\hat{\mathbf{P}}$ is the spectral version of the vector defined in Eq. (7-125). The vector $\hat{\mathbf{P}}$ as defined is analogous to the "probability current" vector of quantum mechanics (see, e.g., Leighton, 1959, Section 2-4E).

As with the Poynting vector, it is customary in optics to define the optical intensity in the scalar domain by

$$I = |\langle\mathbf{P}\rangle|. \tag{7-127}$$

If the optical intensity is customarily defined this way, then why was it defined in terms of the vector **E** or the scalar ψ as in Eq. (2-3) and used in this way throughout the book? We shall discuss this in the next subsection by way of an example.

Irradiance Revisited

For a single plane wave it is well known that the Poynting vector may be expressed by

$$|\mathbf{S}| = \frac{c}{8\pi}\left(\frac{\varepsilon}{\mu}\right)^{1/2}(\mathbf{E}\cdot\mathbf{E}^*) = \frac{c}{8\pi}\left(\frac{\mu}{\varepsilon}\right)^{1/2}(\mathbf{H}\cdot\mathbf{H}^*). \tag{7-128}$$

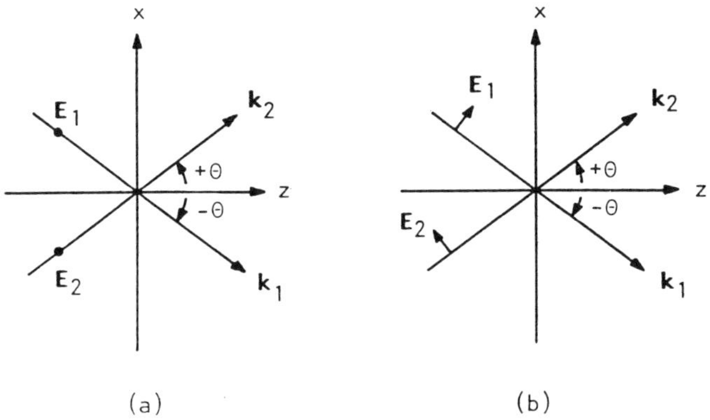

Fig. 7-4. Interference of two plane waves. (*a*) Electric vectors $\mathbf{E}_1$ and $\mathbf{E}_2$ are perpendicular to the *x-z* plane. (*b*) Electric vectors $\mathbf{E}_1$ and $\mathbf{E}_2$ are parallel to the *x-z* plane.

Frequently, in optical experiments, one compares optical intensities in the same medium, and therefore the quantity $\mathbf{E} \cdot \mathbf{E}^*$ in vector theory or VV^* in scalar theory is taken as a measure of optical intensity. From that point on, the quantity $\mathbf{E} \cdot \mathbf{E}^*$ is what is calculated for a wide variety of applications involving plane waves or otherwise. This situation bears closer examination.

Let us consider the interference of two plane waves with electric vectors $\mathbf{E}_1$ and $\mathbf{E}_2$ with the respective propagation vectors $\mathbf{k}_1$ and $\mathbf{k}_2$. The point of discussion concerns the comparison of the Poynting vector and $\mathbf{E} \cdot \mathbf{E}^*$ for the region of interference.

As shown in Fig. 7-4, the *x-z* plane is chosen to be coincident with the plane defined by the vectors $\mathbf{k}_1$ and $\mathbf{k}_2$, and the *z* axis is oriented to bisect the angle 2Θ made by $\mathbf{k}_1$ and $\mathbf{k}_2$. We distinguish between two cases: In Fig. 7-4*a* the electric vectors $\mathbf{E}_1$ and $\mathbf{E}_2$ are perpendicular to the plane of incidence, namely, the *x-z* plane, while in Fig. 7-4*b* they are parallel to the plane of incidence.

With $\hat{\mathbf{i}}$, $\hat{\mathbf{j}}$, and $\hat{\mathbf{k}}$ as the unit vectors along the *x*, *y*, and *z* axes, respectively, the propagation vectors $\mathbf{k}_1$ and $\mathbf{k}_2$ are

$$\begin{aligned} \mathbf{k}_1 &= k(-\hat{\mathbf{i}}\sin\Theta + \hat{\mathbf{k}}\cos\Theta), \\ \mathbf{k}_2 &= k(+\hat{\mathbf{i}}\sin\Theta + \hat{\mathbf{k}}\cos\Theta), \end{aligned} \tag{7-129}$$

where $k = 2\pi/\lambda$ is the magnitude. The electric vectors of Fig. 7-4*a* are parallel to the *y* axis and are given by

$$\begin{aligned} \mathbf{E}_1 &= E_0\hat{\mathbf{j}}\exp(-i\omega t - ikx\sin\Theta + ikz\cos\Theta), \\ \mathbf{E}_2 &= E_0\hat{\mathbf{j}}\exp(-i\omega t + ikx\sin\Theta + ikz\cos\Theta), \end{aligned} \tag{7-130}$$

where E_0 is the magnitude. The corresponding magnetic vectors are parallel to the plane of incidence, and they may be calculated by use of the formula

$$\mathbf{H} = \left(\frac{\varepsilon}{\mu}\right)^{1/2} \mathbf{k} \times \mathbf{E}$$

with subscripts 1 and 2. The expressions so obtained are

$$\begin{aligned}\mathbf{H}_1 &= \left(\frac{\varepsilon}{\mu}\right)^{1/2} E_0(-\hat{\mathbf{i}}\cos\Theta - \hat{\mathbf{k}}\sin\Theta) \\ &\quad \times \exp(-i\omega t - ikx\sin\Theta + ikz\cos\Theta), \qquad (7\text{-}131) \\ \mathbf{H}_2 &= \left(\frac{\varepsilon}{\mu}\right)^{1/2} E_0(-\hat{\mathbf{i}}\cos\Theta + \hat{\mathbf{k}}\sin\Theta) \\ &\quad \times \exp(-i\omega t + ikx\sin\Theta + ikz\cos\Theta).\end{aligned}$$

Observe that the angle between $\mathbf{H}_1$ and $\mathbf{H}_2$ is 2Θ.

For Fig. 7-4*b*, $\mathbf{k}_1$ and $\mathbf{k}_2$ remain the same as in Eq. (7-129) but the electric vectors are now parallel to the plane of incidence and have the expressions

$$\begin{aligned}\mathbf{E}_1 &= E_0(+\hat{\mathbf{i}}\cos\Theta + \hat{\mathbf{k}}\sin\Theta) \\ &\quad \times \exp(-i\omega t - ikx\sin\Theta + ikz\cos\Theta), \\ \mathbf{E}_2 &= E_0(+\hat{\mathbf{i}}\cos\Theta - \hat{\mathbf{k}}\sin\Theta) \\ &\quad \times \exp(-i\omega t + ikx\sin\Theta + ikz\cos\Theta). \qquad (7\text{-}132)\end{aligned}$$

The angle between these two vectors in 2Θ. The corresponding magnetic vectors are parallel to the y axis and are given by

$$\begin{aligned}\mathbf{H}_1 &= \left(\frac{\varepsilon}{\mu}\right)^{1/2} \hat{\mathbf{j}} k E_0 \exp(-i\omega t - ikx\sin\Theta + ikz\cos\Theta), \\ \mathbf{H}_2 &= \left(\frac{\varepsilon}{\mu}\right)^{1/2} \hat{\mathbf{j}} k E_0 \exp(-i\omega t + ikx\sin\Theta + ikz\cos\Theta). \qquad (7\text{-}133)\end{aligned}$$

In an interference experiment we need the linear suppositions

$$\mathbf{E} = \mathbf{E}_1 + \mathbf{E}_2 \quad \text{and} \quad \mathbf{H} = \mathbf{H}_1 + \mathbf{H}_2. \qquad (7\text{-}134)$$

The expressions for $\mathbf{E} \cdot \mathbf{E}^*$ are calculated by using Eqs. (7-130) and (7-132).

For the experiment of Fig. 7-4*a* we get

$$\mathbf{E}\cdot\mathbf{E}^* = 2|E_0|^2[1 + \cos(2kx\sin\Theta)] \tag{7-135}$$

while for that in Fig. 7-4*b* we get

$$\mathbf{E}\cdot\mathbf{E}^* = 2|E_0|^2[1 + \cos 2\Theta\cos(2kx\sin\Theta)]. \tag{7-136}$$

Since $\mathbf{E}_1$ and $\mathbf{E}_2$ are parallel in case (a), they continue to interfere at all Θ larger than zero and smaller than 90°. At 90° there results a standing wave. In case (b), due to the angle 2Θ between $\mathbf{E}_1$ and $\mathbf{E}_2$, the interference term acquires a factor $\cos 2\Theta$.

Now, the time-average Poynting vector may be calculated by using Eq. (7-121) for the superposed fields of Eq. (7-134). For both cases (a) and (b) the Poynting vector is the same and has the expression

$$\mathbf{S} = \frac{\omega}{4\pi}\left(\frac{\varepsilon}{\mu}\right)^{1/2}|E_0|^2\cos\Theta[1 + \cos(2kx\sin\Theta)]\hat{\mathbf{k}}. \tag{7-137}$$

The expression is common for both cases, as is expected, since it describes the energy flow, the amount and direction of which remain the same whether $\mathbf{E}$ is polarized perpendicular or parallel to the plane of incidence.

Consider a photographic plate placed perpendicular to the z axis to record the interference pattern in the region of interference of the two plane waves. In case (a) the expressions for $\mathbf{E}\cdot\mathbf{E}^*$ and $\mathbf{S}$ both predict fringes with period $\lambda/2\sin\Theta$ and visibility unity. In case (b), although both $\mathbf{E}\cdot\mathbf{E}^*$ and $\mathbf{S}$ predict fringes, the visibility predicted by $\mathbf{E}\cdot\mathbf{E}^*$ is $\cos 2\Theta$ but that due to the $\mathbf{S}$ calculation remains unity. The difference in the two calculations for case (b) is extreme when $\Theta = 45°$. In this situation $\mathbf{E}_1$ is perpendicular to $\mathbf{E}_2$; $\mathbf{E}\cdot\mathbf{E}^*$ predicts *absence* of fringes, whereas the $\mathbf{S}$ calculation yields fringes of unit visibility.

Only *one* of the expressions, either $\mathbf{E}\cdot\mathbf{E}^*$ or $\mathbf{S}$, applies to the irradiance measured by the detector (the photographic plate). It is an experimental fact that *no* interference is observed when the electric vector of one wave is perpendicular to that of the other wave. Thus, we conclude that the photographic plate measures $\mathbf{E}\cdot\mathbf{E}^*$ and *not* the Poynting vector.

Wiener's Experiments

These conclusions agree with the experimental results of Wiener (1890), which are reported in many textbooks (see, in particular, Longhurst, 1967,

Section 21.10; and also Born and Wolf, 1970, Section 7.4, and Jenkins and White, 1976, Section 25.12). It is apropos to discuss them briefly here.

In the first of Wiener's experiments a plane wave is incident normally to a plane reflecting surface. The incident and reflected waves form a standing wave pattern. A photographic plate inclined to the normal shows blackening at the electric antinodes.

In a second set of experiments a plane-polarized beam of light is incident at 45° to a reflecting surface. In one experiment the **E** vector is perpendicular to the plane of incidence, and in the other it is parallel to the plane of incidence. These two cases are similar to the situations in Fig. 7-4. The interference between the incident, $\mathbf{E}_{\text{inc}}$, and reflected, $\mathbf{E}_{\text{ref}}$, waves is recorded on an inclined photographic plate. At $\Theta = 45°$, when the **E** vectors are parallel to the plane of incidence, $\mathbf{E}_{\text{inc}} \cdot \mathbf{E}^*_{\text{ref}}$ equals zero, and hence the superposition contains no interference term, while the **H** vectors do contain such a term. In this situation no fringes are recorded on the photographic plate. On the other hand, for the same angle of incidence, when the **E** vectors are perpendicular to the plane of incidence, the interference term is nonzero, while the **H** vectors, which are in this case parallel to the plane of incidence, do not contain an interference term. In this situation fringes are recorded on the photographic plate. In general, the photographic plate shows fringes only when the scalar product $\mathbf{E}_{\text{inc}} \cdot \mathbf{E}^*_{\text{ref}} \neq 0$.

Thus, the observable fringes are caused by the **E** vector and not by the **H** vector. The photographic plate may be replaced by fluorescent screens or other detecting surfaces; the conclusions remain the same. Thus, the output record of an optical detector is in accordance with $\mathbf{E} \cdot \mathbf{E}^*$ and *not* that dictated by the **S** vector calculation.

To complete the story we need a scalar analog of the above experiments of Fig. 7-4. Let V_1 and V_2 be the monochromatic optical scalar fields with propagation vectors as in Fig. 7-4. The superposition $V = V_1 + V_2$ leads to the scalar interference pattern,

$$VV^* = 2A^2[1 + \cos(2kx \sin\Theta)], \tag{7-138}$$

in which the two waves are in phase and have the same absolute value A. The fringe periodicitiy is $\lambda/2\sin\Theta$ as in the vector case. The flow vector **P**, calculated according to Eq. (7-125), has the form

$$\mathbf{P} = (2\omega c^2 k \cos\Theta) VV^* \hat{\mathbf{k}}. \tag{7-139}$$

Equations (7-138) and (7-139) are similar to Eqs. (7-135) and (7-137), where the electric vectors of the interfering waves are mutually parallel and both are perpendicular to the plane of incidence.

Finally, to complete the comparison, the energy density in the scalar case, Eq. (7-124), is made up of two terms, of which the detector responds to the V field, not to its gradient ∇V. This is analogous to the situation in the vector case, where the detector responds to the **E** field and not to the **H** field.

The considerations in this subsection indicate that it would be useful to define a term *measured irradiance* E_m related to $\mathbf{E} \cdot \mathbf{E}^*$ and distinguish it from what may be called *Poynting irradiance* E_P related to the **S** vector. It also appears that it is inappropriate to define optical intensity as the absolute value of the time average of the **S** vector, Eq. (7-122), or of the **P** vector as in Eq. (7-127) (see Webb, 1969, Section 4.2).

Thus, we shall define the measured irradiance

$$\mathsf{E}_m = \left(\frac{\varepsilon}{\mu}\right)^{1/2} \mathbf{E} \cdot \mathbf{E}^* \tag{7-140a}$$

for the vector field and use [see, e.g., Eq. (2-36)]

$$\mathsf{E}_m = \mathsf{C} V V^* \tag{7-140b}$$

for the scalar field. Furthermore, the Poynting irradiance is defined by

$$\mathsf{E}_P = |\langle \mathbf{S} \rangle| \tag{7-141a}$$

for the vector field and

$$\mathsf{E}_P = |\langle \mathbf{P} \rangle| \tag{7-141b}$$

for the scalar field. The terms E_m and E_P may not be used interchangeably.

The Poynting vector and flow vector fulfill the equations of continuity, Eqs. (7-119) and (7-123), respectively. No such conservation law is satisfied by E_m. The conservation law is important to balance the account books of energy and it describes how the field energy is transported. To study the wave theoretic basis of Chandrasekhar's (1960) *radiative transfer*, one must begin with the Poynting vector or the flow vector (see, e.g., Fante, 1981). Nevertheless, the measured irradiance will have to be calculated ultimately to obtain the outcome of optical detection.

WALTHER, MARCHAND, AND WOLF APPROACH

A formulation of generalized radiometry based on the Poynting irradiance was first proposed by Walther (1968). His proposal was developed further by Marchand and Wolf (1974). We shall refer to this generalized radiometry as the Walther, Marchand, and Wolf (WMW) approach. Since 1974 this approach has been widely used and applied by many researchers; we shall refer only to the review articles by Baltes (1977), Wolf (1978), and Baltes et al. (1978).

In what is to follow we shall briefly present the WMW approach by using the notation of the present book, but the symbols for the radiometric functions will be as used by WMW (and will be represented here without "sans serif"). In this way, we shall show how some of their important results follow. For details, the reader is referred to Appendix 7.2, wherein the complete paper by Marchand and Wolf is reproduced for ease of reference.

Our first order of business is to obtain an expression for the spectral radiant power $\hat{\Phi}$. It is found by integrating the normal component of the spectral flow vector $\hat{\mathbf{P}}$ over the surface area of a hemisphere of large radius r,

$$\hat{\Phi} = \int \hat{\mathbf{P}} \cdot \hat{\mathbf{r}}\, r^2\, d\Omega, \tag{7-142}$$

where $\hat{\mathbf{r}}$ is the unit vector normal to the surface of the hemisphere and $r^2\, d\Omega$ is its area element. For this purpose we need the asymptotic form of $\hat{\mathbf{P}}$ for large r.

We begin with the field at a point $\mathbf{r}$ and time t in the form

$$V(\mathbf{r}, t) \to \hat{V}(\mathbf{r}, \nu) \exp(-i\omega t),$$

where $\hat{V}$ is the single Fourier frequency component, and the circular frequency ω is $2\pi\nu$. The position vector $\mathbf{r}$ has components x, y, z. We may express the argument of the field in one of several ways: $\hat{V}(r\hat{\mathbf{n}}, \nu)$, $\hat{V}(x, y, z, \nu)$, or $\hat{V}(\mathbf{x}, z, \nu)$, where $\hat{\mathbf{n}}$ is a unit vector along $\mathbf{r}$.

Consider the two-dimensional spatial Fourier transform

$$\hat{V}(x, y, z, \nu) = \int\int \mathring{V}(\kappa p', \kappa q', 0, \nu) \exp[+i2\pi\kappa(p'x + q'y + m'z)] \times d(\kappa p')\, d(\kappa q').$$

To find the asymptotic form of $\hat{V}$ we put $\mathbf{r} = r\hat{\mathbf{n}} = r(\hat{\mathbf{i}} p + \hat{\mathbf{j}} q + \hat{\mathbf{k}} m)$ and invoke the stationary phase argument on the exponent for large r (see Walther, 1967). The asymptotic form of $\hat{V}$ is found to be

$$\hat{V}(r\hat{\mathbf{n}}, \nu) \simeq -i2\pi m\kappa^2 \mathring{V}(\kappa\mathbf{p}, 0, \nu)\left[\frac{\exp(ikr)}{kr}\right]. \tag{7-143}$$

It differs by a factor of κ^2 from the one given by Walther because of our definition of $\mathring{V}$. It has the form of an expanding spherical wave with the angle dependence dictated by the angular spectrum at $z = 0$.

The asymptotic form of the spectral flow vector is calculated in spherical polar coordinates by use of the asymptotic form of $\hat{V}$. For use in Eq. (7-142) we need only the $\hat{\mathbf{r}}$ component of $\hat{\mathbf{P}}$, which may be expressed in one of two ways:

$$\hat{\mathbf{P}} \cdot \hat{\mathbf{r}} = 2c^3k^2 |\hat{V}|^2 = 2c^3k^2 \frac{m^2\kappa^2}{r^2} |\mathring{V}|^2. \tag{7-144}$$

Now from the point of view of coherence theory, it is the ensemble average of $|\hat{V}|^2$, namely,

$$E[\hat{V}(\mathbf{r}, \nu)\hat{V}^*(\mathbf{r}, \nu)] = \hat{\Gamma}(\mathbf{r}, \mathbf{r}, \nu), \tag{7-145}$$

that enters the calculation. By performing the ensemble average of $\hat{\mathbf{P}} \cdot \hat{\mathbf{r}}$, we get for the spectral radiant power of Eq. (7-142) the following:

$$\hat{\Phi}(r, \nu) = 2c^3k^2 \int_{1/2} \hat{\Gamma}(r\hat{\mathbf{n}}, r\hat{\mathbf{n}}, \nu) r^2 \, d\Omega, \tag{7-146a}$$

which in the spatial frequency domain takes the form

$$\hat{\Phi}(r, \nu) = 2c^3k^2 \int_{1/2} \mathring{\Gamma}(\kappa\mathbf{p}, \kappa\mathbf{p}, z, \nu) m^2\kappa^2 \, d\Omega. \tag{7-146b}$$

It describes the power on a hemisphere of large radius r. Observe that $\mathring{\Gamma}$ appears with its arguments equal: $\mathbf{p}_1 = \mathbf{p}_2 = \mathbf{p}$. In this situation the angular spectrum description of Eq. (5-5) suggests that for real waves the above equation may be replaced by

$$\hat{\Phi}(r, \nu) = 2c^3k^2 \int_{1/2} \mathring{\Gamma}(\kappa\mathbf{p}, \kappa\mathbf{p}, 0, \nu) m^2\kappa^2 \, d\Omega. \tag{7-146c}$$

The approximation of neglecting the evanescent waves was also used in establishing the power relationship of Eq. (5-29). The appearance of $\mathring{\Gamma}$, which belongs to the plane $z = 0$, permits us to express the spectral radiant power in terms of the MSDF in that plane; thus

$$\hat{\Phi}(r, \nu) = 2c^3k^2 \int_{1/2} m^2\kappa^2 \, d\Omega \iint_{-\infty}^{\infty} d\mathbf{x}_1 \, d\mathbf{x}_2 \, \hat{\Gamma}(\mathbf{x}_1, \mathbf{x}_2, 0, \nu)$$
$$\times \exp[-i2\pi\kappa\mathbf{p} \cdot (\mathbf{x}_1 - \mathbf{x}_2)]. \tag{7-147}$$

Now our goal is to express $\hat{\Phi}$ as an area integral and a projected solid angle integral to enable us to isolate an entity that may be identified as a radiance function. Specifically, we wish to cast the above equation in a form similar to Eq. (7-6). This objective may be achieved by following an ingenious procedure due to Walther. Note that $\mathbf{x}_1$ and $\mathbf{x}_2$ are two different points in the plane $z = 0$. Walther introduces the average and difference variables,

$$\mathbf{x} = \tfrac{1}{2}(\mathbf{x}_1 + \mathbf{x}_2) \quad \text{and} \quad \mathbf{x}_{12} = \mathbf{x}_1 - \mathbf{x}_2, \tag{7-148}$$

with the property that $d\mathbf{x}_1\, d\mathbf{x}_2 = d\mathbf{x}\, d\mathbf{x}_{12}$. The expression of $\hat{\Phi}$ may be rewritten in the form

$$\hat{\Phi}(r, \nu) = \int d\mathbf{x} \int_{1/2} m\, d\Omega \kappa^2 \Big\{ 2c^3 k^2 m \int \hat{\Gamma}\big(\mathbf{x} + \tfrac{1}{2}\mathbf{x}_{12}, \mathbf{x} - \tfrac{1}{2}\mathbf{x}_{12}, 0, \nu\big) \times \exp(-i2\pi\kappa\mathbf{p} \cdot \mathbf{x}_{12})\, d\mathbf{x}_{12} \Big\}.$$

The quantity in braces may be identified with generalized spectral radiance. Following WMW, we denote it by

$$B_\omega(\mathbf{r}, \mathbf{s}) = 2c^3 k^2 m \int \hat{\Gamma}\big(\mathbf{x} + \tfrac{1}{2}\mathbf{x}_{12}, \mathbf{x} - \tfrac{1}{2}\mathbf{x}_{12}, 0, \nu\big) \times \exp(-i2\pi\kappa\mathbf{p} \cdot \mathbf{x}_{12})\, d\mathbf{x}_{12}, \tag{7-149}$$

where we have used the vector

$$\mathbf{r} = \hat{\mathbf{i}}x + \hat{\mathbf{j}}y, \qquad z = 0,$$

and the unit vector $\mathbf{s}$,

$$\mathbf{s} = \hat{\mathbf{i}}p + \hat{\mathbf{j}}q + \hat{\mathbf{k}}m.$$

Observe that $\mathbf{s}$ with the independent components p, q is Fourier conjugate to $\mathbf{x}_{12}$; see, for example, Eq. (7-54). Thus, the arguments $\mathbf{r}$ and $\mathbf{s}$ of B_ω are *not* Fourier conjugate, which makes B_ω a valid function in wave theory. The spectral radiant power $\hat{\Phi}$ thus takes the familiar form

$$\hat{\Phi} = \frac{1}{\lambda^2} \int_{z=0} \int_{1/2} B_\omega(\mathbf{r}, \mathbf{s}) m\, d\Omega\, d\mathbf{x}. \tag{7-150}$$

The angle integration is restricted to half the total solid angle.

Now it is a simple matter to define the rest of the radiometric quantities. Marchand and Wolf define the spectral radiant emittance† by

$$E_\omega(\mathbf{r}) = \int_{1/2} B_\omega(\mathbf{r}, \mathbf{s})\, m\, d\Omega \tag{7-151}$$

so that

$$\hat{\Phi} = \frac{1}{\lambda^2} \int_{z=0} E_\omega(\mathbf{r})\, d\mathbf{x}. \tag{7-152}$$

On the other hand, the spectral radiant intensity is defined by

$$J_\omega(\mathbf{s}) = m \int_{z=0} B_\omega(\mathbf{r}, \mathbf{s})\, d\mathbf{x}. \tag{7-153}$$

Then

$$\hat{\Phi} = \frac{1}{\lambda^2} \int_{1/2} J_\omega(\mathbf{s})\, d\Omega. \tag{7-154}$$

This is the structure of generalized radiometry as obtained with the flow vector approach. We shall apply this to a few special cases of interest.

First, let us consider the expression of spectral radiant emittance, Eq. (7-151), and express it in terms of $\hat{\Gamma}$ by using Eq. (7-149); thus

$$E_\omega(\mathbf{r}) = 8\pi^2 c^3 \int d\mathbf{x}_{12}\, \hat{\Gamma}\left(\mathbf{x} + \tfrac{1}{2}\mathbf{x}_{12}, \mathbf{x} - \tfrac{1}{2}\mathbf{x}_{12}, 0, \nu\right)$$
$$\times \int_{1/2} d(\kappa\mathbf{p})\, m \exp(-i2\pi\kappa\mathbf{p} \cdot \mathbf{x}_{12}) \tag{7-155}$$

If $\hat{\Gamma}$ is known for the field, it gives E_ω, which describes the spectral radiant incidence on the plane $z = 0$. If $\hat{\Gamma}$ is known for the source, it gives the source emittance. In either case E_ω is *not* simply related to $\hat{\Gamma}$ as, for example, in Eq. (7-27) or as with the measured irradiance of Eq. (7-140b). The angle integral contained in it leads to spherical Bessel functions as shown by

†In the radiometric nomenclature, the quantity in Eq. (7-151) is called the spectral radiant exitance with reference to a source or the spectral irradiance in relation to a receiving plane.

Marchand and Wolf in their Eq. (39) of Appendix 7.2; thus

$$E_{\omega} = c^3k^2(2\pi)^{3/2}\int d\mathbf{x}_{12}\,\hat{\Gamma}(\mathbf{x}+\tfrac{1}{2}\mathbf{x}_{12},\mathbf{x}-\tfrac{1}{2}\mathbf{x}_{12},0,\nu)\left[\frac{J_{3/2}(kr_{12})}{(kr_{12})^{3/2}}\right], \tag{7-156}$$

where $r_{12} = (x_{12}^2 + y_{12}^2)^{1/2}$.For example, for a Carter–Wolf source, Eq. (7-84), the $\mathbf{x}_{12}$ integral reduces to a constant, making E_{ω} proportional to the optical spectral intensity, in agreement with the customary definition. Such is not the case in general. The MSDF $\hat{\Gamma}$ for a spatially nonstationary source does *not* factor. Equation (7-156) suggests the presence of a wave theoretic microstructure that makes E_{ω} quite different from the customarily defined optical spectral intensity. The spherical Bessel function in the square brackets is slightly broader than, for example, the familiar $[\sin(kr_{12})]/kr_{12}$. For comparison both of these functions are plotted in Fig. 7-5.

In the next two subsections we shall study some of the important results that follow from this theory.

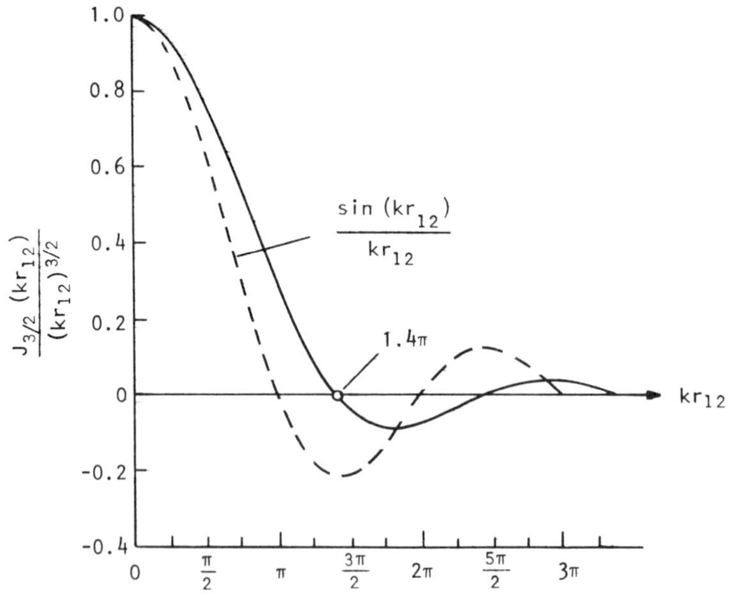

Fig. 7-5. Plot of the function $J_{3/2}(kr_{12})/(kr_{12})^{3/2}$. For comparison, the familiar function $[\sin(kr_{12})]/kr_{12}$ is also plotted on the same scale.

Blackbody Radiation

The spatial coherence function $\hat{g}_{BB}(\mathbf{x}_{12}, \nu)$ of blackbody radiation is given in Eq. (7-115). Its two-dimensional spatial Fourier transform $\mathring{g}_{BB}(\kappa\mathbf{p}, \nu)$ is given in Eqs. (7-117) and (7-118). Now the WMW spectral radiance $B_\omega(\mathbf{r}, \mathbf{s})$ of Eq. (7-149) contains a two-dimensional spatial Fourier transform over the difference variable $\mathbf{x}_{12}$. Therefore, without detailed calculation we see that $B_\omega(\mathbf{r}, \mathbf{s})$ is independent of angle or the variable $\mathbf{s}$ for blackbody radiation. It also follows, from Eq. (7-153), that the spectral radiant intensity J_ω is proportional to m; that is,

$$J_\omega \propto m = \cos\theta. \tag{7-157}$$

These are the well-known properties of a Lambertian source. Indeed, we expect this to be so for a completely randomized cavity radiation. Without taxing the reader's imagination, it follows that the converse is also true, namely, that the spectral radiance is a constant independent of angle only if the spatial Fourier transform of $\hat{\Gamma}$ has a $1/m$ dependence for real waves. It is *not* a constant SFS; the weighting $1/m$ is necessary to obtain a Lambertian behavior. This implies that some partial coherence is necessary to exhibit Lambertian properties.

On the other hand, the requirement of constant SFS (uniform weighting) was found to be essential in establishing the results listed in Table 7-3 for a noncoherent source. There it was shown that the noncoherent source is Lambertian.

Noncoherent Source

The noncoherent source has a delta-correlated MSDF $\hat{\Gamma}$. The spatial Fourier transform operation contained in Eq. (7-149) immediately shows that the spectral radiance is proportional to m; that is,

$$B_\omega(\mathbf{r}, \mathbf{s}) \propto m = \cos\theta. \tag{7-158}$$

It follows from Eq. (7-153) that the angular dependence of the spectral radiant intensity is

$$J_\omega(\mathbf{s}) \propto \cos^2\theta. \tag{7-159}$$

Thus, a source that has a constant or uniform SFS is *not* Lambertian in the WMW approach. The optical interpretation of a constant SFS is that the source is uniform in angle but, according to Eq. (7-159), such a source will appear darker when viewed at larger angles from the normal.

SUMMARY AND CONCLUSIONS

We have presented two approaches to generalized radiometry. They differ mainly in the choice of the field quantity that is relevant to the detection of radiation. In the first approach the irradiance was defined in terms of the measured irradiance of Eq. (7-140); that is, it is directly related to the square of the absolute value of the scalar field variable.

The functions of generalized radiometry and their interrelationships are given in Table 7-2. In this scheme the generalized radiance is defined as a complex cross correlation that is a function of two space points and two directions.

This formalism was applied to the case of a uniform noncoherent source. It has a constant spatial frequency spectrum. It was shown that if the coherence in the field is neglected, then generalized radiometry reduces to conventional radiometry; all the functions then become real and nonnegative (see Table 7-3). It was shown that the uniform noncoherent source is Lambertian. The radiometric functions in this case are independent of position and direction.

The formalism of generalized radiometry was applied to a partially coherent source case. Contact with conventional radiometry was established by defining a reduced radiance function and by using the approximation of neglecting the coherence in the field. It was found that the reduced radiance function is real but not necessarily nonnegative. However, the reduced radiant intensity function was found to be real and nonnegative. The results are given in Table 7-4. The radiometric functions for the partially coherent case depend on position as well as direction.

The special case of the Carter–Wolf quasihomogeneous, partially coherent source was also studied with the formalism of generalized radiometry. The results are given in Table 7-5. In this case the reduced spectral radiance and the reduced spectral radiant intensity are both real as well as nonnegative.

The formalism of generalized radiometry was applied to the case of a completely coherent source. The resulting relationships [see Eqs. (7-92)–(7-98)] are found to be unique in the sense that they have no analog in conventional radiometry.

To study the radiation from a blackbody source, the mutual spectral density function was constructed by following the Carter–Wolf source model. The coherence function was taken to be $[\sin(kr_{12})]/kr_{12}$ as dictated by the theory of cavity radiation. Although this coherence function is very narrow, the radiation is partially coherent since its spatial frequency spectrum is not "white;" that is, it is not constant. Its form is $1/\cos\theta$; it grows to ∞ as the viewing angle θ approaches $\pi/2$ from the normal to the source.

The application of generalized radiometry to this case yielded nonphysical results.

In the second approach, namely, the one due to Walther, Marchand, and Wolf, the irradiance is defined in terms of the Poynting irradiance of Eq. (7-141). In this approach, the spectral radiant emittance is found to be proportional to the integral of the product of $\hat{\Gamma}$ and $J_{3/2}(kr_{12})/(kr_{12})^{3/2}$ [see Eq. (7-156)].

One begins with the flow vector and defines the spectral radiance function as in Eq. (7-149). The other radiometric quantities were defined in Eqs. (7-151)–(7-154). In this approach, blackbody radiation whose coherence function has a nonuniform spatial frequency spectrum, of the form $1/\cos\theta$, exhibits Lambertian properties.

The noncoherent source, characterized by a uniform spatial frequency spectrum over the real waves, is found to have properties quite different from Lambertian; in particular, the spectral radiant intensity has $\cos^2\theta$ angular dependence. It implies that a uniform noncoherent source will appear darker at larger viewing angles from its normal. In this approach one is compelled to abandon the physical–optical interpretation of the spatial frequency spectrum, for a nonuniform $(1/\cos\theta)$ weighting is the only one that will lead to uniform radiation (Lambertian) characteristics.

These are the two approaches to generalized radiometry and some of the elementary results that readily follow from them. They are gathered together for the perusal of the reader.

REFERENCES

Aharoni, J. (1959). *The Special Theory of Relativity*, Clarendon Press, Oxford, England, 331 pp.

Baltes, H. P. (1977). Coherence and radiation laws, *Appl. Phys.* **12**:221–224.

Baltes, H. P., J. Geist, and A. Walther (1978). Radiometry and coherence, in *Inverse Source Problems in Optics*, Vol. 9, H. P. Baltes, Ed., Springer-Verlag, Berlin.

Bartell, F. O., and W. L. Wolfe (1973). Polythermal blackbody simulators (Abstr.), *J. Opt. Soc. Am.* **63** (10):1286.

Bartell, F. O., and W. L. Wolfe (1976). Cavity radiation theory, *Infrared Phys.* **16**:13–24.

Born, M., and E. Wolf (1970). *Principles of Optics*, 4th ed., Pergamon, New York, 808 pp.

Bouguer, Pierre (1760). *Optical Treatise on the Gradation of Light*, translation by W. E. Knowles Middleton (1961), University of Toronto Press, Toronto.

Carter, W. H., and E. Wolf (1977). Coherence and radiometry with quasihomogeneous planar sources, *J. Opt. Soc. Am.* **67** (6):785–796.

Chandrasekhar, S. (1960). *Radiative Transfer*, Dover, New York, 393 pp.

Eyges, L. (1972). *The Classical Electromagnetic Field*, Addison-Wesley, Reading, MA, Sec. 11.9, particularly the discussion after Eq. (11.74).

Fante, R. L. (1981). Relationship between radiative-transport theory and Maxwell's equations in dielectric media, *J. Opt. Soc. Am.* **71** (4):460–468.

Feynman, R. P., R. B. Leighton, and M. Sands (1964). *The Feynman Lectures on Physics*, Addison-Wesley, Reading, MA, Vol. II, Sec. 27-4.

Glauber, R. J. (1963). Coherent and incoherent states of the radiation field, *Phys. Rev.* **131** (6):2766–2788.

Gradshteyn, I. S., and Ryzhik, I. M. (1965). *Tables of Series, Products, and Integrals*, 4th ed., Academic Press, New York, 1086 pp.

Grum, F., and R. J. Becherer (1979). *Optical Radiation Measurements: Volume 1, Radiometry*, Academic Press, New York.

Hecht, E., and A. Zajac (1976). *Optics*, Addison-Wesley, Reading, MA, 565 pp.

Illuminating Engineering Society (1968). *USA Standard Nomenclature and Definitions for Illuminating Engineering*, RP-16, Illuminating Engineering Society, New York.

Jackson, J. D. (1963). *Classical Electrodynamics*, Wiley, New York, Sec. 6.8.

Jenkins, F. A., and H. E. White (1976). *Fundamentals of Optics*, McGraw-Hill, New York, Sec. 25.12.

Jones, R. C. (1963). Terminology in photometry and radiometry, *J. Opt. Soc. Am.* **53** (11):1314–1315.

Kano, Y., and E. Wolf (1962). Temporal coherence of blackbody radiation, *Proc. Phys. Soc. Lond.* **80** (6):1273–1276.

Klauder, J., and E. C. G. Sudarshan (1968). *Fundamentals of Quantum Optics*, Benjamin, New York, 279 pp.

Klein, M. V. (1970). *Optics*, Wiley, New York, 647 pp.

Leighton, R. B. (1959). *Principles of Modern Physics*, McGraw-Hill, New York, 795 pp.

Longhurst, R. S. (1967). *Geometrical and Physical Optics*, 2nd ed., Wiley, New York, 534 pp.

MacAdam, D. L. (1967). Nomenclature and symbols for radiometry and photometry, *J. Opt. Soc. Am.* **57** (6):854.

Mandel, L., and E. Wolf (1965). Coherence properties of optical fields, *Revs. Mod. Phys.* **37** (2):231–287.

Marathay, A. S. (1975). Diffraction of light from an aperture on a spherical surface, *J. Opt. Soc. Am.* **65** (8):909–913.

Marathay, A. S. (1976a). Radiometry of partially coherent fields, I, *Opt. Acta* **23** (10):785–794.

Marathay, A. S. (1976b). Radiometry of partially coherent fields, II, *Opt. Acta* **23** (10):795–798.

Marathay, A. S., and G. B. Parrent, Jr. (1970). Use of scalar theory in optics, *J. Opt. Soc. Am.* **60** (2):243–245.

Marathay, A. S., and S. Prasad (1980). Rayleigh–Sommerfeld diffraction theory and Lambert's law, *Pramana* **14** (2):103–111.

Marchand, E. W., and E. Wolf (1974). Radiometry with sources of any state of coherence, *J. Opt. Soc. Am.* **64** (9):1219–1226.

Mechtly, E. A. (1969). The International System of Units, Rep. NASA P-7102, p. 44, Office of Tech. Util., NASA, Washington, D. C.

Mehta, C. L. (1963). Coherence-time and effective bandwidth of blackbody radiation, *Nuovo Cimento* **28** (2):401–408.

Mehta, C. L., and E. Wolf (1964a). Coherence properties of blackbody radiation. I. Correlation tensors of the classical field, *Phys. Rev. A* **134** (5):1143–1149.

Mehta, C. L., and E. Wolf (1964b). Coherence properties of blackbody radiation. II. Correlation tensors of the quantized field, *Phys. Rev. A* **134** (5):1149–1153.

Nicodemus, F. E., ed. (1976). *Self-Study Manual of Optical Radiation Measurements*, U.S. Department of Commerce, National Bureau of Standards, available from Superintendent of Documents, U.S. Government Printing Office, Washington, D.C.

Palmer, J. M. (1979). Spectral radiant exitance of a nonisothermal cavity radiator, *Appl. Opt.* **18** (6):758–760.

Roman, P. (1965). *Advanced Quantum Theory*, Addison-Wesley, Reading, MA, Sec. 1-7.

Skinner, T. J. (1965). Energy Considerations, Propagation in a Random Medium and Imaging in Scalar Coherence, Ph.D. Dissertation, Boston University.

Walsh, J. W. T. (1965). *Photometry*, 3rd ed., Dover, New York, p. 1.

Walther, A. (1967). Systematic approach to teaching of lens theory, *Am. J. Phys.* **35** (9):808–816.

Walther, A. (1968). Radiometry and coherence, *J. Opt. Soc. Am.* **58** (9):1256–1259.

Webb, R. H. (1969). *Elementary Wave Optics*, Academic Press, New York.

Wiener, O. (1890). Stehende Lichtwellen und die Schwingungsrichtung Polarisirten Lichtes, *Ann. Phys. (Leipz.)* **40**:203–243.

Wolf, E. (1978). Coherence and radiometry, *J. Opt. Soc. Am.* **68** (1):6–17.

Wolfe, W. L. (1965). *Handbook of Military Infrared Technology*, Office of Naval Research, Department of the Navy, available from Superintendent of Documents, U.S. Government Printing Office, Washington, D.C.

Wolfe, W. L., and G. J. Zissis (1978). *The Infrared Handbook*, Office of Naval Research, Department of the Navy, available from Superintendent of Documents, U.S. Government Printing Office, Washington, D.C.

Wyatt, C. L. (1978). *Radiometric Calibration: Theory and Methods*, Academic Press, New York, 200 pp.

APPENDIX 7.1. COMPARISON OF THE RAYLEIGH-SOMMERFELD THEORY FOR A PLANE AND SPHERICAL GEOMETRY

In the Rayleigh–Sommerfeld diffraction theory, a Green's function is constructed so that it vanishes on a surface containing the diffracting aperture. For the aperture in a plane, one specifies the complex amplitude transmittance and the incident wave amplitude at each point in the aperture. The sources are to the left of this plane and the diffracted amplitude is calculated in the right half-space, $z > 0$. In the case of an aperture on a sphere, the complex amplitude transmittance and the incident wave amplitude are specified at points on the surface of the sphere within the domain $\mathcal{A}$ of the aperture. It is assumed to be zero outside $\mathcal{A}$. The sources are assumed to be inside the sphere and the diffracted amplitude is calculated in the region outside the sphere. The Green's function is constructed in such a way that it is zero on the surface of the sphere containing the aperture $\mathcal{A}$.

TABLE A7-1. COMPARISON OF THE RAYLEIGH–SOMMERFELD THEORY FOR PLANE AND SPHERICAL GEOMETRY

Plane Geometry	Spherical Geometry
Aperture $\mathcal{Q}$ on the plane defined by $z = z_1$	Aperture $\mathcal{Q}$ on the sphere of radius r_1
Green's function $G = 0$ on $z = z_1$	$G = 0$ on $r = r_1$
Diffracted field:	Diffracted field:
$\psi(x_2, y_2, z_2) = \iint_{\mathcal{Q}} dx_1\, dy_1 \left[\psi_{\mathcal{Q}}(x_1, y_1, z_1) \times \left\{ -\frac{\partial G}{\partial n}(x_2 - x_1, y_2 - y_1, z_2 - z_1) \right\} \right]$	$\psi(r_2, \theta_2, \phi_2) = \iint_{\mathcal{Q}} r_1^2\, d\Omega_1 \left[\psi_{\mathcal{Q}}(r_1, \theta_1, \phi_1) \times \left\{ -\frac{\partial G}{\partial n}(\mathbf{r}_2, \mathbf{r}_1) \right\} \right]$
The eigenfunctions for this geometry are plane waves, hence use (spatial) Fourier transform:	The eigenfunctions for this geometry are spherical harmonics, hence use Laplaces series:
$\tilde{\psi}(\kappa p, \kappa q, z_2) = \tilde{\psi}_{\mathcal{Q}}(\kappa p, \kappa q, z_1) \dfrac{\exp(ikmz_2)}{\exp(ikmz_1)}$	$\psi_{lm}(r_2) = \psi_{\mathcal{Q}lm}(r_1) \dfrac{h_l^{(1)}(kr_2)}{h_l^{(1)}(kr_1)}$
$\exp(ikmz)$ = z-dependent solution of the Hemholtz equation	$h_l^{(1)}(kr)$ = radial solution of the Hemholtz equation
Amplitude ψ or intensity $\lvert\psi\rvert^2$ is defined over *planes* of constant z which are parallel to the $z = z_1$ plane over which $G = 0$	ψ or $\lvert\psi\rvert^2$ is defined over *spheres* of radius r_2 concentric to the sphere of radius r_1, on which $G = 0$

A comparison of the formulation of the plane and spherical geometry is summarized in Table A7-1. The material for this table is from Marathay and Prasad (1980) published in *Pramana* **14** (2):103–111.

APPENDIX 7.2. MARCHAND AND WOLF PAPER

In this appendix, the entire paper by Marchand and Wolf entitled "Radiometry with Sources of Any State of Coherence" is reproduced for ready reference. This paper was published in the *Journal of the Optical Society of America*, Vol. 64, No. 9, Sept. 1974, pp. 1219–1226. Permission to reproduce the paper was kindly granted by Dr. E. W. Marchand and Prof. E. Wolf and by the Optical Society of America.

Radiometry with sources of any state of coherence*

E. W. Marchand

Research Laboratories, Eastman Kodak Co., Rochester, New York 14650

E. Wolf

Department of Physics and Astronomy, University of Rochester, Rochester, New York 14627

(Received 20 February 1974)

The basic laws of radiometry are generalized to fields generated by a two-dimensional stationary source of any state of coherence. Important in this analysis is the concept of the generalized radiance function, introduced by Walther in 1968. The concepts of generalized radiant emittance and of generalized radiant intensity are introduced and it is shown how all these quantities may be expressed in terms of coherence functions of the source. Both the generalized radiance of Walther and the generalized radiant emittance may take on negative values, indicating that these quantities have, in general, a less-direct physical meaning than have the corresponding quantities of traditional radiometry (which presumably represents the incoherent limit of the present theory). The generalized radiant intensity is, however, always found to be non-negative and, just as in the incoherent limit, represents the angular distribution of the energy flux in the far zone.

Index Headings: Radiometry; Coherence.

Although the concepts of radiometry are familiar to anyone who has even the most rudimentary knowledge of optics, it does not appear to be generally realized that the fundamental laws of radiometry have never been systematically derived from the basic theories of light. Radiometry appears to have been systematized around the turn of the century in connection with the theory of heat radiation, and a well-known book by Max Planck,[1] the first edition of which appeared in 1906, remains probably the most comprehensive account to this day of the fundamentals of the subject. It is taken for granted that traditional radiometry describes, in some approximation, the behavior of radiation fields created by incoherent sources, and it is known that blackbody sources, with which the early investigations on heat radiation were chiefly concerned, are among the most incoherent ones found in nature. Yet modern researches in coherence theory have shown that even such sources can generate fields that are highly correlated (i.e., spatially highly coherent) throughout arbitrarily large regions of space, a result expressed quantitatively by the so-called VanCittert–Zernike theorem.[2] This fact raises serious questions as to the range of validity of even the most basic laws of radiometry and underscores the need for clarifying how, and under what conditions, these laws can be deduced from the presently accepted theories of light. Closely related to these questions is the problem, of considerable practical importance at the present time, of the possibility of extending the concepts and the laws of traditional radiometry to fields produced by laser sources and, more generally, by sources of arbitrary states of coherence.

In the present paper, we take a step towards solving these problems and present a generalization of some of the basic concepts of radiometry to fields generated by any statistically stationary two-dimensional source.[3] Basic in our treatment is the concept of generalized radiance for a source of any state of coherence, introduced a few years ago by Walther[4] in an important investigation. In terms of it, we define generalizations of the concepts of radiant emittance and of radiant intensity and examine these quantities in some detail. We show how the generalized radiance, the generalized radiant emittance, and the generalized radiant intensity may be expressed in terms of various second-order correlation functions of the fluctuating source. A result emerges that may appear somewhat surprising without closer examination: Both the generalized radiance of Walther and our generalized radiant emittance may take on negative values, and they also exhibit some other apparently unphysical features. These facts indicate that neither of these two quantities is, in general, measurable; but it will become clear that they are nevertheless useful mathematical constructs in terms of which considerable insight can be obtained into the behavior of partially coherent fields. The generalized radiant intensity is found to be always non-negative and to represent correctly the (measurable) angular distribution of the energy flux in the far zone. We also examine the incoherent and the coherent limits of our theory. In the limiting case of incoherence both the radiance and the radiant emittance are always found to be non-negative, and this limit obviously represents traditional radiometry. However, we also find that a strictly incoherent finite source cannot radiate in accordance with Lambert's law.

Only the basic mathematical structure of the generalized radiometry is discussed in this paper. We plan to present, in a subsequent paper, numerical results illustrating radiation properties of sources of different geometries and different states of coherence.

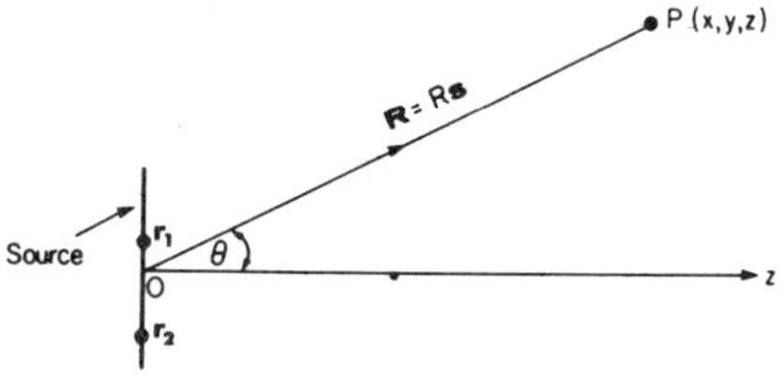

FIG. 1. Illustration of the notation.

I. GENERALIZATION OF BASIC CONCEPTS OF RADIOMETRY TO FIELDS CREATED BY SOURCES OF ANY STATE OF COHERENCE

We consider an optical field generated by a source occupying a portion of the plane $z=0$, around the origin O of our coordinate system. The source is assumed to be statistically stationary but of any state of coherence and to emit light into the half-space $z>0$. We have shown in an earlier paper[5] that the intensity $I(\mathbf{R})$ at a point $P(x,y,z)$ in the far zone $(k|\mathbf{R}|\to\infty)$ may be expressed in the form (see Fig. 1)

$$I(\mathbf{R})=\int_0^\infty I_\omega(\mathbf{R})\,d\omega, \tag{1}$$

with

$$I_\omega(\mathbf{R})\sim\left(\frac{2\pi}{kR}\right)^2\cos^2\theta\,\mathcal{A}(\mathbf{s},\mathbf{s};\omega). \tag{2}$$

Here

$$k=\omega/c=2\pi/\lambda \tag{3}$$

is the wave number, ω the frequency, λ the wavelength, c the vacuum velocity of light, and $\mathcal{A}(\mathbf{s},\mathbf{s};\omega)$ is the angular self-correlation function of the field in the half-space $z>0$ [cf. Eq. (12) below]. In Eq. (2), $\mathbf{s}$ is the real unit vector in the $\mathbf{R}$ direction,

$$\mathbf{s}=\frac{\mathbf{R}}{R}, \tag{4}$$

with components

$$s_x=\frac{x}{R},\quad s_y=\frac{y}{R},\quad s_z=\frac{z}{R}=(1-s_x^2-s_y^2)^{\frac{1}{2}}=\cos\theta, \tag{5}$$

where θ is the angle between the position vector $\mathbf{R}$ and the z axis.

The intensity I appearing in Eq. (1) was defined in the manner customary in physical optics, i.e., as the average of the square of the absolute value of a complex scalar field variable. It is usually assumed (Ref. 2, §8.4) that this quantity is equal, in appropriate units, to the absolute value of the average Poynting vector $\mathbf{S}$, i.e.,

$$I(\mathbf{R})=|\langle\mathbf{S}(\mathbf{R})\rangle|. \tag{6}$$

However, in radiometry, the basic quantity is the radiant flux. Now a surface element $d\sigma$ around a point P in the far zone, at distance R from the origin, at right angles to the direction OP may be expressed in the form

$$d\sigma=R^2\,d\Omega, \tag{7}$$

where

$$d\Omega=\frac{ds_x\,ds_y}{s_z} \tag{8}$$

is the element of the solid angle that the surface element subtends at the origin O. Hence, from the physical significance of the Poynting vector (Ref. 2, p. 9) and from the relations (6) and (7), it follows that the flux traversing the area element $d\sigma$ and the intensity I are related by the formula

$$dF=IR^2\,d\Omega. \tag{9}$$

Hence Eqs. (1) and (2) may be rewritten in the following form, which is more appropriate to the purposes of radiometry,

$$dF=\int_0^\infty dF_\omega\,d\omega, \tag{10}$$

where

$$dF_\omega\sim\left(\frac{2\pi}{k}\right)^2\cos^2\theta\,\mathcal{A}(\mathbf{s},\mathbf{s};\omega)\,d\Omega. \tag{11}$$

We take Eqs. (10) and (11) as the starting point of our analysis, bearing in mind that we are considering light created by a stationary plane source of any state of coherence.

We will now formally express the angular self-correlation function $\mathcal{A}(\mathbf{s},\mathbf{s};\omega)$ that occurs in Eq. (11) as an integral over the source plane. According to Eq. (17) of Ref. 5 and the Fourier inverse of Eq. (15) of Ref. 5,

$$\mathcal{A}(\mathbf{s},\mathbf{s};\omega)=\left(\frac{k}{2\pi}\right)^4\iint_{(z=0)}W(\mathbf{r}_1,\mathbf{r}_2;\omega)\times\exp\{-ik[(\mathbf{r}_1-\mathbf{r}_2)\cdot\mathbf{s}]\}\,d^2\mathbf{r}_1\,d^2\mathbf{r}_2, \tag{12}$$

where $W(\mathbf{r}_1,\mathbf{r}_2;\omega)$ is the cross-spectral density function for the pair of points $\mathbf{r}_1$ and $\mathbf{r}_2$ in the source plane $z=0$. In the double integral on the right-hand side of Eq. (12), it is to be understood that the points $\mathbf{r}_1$ and $\mathbf{r}_2$ take independently all possible positions in the source plane $z=0$. Let us introduce new variables of integration in Eq. (12) by means of the formulas

$$(\mathbf{r}_1-\mathbf{r}_2)=\mathbf{r}',\quad \tfrac{1}{2}(\mathbf{r}_1+\mathbf{r}_2)=\mathbf{r}. \tag{13}$$

Then Eq. (12) can readily be shown to be expressible in the form

$$\mathcal{A}(\mathbf{s},\mathbf{s};\omega)=\left(\frac{k}{2\pi}\right)^2\frac{1}{\cos\theta}\int_{(z=0)}B_\omega(\mathbf{r},\mathbf{s})\,d^2\mathbf{r}, \tag{14}$$

where

$$B_\omega(\mathbf{r},\mathbf{s})=\left(\frac{k}{2\pi}\right)^2\cos\theta\int_{(z'=0)} W(\mathbf{r}+\tfrac{1}{2}\mathbf{r}',\mathbf{r}-\tfrac{1}{2}\mathbf{r}';\omega)\times\exp(-ik\mathbf{s}\cdot\mathbf{r}')\,d^2\mathbf{r}' \quad (15)$$

and the integrals in Eqs. (14) and (15) extend over the plane $z=0$ and $z'=0$, respectively. On substituting from Eq. (14) into (11) and recalling that according to Eq. (5) $s_z=\cos\theta$, we find that

$$dF_\omega\sim\cos\theta\,d\Omega\int_{(z=0)} B_\omega(\mathbf{r},\mathbf{s})\,d^2\mathbf{r}, \quad (16)$$

so that the flux at frequency ω, radiated by the source across the hemisphere at infinity in the half-space $z>0$, is given by

$$F_\omega\sim\int_{(2\pi)} d\Omega\cos\theta\int_{(z=0)} B_\omega(\mathbf{r},\mathbf{s})\,d^2\mathbf{r}, \quad (17)$$

where the integration extends over the source plane $z=0$ and over the solid angle subtended by the hemisphere in the half-space $z>0$ and centered at the origin.

Formally Eq. (17) is of the same mathematical form as a basic law of radiometry, for the flux generated by a spatially incoherent source with radiance function $B_\omega(\mathbf{r},\mathbf{s})$. However, Eq. (17), with $B_\omega(\mathbf{r},\mathbf{s})$ defined by Eq. (15), now gives the flux generated by a stationary source of any state of coherence. We shall refer to the function $B_\omega(\mathbf{r},\mathbf{s})$, defined by Eq. (15), as *generalized radiance*. It is the same function as that introduced not long ago by Walther.[4] However, this quantity must be interpreted with some caution, because, as we shall see later, it does not possess all the properties normally attributed to radiance. In particular, as we show in Sec. VI, the generalized radiance may be negative for some values of its arguments and does not, in general, vanish for all points $\mathbf{r}$ in the source plane that lie outside the domain occupied by the source. Nevertheless, as we shall see, the generalized radiance is a useful quantity, because, in terms of it, the concepts of radiant emittance and of radiant intensity can readily be generalized to fields created by sources of any state of coherence. For this purpose, we need only to rewrite Eq. (17) in two different ways,

$$F_\omega=\int_{(z=0)} E_\omega(\mathbf{r})\,d^2\mathbf{r} \quad (18)$$

and

$$F_\omega=\int_{(2\pi)} J_\omega(\mathbf{s})\,d\Omega, \quad (19)$$

where

$$E_\omega(\mathbf{r})=\int_{(2\pi)} B_\omega(\mathbf{r},\mathbf{s})\cos\theta\,d\Omega \quad (20)$$

and

$$J_\omega(\mathbf{s})=\cos\theta\int_{(z=0)} B_\omega(\mathbf{r},\mathbf{s})\,d^2\mathbf{r}. \quad (21)$$

The expressions (20) and (21) are formally identical with the usual expressions of traditional radiometry for the radiant emittance and the radiant intensity, respectively, from a spatially incoherent source. Hence we shall refer to the quantity $E_\omega(\mathbf{r})$, defined by Eq. (20), as the *generalized radiant emittance* (at frequency ω) at the point $\mathbf{r}$ of the source and to the quantity $J_\omega(\mathbf{s})$, defined by Eq. (21), as the *generalized radiant intensity* (at frequency ω) in the $\mathbf{s}$ direction. To avoid possible misunderstanding, we stress that the radiometric term "radiant intensity" [defined by Eq. (21)] and the physical-optics term "intensity" [defined by Eq. (6)] have different meanings.

We now examine briefly some of the properties of these quantities and show that they may be expressed in several alternative forms involving various correlation functions of the optical field.

II. THE GENERALIZED RADIANCE

We defined the generalized radiance by the formula (15), viz.,

$$B_\omega(\mathbf{r},\mathbf{s})=\left(\frac{k}{2\pi}\right)^2\cos\theta\int_{(z'=0)} W(\mathbf{r}+\tfrac{1}{2}\mathbf{r}',\mathbf{r}-\tfrac{1}{2}\mathbf{r}';\omega)\times\exp(-ik\mathbf{s}\cdot\mathbf{r}')\,d^2\mathbf{r}', \quad (22)$$

W being the cross-spectral density function. We show in Appendix A that B_ω may also be expressed in the form

$$B_\omega(\mathbf{r},\mathbf{s})=k^2\cos\theta\int_{(\mathbf{f}\text{ plane})} \hat{W}(k\mathbf{s}+\tfrac{1}{2}\mathbf{f},-k\mathbf{s}+\tfrac{1}{2}\mathbf{f};\omega)\times\exp(i\mathbf{f}\cdot\mathbf{r})\,d^2\mathbf{f}, \quad (23)$$

where

$$\hat{W}(\mathbf{f}_1,\mathbf{f}_2;\omega)=\frac{1}{(2\pi)^4}\iint_{(z=0)} W(\mathbf{r}_1,\mathbf{r}_2;\omega)\times\exp\{-i(\mathbf{f}_1\cdot\mathbf{r}_1+\mathbf{f}_2\cdot\mathbf{r}_2)\}\,d^2\mathbf{r}_1\,d^2\mathbf{r}_2 \quad (24)$$

is the four-dimensional spatial Fourier transform of the cross-spectral density function $W(\mathbf{r}_1,\mathbf{r}_2;\omega)$.

Alternatively, if we recall that, according to Eq. (17) of Ref. 5, $\hat{W}$ and the angular correlation function $\mathfrak{A}$ are connected by the relation

$$\mathfrak{A}(\mathbf{s}_1,\mathbf{s}_2;\omega)=k^4\hat{W}(k\mathbf{s}_1,-k\mathbf{s}_2;\omega), \quad (25)$$

we see from Eqs. (23) and (25), if we change the

variables of integration from $\mathbf{f}$ to $\mathbf{s}'=\mathbf{f}/k$, that the generalized radiance may also be expressed in the form

$$B_\omega(\mathbf{r},\mathbf{s})=\cos\theta\iint_{-\infty}^{\infty}\mathcal{A}(\mathbf{s}+\tfrac{1}{2}\mathbf{s}',\ \mathbf{s}-\tfrac{1}{2}\mathbf{s}';\omega)\times\exp(ik\mathbf{s}'\cdot\mathbf{r})\,ds_x'\,ds_y'. \quad (26)$$

The formal similarity in the structure of the integrals (22) and (26) should be noted. Equation (22) expresses the generalized radiance in terms of the cross-spectral density function $W(\mathbf{r}_1,\mathbf{r}_2;\omega)$, i.e., in terms of correlations of the field at pairs of points on the source plane. Equation (26) expresses the generalized radiance in terms of the angular correlation function $\mathcal{A}(\mathbf{s}_1,\mathbf{s}_2;\omega)$, i.e., in terms of correlations of the plane waves in the angular spectrum representation that are propagated in pairs of (possibly complex) directions.

Although for each point $\mathbf{r}$ the domain of integration of the right-hand side of Eq. (22) is formally infinite, the integration extends, actually, over a finite domain of the $\mathbf{r}'$ plane. This follows from the fact that the source was assumed to be finite and hence the integrand in Eq. (22) vanishes when either or both of the points $(\mathbf{r}+\frac{1}{2}\mathbf{r}')$ and $(\mathbf{r}-\frac{1}{2}\mathbf{r}')$ falls outside the area of the plane $z=0$ occupied by the source.

In Appendix B we show that the generalized radiance $B_\omega(\mathbf{r},\mathbf{s})$ is always real. However, we shall see later (end of Sec. VI) that it may take on negative values in some cases. Hence the generalized radiance cannot, in general, be interpreted as a true flux density. This fact, which may appear somewhat surprising at first, can be appreciated readily if we recall that the generalized radiance was introduced by means of the relation [Eq. (16)]

$$dF_\omega\sim\cos\theta\,d\Omega\int_{(z=0)}B_\omega(\mathbf{r},\mathbf{s})\,d^2\mathbf{r}, \quad (27)$$

where dF_ω represents the flux at frequency ω, radiated by the source into the solid angle $d\Omega$ about the direction $\mathbf{s}$. Thus, although dF_ω is necessarily non-negative, there is no reason to expect that B_ω will also have this property. In fact, because B_ω is a function of conjugate Fourier variables, namely, $\mathbf{r}$ and $\mathbf{s}$, we cannot prescribe the values of B_ω in Eq. (16) arbitrarily for all values of its arguments. This situation is strictly analogous to that encountered in connection with the Wigner distribution function in quantum statistics.[6] The Wigner distribution function is also a function of conjugate Fourier variables and is well known to take on negative values, in general. Nevertheless, this distribution function is a useful mathematical tool, as it allows calculation of various expectation values to be carried out by means of formulas that are strictly similar to those employed in ordinary probability theory. We might expect the generalized radiance to share this property with the Wigner distribution function, but we shall not study this question in the present paper.

Although, as already pointed out, the generalized radiance can take on negative values, it can be shown to be non-negative for a spatially incoherent source. This result will be established in Sec. VI.

Irrespective of the state of coherence of the source, the following result holds:

$$\int_{(z=0)}B_\omega(\mathbf{r},\mathbf{s})\,d^2\mathbf{r}=(2\pi k)^2\cos\theta\hat{W}(k\mathbf{s},-k\mathbf{s};\omega). \quad (28)$$

This identity follows when the integral of the expression (23) for B_ω is taken with respect to $\mathbf{r}$ over the whole z plane. If the order of the $\mathbf{r}$ and $\mathbf{f}$ integrations is then interchanged and use is made of the relation

$$\int_{(\mathbf{f}\text{ plane})}\exp(i\mathbf{f}\cdot\mathbf{r})\,d^2\mathbf{r}=(2\pi)^2\delta^{(2)}(\mathbf{f}), \quad (29)$$

where $\delta^{(2)}(\mathbf{f})$ is the two-dimensional Dirac δ function, the relation (28) is then immediately obtained.

III. THE GENERALIZED RADIANT EMITTANCE

The generalized radiant emittance was defined by formula (20) as

$$E_\omega(\mathbf{r})=\int_{(2\pi)}B_\omega(\mathbf{r},\mathbf{s})\cos\theta\,d\Omega. \quad (30)$$

If we use expression (15) for B_ω, Eq. (30) becomes

$$E_\omega(\mathbf{r})=\int_{(z'=0)}W(\mathbf{r}+\tfrac{1}{2}\mathbf{r}',\ \mathbf{r}-\tfrac{1}{2}\mathbf{r}';\omega)K_\omega(\mathbf{r}')\,d^2\mathbf{r}', \quad (31)$$

where

$$K_\omega(\mathbf{r}')=\left(\frac{k}{2\pi}\right)^2\iint_{s_x^2+s_y^2\leqslant 1}\cos\theta\exp(-ik\mathbf{s}\cdot\mathbf{r}')\,ds_x\,ds_y. \quad (32)$$

The integral on the right-hand side of Eq. (32) may be evaluated readily. If we change to polar coordinates via the relations

$$s_x=\rho\cos\phi,\qquad s_y=\rho\sin\phi, \quad (33a)$$

$$x'=r'\cos\phi',\qquad y'=r'\sin\phi', \quad (33b)$$

and use the fact that $\cos\theta=(1-s_x^2-s_y^2)^{\frac{1}{2}}$, Eq. (32) becomes

$$K_\omega(\mathbf{r}')=\left(\frac{k}{2\pi}\right)^2\int_0^1\rho(1-\rho^2)^{\frac{1}{2}}\,d\rho\times\int_0^{2\pi}\exp[-ik\rho r'\cos(\phi-\phi')]\,d\phi, \quad (34)$$

$$r'=|\mathbf{r}'|=(x'^2+y'^2)^{\frac{1}{2}}.$$

The integral with respect to ϕ is well known to have the value[7] $2\pi J_0(k\rho r')$ where J_0 is the Bessel function of the first kind and zero order. Hence

$$K_\omega(\mathbf{r}')=\frac{k^2}{2\pi}\int_0^1 \rho(1-\rho^2)^{\frac{1}{2}}J_0(k\rho r')\,d\rho. \tag{35}$$

The integral on the right-hand side may be evaluated in terms of a spherical Bessel function. We have[8]

$$\int_0^1 \rho(1-\rho^2)^{\frac{1}{2}}J_0(k\rho r')\,d\rho=\left(\frac{\pi}{2}\right)^{\frac{1}{2}}\frac{J_{\frac{3}{2}}(kr')}{(kr')^{\frac{3}{2}}}, \tag{36}$$

where $J_{\frac{3}{2}}(x)$ is the Bessel function of the first kind and order $\frac{3}{2}$, whose explicit form, in terms of trigonometric functions, is

$$J_{\frac{3}{2}}(x)=\left(\frac{2}{\pi x}\right)^{\frac{1}{2}}\left[\frac{\sin x}{x}-\cos x\right]. \tag{37}$$

From Eqs. (35) and (36), we obtain the following expression for K,

$$K_\omega(\mathbf{r}')=\frac{k^2}{2(2\pi)^{\frac{1}{2}}}\frac{J_{\frac{3}{2}}(kr')}{(kr')^{\frac{3}{2}}}. \tag{38}$$

Using this result in Eq. (31), we finally obtain the expression for the generalized radiant emittance in terms of the cross-spectral density function W of the source,

$$E_\omega(\mathbf{r})=\frac{k^2}{2(2\pi)^{\frac{1}{2}}}\int_{(z'=0)} W(\mathbf{r}+\tfrac{1}{2}\mathbf{r}',\mathbf{r}-\tfrac{1}{2}\mathbf{r}';\omega)\frac{J_{\frac{3}{2}}(kr')}{(kr')^{\frac{3}{2}}}\,d^2\mathbf{r}'. \tag{39}$$

The similarity, in the mathematical structure, of expression (15) for the generalized radiance B_ω and expression (39) for the generalized radiant emittance E_ω should be noted.

We shall demonstrate in Sec. VI that the generalized radiant emittance, like the generalized radiance, may take on negative values. However, as shown in Sec. V, the generalized radiant emittance is non-negative for a spatially incoherent source.

IV. THE GENERALIZED RADIANT INTENSITY

We introduced the generalized radiant intensity for a two-dimensional source of any state of coherence by formula (21), viz.,

$$J_\omega(\mathbf{s})=\cos\theta\int_{(z=0)} B_\omega(\mathbf{r},\mathbf{s})\,d^2\mathbf{r}, \tag{40}$$

where $B_\omega(\mathbf{r},\mathbf{s})$ is the generalized radiance. We now express J_ω in several alternative forms and show that it represents the angular distribution of the flux density in the far zone.

From Eqs. (40) and (28), we obtain an expression for the generalized radiant intensity in terms of the four-dimensional spatial Fourier transform $\hat{W}$ [Eq. (24)] of the cross-spectral density function in the source plane,

$$J_\omega(\mathbf{s})=(2\pi k)^2\hat{W}(k\mathbf{s},-k\mathbf{s};\omega)\cos^2\theta. \tag{41}$$

Alternatively, we may express J_ω in terms of the spectral density function W of the field in the far zone, using the asymptotic relation (34) of Ref. 5. We then obtain from Eq. (41) the formula

$$J_\omega(\mathbf{s})\simeq R^2W(R\mathbf{s},R\mathbf{s};\omega),\quad kR\gg 1 \tag{42}$$

where, as before, $\mathbf{R}$ is a point in the direction $\mathbf{s}$ in the far zone (see Fig. 1). Because the spectral density function $W(\mathbf{R},\mathbf{R};\omega)$ is necessarily non-negative, so is the generalized radiant intensity $J_\omega(\mathbf{s})$, i.e.,

$$J_\omega(\mathbf{s})\geq 0, \tag{43}$$

irrespective of the state of coherence of the source.

Alternatively, using the relation (25) we may also express J_ω in terms of the angular self-correlation function $\mathcal{A}(\mathbf{s},\mathbf{s};\omega)$. From Eq. (41) we then obtain

$$J_\omega(\mathbf{s})=\left(\frac{2\pi}{k}\right)^2\cos^2\theta\,\mathcal{A}(\mathbf{s},\mathbf{s};\omega). \tag{44}$$

Now, if we compare Eq. (44) with Eq. (2) and also use Eq. (6), we see that

$$J_\omega(\mathbf{s})=R^2|\langle\mathbf{S}_\omega(R\mathbf{s})\rangle|,\quad kR\gg 1 \tag{45}$$

where $\langle\mathbf{S}_\omega(R\mathbf{s})\rangle$ is the averaged Poynting vector at the point $\mathbf{R}=R\mathbf{s}$.

Multiplying both sides of Eq. (45) by the element $d\Omega$ of the solid angle around the $\mathbf{s}$ direction and using formula (7) for the corresponding element $d\sigma$ of the surface of a large sphere of radius R centered at the source, we obtain

$$J_\omega(\mathbf{s})d\Omega\sim|\langle\mathbf{S}_\omega(R\mathbf{s})\rangle|\,d\sigma,\quad kR\gg 1. \tag{46}$$

This formula shows that our generalized radiant intensity represents the flux per unit solid angle around the direction $\mathbf{s}$, emitted by the source into the far zone. Thus, the generalized radiant intensity introduced by Eqs. (21) and (15) for radiation from a source of any state of coherence has the same physical significance as the radiant intensity of traditional radiometry, even though the generalized radiance $B_\omega(\mathbf{r},\mathbf{s})$ and the generalized radiant emittance $E_\omega(\mathbf{r})$ may take on negative values.

V. SPECIAL CASE (A), INCOHERENT SOURCES

In the idealized case when the source is spatially strictly incoherent, the cross-spectral density function W for points in the source plane may be represented by a two-dimensional Dirac δ function,[9]

$$W(\mathbf{r}_1,\mathbf{r}_2;\omega)=w(\mathbf{r}_1,\omega)\delta^{(2)}(\mathbf{r}_1-\mathbf{r}_2), \tag{47}$$

where

$$w(\mathbf{r}_1,\omega)\geq 0, \tag{48}$$

and, of course, $w(\mathbf{r}_1,\omega)\equiv 0$ if the point $\mathbf{r}_1$ is outside the source area. In this case the expression (15) for the generalized radiance reduces to

$$B_\omega(\mathbf{r},\mathbf{s})=\left(\frac{k}{2\pi}\right)^2 \cos\theta\, w(\mathbf{r},\omega). \tag{49}$$

With B_ω given by Eq. (49), the expressions (20) and (21) become

$$E_\omega(\mathbf{r})=\left(\frac{k}{2\pi}\right)^2 w(\mathbf{r},\omega)\int_{(2\pi)} \cos^2\theta\, d\Omega, \tag{50}$$

$$J_\omega(\mathbf{s})=\left(\frac{k}{2\pi}\right)^2 \cos^2\theta \int_{(z=0)} w(\mathbf{r},\omega)\, d^2\mathbf{r}. \tag{51}$$

The integral in Eq. (50) may be evaluated readily by expressing $d\Omega$ in spherical polar coordinates, and is found to have the value $2\pi/3$. Hence

$$E_\omega(\mathbf{r})=\frac{k^2}{6\pi}w(\mathbf{r},\omega). \tag{52}$$

It is useful to express B_ω and J_ω in terms of E_ω rather than in terms of $w(r;\omega)$. From Eqs. (49) and (52), we see that

$$B_\omega(\mathbf{r},\mathbf{s})=\frac{3}{2\pi}E_\omega(\mathbf{r})\cos\theta, \tag{53}$$

and, in view of Eq. (52), Eq. (51) may be expressed in the form

$$J_\omega(\mathbf{s})=\frac{3}{2\pi}\cos^2\theta \int_{(z=0)} E_\omega(\mathbf{r})\, d^2\mathbf{r}. \tag{54}$$

For points of observation in the direction at right angles to the plane of the source, $s_x=s_y=0$ and $\theta=0$, and Eq. (54) gives $J_\omega=J_\omega^{(0)}$, where

$$J_\omega^{(0)}=\frac{3}{2\pi}\int_{(z=0)} E_\omega(\mathbf{r})\, d^2\mathbf{r}. \tag{55}$$

From Eqs. (54) and (55), it follows that

$$J_\omega(\mathbf{s})=J_\omega^{(0)}\cos^2\theta. \tag{56}$$

According to Eq. (56), the radiant intensity now decreases in proportion to $\cos^2\theta$ as θ increases—not in proportion to $\cos\theta$ characteristic of an isotropically radiating source. This result implies that a spatially incoherent plane source of finite extent cannot radiate isotropically into the half-space $z>0$, irrespective of the exact form of its radiant-emittance function $E_\omega(\mathbf{r})$. This result is not contradicted by blackbody sources which, as is well known, radiate isotropically and are often incorrectly considered to be spatially incoherent. A careful analysis shows that blackbody-radiation sources exhibit field correlations over distances of the order of the mean wavelength of the radiation.[10]

We see from Eqs. (49) and (52) that, in view of Eq. (48), the generalized radiance and the generalized radiant emittance of a plane incoherent source are necessarily non-negative. The generalized radiant intensity is, of course, also non-negative.

VI. SPECIAL CASE (B), COHERENT SOURCES

For a coherent source the cross-spectral density function W factorizes in the form[11]

$$W(\mathbf{r}_1,\mathbf{r}_2;\omega)=v(\mathbf{r}_1)v^*(\mathbf{r}_2). \tag{57}$$

Of course, $v(\mathbf{r})=0$ when the point $\mathbf{r}$ lies outside the source area. On substituting from Eq. (57) into Eq. (15), we obtain, for the generalized radiance of a coherent source, the expression

$$B_\omega(\mathbf{r},\mathbf{s})=\left(\frac{k}{2\pi}\right)^2 \cos\theta \int_{(z'=0)} v(\mathbf{r}+\tfrac{1}{2}\mathbf{r}')v^*(\mathbf{r}-\tfrac{1}{2}\mathbf{r}') \times\exp(-ik\mathbf{s}\cdot\mathbf{r}')\, d^2\mathbf{r}'. \tag{58}$$

The generalized radiant emittance of a coherent source may be determined by substituting from Eq. (57) into Eq. (39). We then obtain

$$E_\omega(\mathbf{r})=\frac{k^2}{2(2\pi)^{\frac{1}{2}}}\int_{(z=0)} v(\mathbf{r}+\tfrac{1}{2}\mathbf{r}')v^*(\mathbf{r}-\tfrac{1}{2}\mathbf{r}')\frac{J_{\frac{3}{2}}(kr')}{(kr')^{\frac{3}{2}}}\, d^2\mathbf{r}'. \tag{59}$$

To determine the generalized radiant intensity of a coherent source, we shall use Eq. (41) and recall that, according to Eq. (24), $\hat{W}(\mathbf{f}_1,\mathbf{f}_2;\omega)$ is the four-fold Fourier inverse of the cross-spectral density function $W(\mathbf{r}_1,\mathbf{r}_2;\omega)$. In the present case, when W is given by Eq. (57), $\hat{W}$ clearly has the form

$$\hat{W}(\mathbf{f}_1,\mathbf{f}_2;\omega)=\hat{v}(\mathbf{f}_1)\hat{v}^*(-\mathbf{f}_2), \tag{60}$$

where $\hat{v}(\mathbf{f})$ is the two-fold Fourier transform of $v(\mathbf{r})$, i.e.,

$$\hat{v}(\mathbf{f})=\frac{1}{(2\pi)^2}\int_{(z=0)} v(\mathbf{r})\exp(-i\mathbf{f}\cdot\mathbf{r})\, d^2\mathbf{r}. \tag{61}$$

From Eqs. (41), (60), and (61), it follows that the generalized radiant intensity of a coherent source is given by

$$J_\omega(\mathbf{s})=(2\pi k)^2\cos^2\theta\,|\hat{v}(k\mathbf{s})|^2. \tag{62}$$

From Eqs. (58) and (59), we can readily verify our earlier assertions that both the generalized radiance and the generalized radiant emittance may take on negative values. To see this, consider, for example, the values of these quantities at a point of the source that coincides with the origin $\mathbf{r}=0$ of our coordinate system. The expressions (58) and (59) then reduce to

$$B_\omega(0,\mathbf{s})=\left(\frac{k}{2\pi}\right)^2\cos\theta\int_{(z'=0)} v(\tfrac{1}{2}\mathbf{r}')v^*(-\tfrac{1}{2}\mathbf{r}')\times\exp(-ik\mathbf{s}\cdot\mathbf{r}')\,d^2\mathbf{r}', \quad (63)$$

$$E_\omega(0)=\frac{k^2}{2(2\pi)^{\frac{1}{2}}}\int_{(z'=0)} v(\tfrac{1}{2}\mathbf{r}')v^*(-\tfrac{1}{2}\mathbf{r}')\frac{J_{\frac{1}{2}}(kr')}{(kr')^{\frac{1}{2}}}\,d^2\mathbf{r}'. \quad (64)$$

Suppose, next, that the source is uniform, i.e., $v(\mathbf{r})=v_0=$ const when the point $\mathbf{r}$ is within the area of the plane $z=0$ occupied by the source and is zero elsewhere on that plane. Moreover, suppose that the source has the form of a square of sides a, parallel to the x and y axes and centered at the origin. Equation (63) then may readily be shown to reduce to

$$B_\omega(0,\mathbf{s})=\frac{|v_0|^2(ka)^2}{\pi^2}\left(\frac{\sin(kas_x)}{kas_x}\right)\left(\frac{\sin(kas_y)}{kas_y}\right)\cos\theta. \quad (65)$$

If we now choose the $\mathbf{s}$ direction to be in the xz plane, then $s_y=0$, $s_x=\sin\theta$, and Eq. (65) gives

$$B_\omega(0,\mathbf{s})=\frac{|v_0|^2(ka)^2}{\pi^2}\left[\frac{\sin(ka\sin\theta)}{ka\sin\theta}\right]\cos\theta. \quad (66)$$

Clearly, B_ω will be negative, for example, for directions for which $\pi<ka\sin\theta<2\pi$. If we choose the side of the square $a=10^5\lambda$, we see from Eq. (66) that $B_\omega(0,\mathbf{s})$ will be negative for θ values such that $5\times10^{-6}<\sin\theta<10^{-5}$, i.e., 2.85×10^{-4} degrees$<\theta<5.7\times10^{-4}$ degrees.

An example of a situation where the generalized radiant emittance becomes negative is provided by a uniform coherent source that has the form of an annulus about the origin $\mathbf{r}=0$. If r_1 and r_2 are the inner and outer radii of the annulus, Eq. (64) can readily be shown to reduce to

$$E_\omega(0)=|v_0|^2\left[\frac{\sin(2kr_1)}{2kr_1}-\frac{\sin(2kr_2)}{2kr_2}\right]. \quad (67)$$

This formula shows, incidently, that the generalized radiant emittance at the center of the annulus, i.e., in the source plane outside the domain occupied by the source itself has a nonzero value in general. Moreover, if we choose suitable values for the two radii, the generalized radiant emittance at the origin becomes negative. For example, with the choice $r_1=10^5\lambda$, $r_2=(10^5+\frac{1}{8})\lambda$, Eq. (67) gives $E_\omega(0)=-8\times10^{-7}|v_0|^2$.

The peculiar features of the generalized radiance and of the generalized radiant emittance, illustrated by the two examples just considered, underscore our contention that, in general, these two quantities are not measurable. On the other hand, the generalized radiant intensity defined by Eq. (21) may be shown to be non-negative in both of these cases, in agreement with the general conclusion expressed by Eq. (43) Clearly, the generalized radiant intensity is the basic measurable quantity that characterizes some of the radiation properties of a source of any state of coherence.

APPENDIX A: DERIVATION OF THE EXPRESSION (23) FOR THE GENERALIZED RADIANCE

The generalized radiance was defined by the formula (15), viz.,

$$B_\omega(\mathbf{r},\mathbf{s})=\left(\frac{k}{2\pi}\right)^2\cos\theta\int_{(z'=0)} W(\mathbf{r}+\tfrac{1}{2}\mathbf{r}',\mathbf{r}-\tfrac{1}{2}\mathbf{r}';\omega)\times\exp(-ik\mathbf{s}\cdot\mathbf{r}')\,d^2\mathbf{r}', \quad (\mathrm{A1})$$

where $W(\mathbf{r}_1,\mathbf{r}_2;\omega)$ is the cross-spectral density function.

Let us represent W as a four-dimensional Fourier integral [given by the inverse of Eq. (24)]

$$W(\mathbf{r}_1,\mathbf{r}_2;\omega)=\iint_{(\mathbf{f}_1,\mathbf{f}_2\text{ planes})}\hat{W}(\mathbf{f}_1,\mathbf{f}_2;\omega)\times\exp[i(\mathbf{f}_1\cdot\mathbf{r}_1+\mathbf{f}_2\cdot\mathbf{r}_2)]\,d^2\mathbf{f}_1\,d^2\mathbf{f}_2. \quad (\mathrm{A2})$$

If we substitute from Eq. (A2) into Eq. (A1), rearrange some of the terms in the integrand, and invert the order of integration, we find that

$$B_\omega(\mathbf{r},\mathbf{s})=\left(\frac{k}{2\pi}\right)^2\cos\theta\iint_{(\mathbf{f}_1,\mathbf{f}_2\text{ planes})}\hat{W}(\mathbf{f}_1,\mathbf{f}_2;\omega)\times\exp[i(\mathbf{f}_1+\mathbf{f}_2)\cdot\mathbf{r}]G(\tfrac{1}{2}\mathbf{f}_1-\tfrac{1}{2}\mathbf{f}_2-k\mathbf{s})\,d^2\mathbf{f}_1\,d^2\mathbf{f}_2, \quad (\mathrm{A3})$$

where

$$\begin{aligned}G(\tfrac{1}{2}\mathbf{f}_1-\tfrac{1}{2}\mathbf{f}_2-k\mathbf{s})&=\int_{(z'=0)}\exp[i\mathbf{r}'\cdot(\tfrac{1}{2}\mathbf{f}_1-\tfrac{1}{2}\mathbf{f}_2-k\mathbf{s})]\,d^2\mathbf{r}'\\&=(2\pi)^2\delta^{(2)}(\tfrac{1}{2}\mathbf{f}_1-\tfrac{1}{2}\mathbf{f}_2-k\mathbf{s})\\&=4(2\pi)^2\delta^{(2)}(\mathbf{f}_1-\mathbf{f}_2-2k\mathbf{s}),\end{aligned} \quad (\mathrm{A4})$$

and $\delta^{(2)}$ is again the two-dimensional Dirac δ function. Integration in Eq. (A3) with respect to $\mathbf{f}_2$ is trivial; we obtain for B_ω

$$B_\omega(\mathbf{r},\mathbf{s})=4k^2\cos\theta\int_{(\mathbf{f}\text{ plane})}\hat{W}(\mathbf{f}_1,\mathbf{f}_1-2k\mathbf{s};\omega)\times\exp[2i(\mathbf{f}_1-k\mathbf{s})\cdot\mathbf{r}]\,d^2\mathbf{f}_1. \quad (\mathrm{A5})$$

Let us now introduce in place of $\mathbf{f}_1$ a new variable

$$\mathbf{f}=2(\mathbf{f}_1-k\mathbf{s}). \quad (\mathrm{A6})$$

Formula (A5) then becomes

$$B_\omega(\mathbf{r},\mathbf{s})=k^2\cos\theta \int_{(\mathbf{f}\text{ plane})} \hat{W}(k\mathbf{s}+\tfrac{1}{2}\mathbf{f},\, -k\mathbf{s}+\tfrac{1}{2}\mathbf{f};\omega) \times\exp(i\mathbf{f}\cdot\mathbf{r})\, d^2\mathbf{f}, \quad \text{(A7)}$$

which is Eq. (23).

APPENDIX B: REALITY OF THE GENERALIZED RADIANCE

In this Appendix we show that the generalized radiance, defined by Eq. (15), viz.,

$$B_\omega(\mathbf{r},\mathbf{s})=\left(\frac{k}{2\pi}\right)^2 \cos\theta \int_{(z'=0)} W(\mathbf{r}+\tfrac{1}{2}\mathbf{r}',\, \mathbf{r}-\tfrac{1}{2}\mathbf{r}';\omega) \times\exp(-ik\mathbf{s}\cdot\mathbf{r}')\, d^2\mathbf{r}' \quad \text{(B1)}$$

is real. We first take the complex conjugate (denoted by an asterisk) of Eq. (B1) and obtain

$$B^*_\omega(\mathbf{r},\mathbf{s})=\left(\frac{k}{2\pi}\right)^2 \cos\theta \int_{(z'=0)} W^*(\mathbf{r}+\tfrac{1}{2}\mathbf{r}',\, \mathbf{r}-\tfrac{1}{2}\mathbf{r}';\omega) \times\exp(ik\mathbf{s}\cdot\mathbf{r}')\, d^2\mathbf{r}'. \quad \text{(B2)}$$

Now, from the definition of the cross-spectral density [cf. Eq. (6) of Ref. 5], it follows that

$$W^*(\mathbf{r}_1,\mathbf{r}_2;\omega)=W(\mathbf{r}_2,\mathbf{r}_1;\omega), \quad \text{(B3)}$$

so that Eq. (B2) can be expressed in the form

$$B^*_\omega(\mathbf{r},\mathbf{s})=\left(\frac{k}{2\pi}\right)^2 \cos\theta \int_{(z'=0)} W(\mathbf{r}-\tfrac{1}{2}\mathbf{r}',\, \mathbf{r}+\tfrac{1}{2}\mathbf{r}';\omega) \times\exp(ik\mathbf{s}\cdot\mathbf{r}')\, d^2\mathbf{r}'. \quad \text{(B4)}$$

Next we change the variables of integration on the right-hand side of Eq. (B4) from $\mathbf{r}'$ to $-\mathbf{r}'$ and obtain

$$B^*_\omega(\mathbf{r},\mathbf{s})=\left(\frac{k}{2\pi}\right)^2 \cos\theta \int_{(z'=0)} W(\mathbf{r}+\tfrac{1}{2}\mathbf{r}',\, \mathbf{r}-\tfrac{1}{2}\mathbf{r}';\omega) \times\exp(-ik\mathbf{s}\cdot\mathbf{r}')\, d^2\mathbf{r}'. \quad \text{(B5)}$$

Comparison of Eq. (B5) with Eq. (B1) shows that

$$B^*_\omega(\mathbf{r},\mathbf{s})=B_\omega(\mathbf{r},\mathbf{s}), \quad \text{(B6)}$$

implying that the generalized radiance $B_\omega(\mathbf{r},\mathbf{s})$ is real.

REFERENCES

*Research supported in part by the Air Force Office of Scientific Research.
[1]M. Planck, *The Theory of Heat Radiation*, translation from the Second Edition (Dover, New York, 1959).
[2]M. Born and E. Wolf, *Principles of Optics, 4th ed.* (Pergamon, Oxford and New York, 1970), §10.4.2.
[3]A preliminary account of our main results was published in Opt. Commun. **6**, 305 (1972) and was also presented at a meeting of the Optical Society of America held in Rochester, N. Y., 9–12 October 1973 [J. Opt. Soc. Am. **63**, 1285A (1973)]. In Eq. (12) of the published paper, a factor $(2\pi)^{-2}$ should be omitted.
[4]A. Walther, J. Opt. Soc. Am. **58**, 1256 (1968). In a recent paper [J. Opt. Soc. Am. **63**, 1622 (1973)] Walther modified his original definition of the generalized radiance after asserting that it depends on the choice of the coordinate system. This assertion, however, is misleading in the context of his earlier paper relating to radiation from planar sources. For, as we show in a Letter on p. 1273 in the present issue, the generalized radiance, as originally defined by Walther, is, in fact, invariant with respect to an arbitrary displacement of the origin of coordinates in the source plane and is also invariant with respect to rotation of axes about the normal to the plane of the source. In any case, as will be clear from the discussion in the present paper, it is not the generalized radiance, but rather the generalized radiant intensity that has a direct physical significance.
[5]E. W. Marchand and E. Wolf, J. Opt. Soc. Am. **62**, 379 (1972). In Eq. (10) and in some of the subsequent equations of this reference there is an error: m_2 should be replaced by its complex conjugate m_2^*. This error does not, however, affect the main results.
[6]E. Wigner, Phys. Rev. **40**, 749 (1932). For a good discussion of some of the properties of the Wigner distribution function, see K. Imre, E. Ozizmir, M. Rosenbaum, and P. F. Zweifel, J. Math. Phys. **8**, 1907 (1967).
[7]See, for example, G. N. Watson, *A Treatise on the Theory of Bessel Functions* (Cambridge U. P., 1922), p. 20, Eq. (5) (with an obvious substitution).
[8]I. S. Gradsteyn and I. M. Ryzhik, *Tables of Integrals, Series and Products* (Academic, New York, 1965), p. 688, formula 1 of §6.567, with $\nu=0$, $\mu=\frac{1}{2}$.
[9]Alternative representations of spatially incoherent sources are discussed in a paper by M. Beran and G. Parrent, Nuovo Cimento **27**, 1049 (1963).
[10](a) C. L. Mehta, Nuovo Cimento **28**, 401 (1963); (b) R. C. Bourret, Nuovo Cimento **18**, 347 (1960); (c) C. L. Mehta and E. Wolf, Phys. Rev. **134**, A1143 (1964); Phys. Rev. **134**, A1149 (1964).
[11]J. Peřina, *Coherence of Light* (Van Nostrand, London, 1972), §4.2.

Author Index

Subject Index